DIALOGUE AS A COLLECTIVE MEANS OF DESIGN CONVERSATION

DIALOGUE AS A COLLECTIVE MEANS OF DESIGN CONVERSATION

Edited by

Patrick M. Jenlink
Stephen F. Austin State University
Nacogdoches, Texas and
International Systems Institute
Carmel, California

Bela H. Banathy
Saybrook Graduate School and Research Center
San Francisco, California and
International Systems Institute
Carmel, California

Springer

Patrick M. Jenlink
Stephen F. Austin State University
Nacogdoches, TX, USA

Bela H. Banathy
Late of Saybrook Graduate School
San Francisco, CA, USA

Library of Congress Control Number: 2007938313

ISBN-13: 978-0-387-75842-8 e-ISBN-13: 978-0-387-75843-5

Printed on acid-free paper.

9 8 7 6 5 4 3 2 1

springer.com

Memoriam—Bela H. Banathy

Originally, this compendium began as a conversation between Bela and myself, and was shaped by our belief that dialogue offered a means for humankind to collectively work together toward a future society that was more civil, and hopefully more concerned with its own evolution. Bela and I visited, frequently, about the importance of dialogue as collective means of communication, and the dialogic nature of design conversation as a collective means of design in social systems. A means of communicative action that would animate a self-guided evolution of humankind through the creation new systems. We hoped that our species, *Homo Sapiens Sapiens*, would overcome its destructive capabilities and foster new creative possibilities for the future generations of humankind that would follow.

Unfortunately, Bela was not to see the compendium in its completed form. On September 4, 2003 Bela passed away, leaving the world a better place for his presence, and our lives richer for having known him.

Bela found inspiration in the works of William Blake (1991), in particular the illuminated work *Jerusalem: The Emancipation of the Great Albion*. It is from this work that I quote the following passage, in memory of Bela. I believe it illuminates his life and his work as a systems scholar and practitioner (Blake, pl. 10, l. 20):

> I must create a system,
> Or be enslaved by another Man's;
> I will not Reason and Compare,
> My business is to Create.

Bela's life was one of creating. He was a source of great energy in the ebb and flow of humanity. Bela was a systems scholar and practitioner concerned with creating systems that would make the world a better place for humankind. Bela's life was lived with purpose, and his legacy of systems knowledge and ideals serve to guide the work ahead for all of us that care for the future of humankind. He will be missed.

With this compendium, the journey Bela and I set upon several years ago is now complete. However, the work set in motion by Bela is far from done. I trust that those who knew Bela will continue his work, hopefully toward the betterment of humankind. Bela would have liked that.

Patrick M. Jenlink
June 25, 2007

Dedication

To the Family of Bela H. Banathy
and
To the Past, Present, and Future Members of the
International Systems Institute

Acknowledgements

I wish to thank Gary Folven, Senior Editor, Operations Research/Management Science & Information Systems, Springer USA for his assistance in completing the publishing journey that Bela and I began together. As well, I wish to thank Ken Derham, Editorial Director at Springer UK and friend to Bela for his assistance and support to bring this project to fruition. Bela would have liked this very much.

I want to thank Bela's family for their assistance and their strength over the past several years. Bela so loved his family and I know he would want to acknowledge your support. In particular I want to thank Bela A. Banathy for his friendship and continued support to see his father's work to completion.

On behalf of Bela and myself, I want to thank the all the authors for both this volume and the volume that preceded for their steadfast support and commitment to Bela's work and for their contribution this Bela's living legacy.

Finally, I want to thank Guenther Ossmitz for the original photograph taken at the Fuschl Conversation, Fuschl Austria and Doug Walton for his assistance in creating the photographic effects for the cover art.

Preface

We are, as a social species–*Homo Sapien Sapien*–communicative by nature. Communication presupposes community, which in turn means a communion of consciousness of the persons in the community. That we are also sentient beings places within our reach the capability to be equally creative and destructive, whether through discourse or social action. In our evolution as a species we have demonstrated our capacity for terrible acts of destruction—the most horrible of nightmares. The Holocaust, Hiroshima and Nagasaki, September 11, 2001, the Darfur Conflict stand as examples of such nightmares. Yet, as a species, humankind has demonstrated its capacity for the beauty of its creative potential; we have demonstrated our capacity to come together in times of great tragedy as a communityto communicate in such ways as to create solutions and foster new hope for the future.

When we consider, in relation to being a species characterized as communicative and sentient, that we are also an extremely diverse species, we are filled with great potential. Such potential in our diversity, when acted upon in the service of humankind, replenishes and renews. In contrast, when our diversity is not acted upon, the spirit of humankind is lessened.

The fact that we display great diversity–that we are different culturally, linguistically, ethnically, politically–figures largely into our potential for creative actions, and holds promise for present and future generations to overcome the destructive nature of our species that all-to-often marks our evolutionary history. The realization of our potentialities as a species, rests in no small measure, in our capacity for collective creativity and in our capacity to achieve new levels of consciousness. The significant problems we face today in our cities, nation states, and global society cannot be solved at the same level of consciousness with which we created them.

That we have the capacity, as a species, to transcend existing levels of consciousness, while important, is not significant unless we act on our capacity so as to benefit humankind. Such transcendence is understood as a liberating process, seeking to release humanity from existing societal structures and cultural patterns that are oppressive and which seek to disadvantage some individuals while privileging others. The nature of design conversation as communicative and emancipatory action relies on discourses that are democratizing and authentically participative. Design conversation provides a public sphere, in which participants' voices are valued, listened to, and have a primary role in determining the conceptions and actions necessary for designing a new system.

It is in the recognition that our species is ultimately responsible for its future that we also recognize the necessity for fostering new forms of communication—communicative action—that serve the collective needs of our species; the needs of designing social systems, fostering new levels of consciousness, and creating alternative possible futures that transcend the conscious destructive acts that have defined our species. If we are to embrace our responsibilities in the global society, we must engage in designing new systems that will guide our journey to the future. It is incumbent of us to create conditions favorable for communicative action, for us to understand each other, for peace and non-violence and the betterment of all humankind to become markers that define our species. That such communicative action is possible–for each person to stand where the other is standing–will require much of our species.

In this Compendium, the contributing authors set forth their ideas, experiences, and perspectives as the path of a *learning journey*—a journey of new meaning, of new understanding, and of becoming self-aware of design conversation as future creating and consciousness evolving.

The Compendium is organized by five themes. *Section I* examines foundational perspectives of design conversation. This examination helps to create a foundation for a deeper study of the emergent and salient aspects of conversation in relation to social creativity and the evolution of human consciousness. Authors examine design conversation from philosophical, cultural, spiritual, and historical perspectives. *Sections II-IV* examine the philosophical and theoretical perspectives as well as methodological ideas related to conversation. These writings also explore different modalities of conversation and the application of design conversation within and across various types of design settings and human experiences. Also examined is the importance of capacity building for engaging in conversation, as well as providing insight into how to build capacity and develop the capability of the human system for conversation. In *Section V* the editor reflectively examines the contributions to the book and presents his own thoughts on the next steps in the evolutionary relationship of conversation, human systems, and systems design.

Contents

INTRODUCTION

Chapter 1

DESIGN CONVERSATION
Future Building and Consciousness Evolving

PATRICK M. JENLINK
Stephen F. Austin State University and International Systems Institute

1. INTRODUCTION

In this opening chapter, first, I review the purpose of this Compendium and provide an overview of the *learning journey* presented by the editors and authors. The chapter is presented in two parts. *Part I* begins with an exploration of the meaning of design conversation. Then, I examine the relationship of design conversation to future building and consciousness evolving within society. Design conversation is introduced as a form of conversation that enables our species to transcend existing systems through communicative and emancipatory action. As such, design conversation relies on discourses that are democratizing and authentically participative. Design conversation creates social spaces, within which participants' voices are valued, listened to, and have an influence on the conceptions and actions necessary for designing a new system. The chapter will examine why design conversation is important as well as how it may be used to create a collective evolutionary consciousness essential to designing our own future. *Part II* overviews how the compendium is framed into five themes, which present a reflective context for exploring dialogue conversation.

2. AN OVERVIEW OF THE LEARNING JOURNEY

A two-pronged *purpose* has guided the development of this Compendium. The first purpose was to introduce the learner to design conversation as the means of collective communication for which we must be both students and consumers as members of a changing, global society. The second purpose of the Compendium is to demonstrate–and develop an appreciation for–the empowering and liberating quality of conversation as a medium and means of communication for collective creativity and societal change. The Compendium offers a rich set of perspectives and experiences related to design conversation as future creating and consciousness evolving means of collective communication.

The path of the *learning journey* begins by exploring design conversation. This exploration sets in place a historical perspective of design conversation, providing an examination of the emergent and developing characteristics of disciplined conversation. We next explore the different perspectives of design conversation that are sources of methodological ideas and practices of dialogue. We examine different modalities of design conversation in a variety of settings, and provide examples of conversation events. We also provide practical applications of design conversation that illuminate different approaches to systems design that enable individuals, groups, and communities to initiate, engage in, and guide the disciplined inquiry of design.

In *Part One*, I explore the meaning of design conversation and its role in future creating and consciousness evolving. First, I examine the meaning of design conversation, exploring the etymological roots and historical origins. Then I explore dialogue as a method and means of collective communication in the larger context of culture and society. In *Part Two*, I provide an overview of the five organizing themes for the Compendium, and briefly examine author contributions that set the path for our *learning journey* towards dialogue as a collective means of communication.

PART ONE: DESIGN CONVERSATION AND ITS ROLE IN FUTURE CREATING AND CONSCIOUSNESS EVOLVING

In this *Part*, first, I explore conversation as design medium, then follow with an examination of design conversation and its distinguishing features. Then I explore design conversation as future creating and a means of collective communication in systems design. Design conversation as consciousness evolving is then examined in relation to transforming society.

3. CONVERSATION AS DESIGN MEDIUM

Social systems design, as Banathy (1996) explains, "is a process that carries a stream of shared meaning by a free flow of discourse among the stakeholders who seek to create a new system" (p. 213). In order to understand the communicative nature and mediational importance of the design conversation, "various modes of social discourse are explored to search for the mode that is the most appropriate to systems design" (p. 213). As such, these different modes of social discourse serve as a medium for various design activities within the design community. Examples of the different modes of discourse include inquiry discourse (Banathy, 1996; Burbules, 1993; Isaacs, 1996), critical discourse (Burbles, 1993; Carr, 1997; Giroux & McLaren, 1986; Lambert, 1995), sustaining discourse (Hollingsworth, 1994; Lambert, 1995), debriefing conversations (Jenlink, Reigeluth, Carr, & Nelson, 1996), conversations with contexts (Banathy, 1996; Jenlink, 1995) and dialogue conversation which serves as a medium for emotional and cultural change (de Mare, 1991).

The root of the word conversation means, "to turn together." Conversation, then, is people speaking together, engaged in reflection and deliberation. This reflection and deliberation enables participants to "consciously and unconsciously weigh out different views, finding some with which they agree, and others that they dislike" (Isaacs, 1993, p. 35). As Berger and Luckman (1966) suggest, people engage in conversation for the purpose of coordinating their separate interpretations of experience to establish shared meanings, therefore

> the most important vehicle of reality-maintenance is conversation. . . . At the same time that the conversational apparatus ongoingly maintains reality, it ongoingly *modifies* it. . . . Thus the fundamental reality-maintaining fact is the continuing use of the same language to objectify unfolding biographical experience. In the widest sense, all who employ this same language are reality-maintaining others. (pp. 172-173)

Conversation, by its very nature, is relational. Whether it is dialogue, discussion, debate, or design discourse (Banathy, 1996; Jenlink & Carr, 1996), it embodies relationships as a foundation of the interaction between one individual and another. In this sense, relation is the foundation of conversation. Likewise, relation is the foundation of an ethic of care. An ethic of care "is fundamentally concerned with how human beings meet and treat one another. It is not unconcerned with individual rights, the common good, or community traditions, but it de-emphasizes these concepts and recasts them in terms of relations" (Noddings, 1993, p. 45).

Martin Buber (1970) in his explication of the "principle of relatedness" acknowledged two modes of encountering or relating to other entities. He identified these as the I-Thou mode, which is the way of relation, and the I-It mode in which we observe others or listen to what they have to say by assimilating it to prestructures, or preselected ways of interpreting and understanding the phenomena we experience. The I-It mode is instrumental in the sense that as we encounter, through conversation and experience, events and phenomena, we appropriate materials and ideas to be used for our purposes. However, the I-thou mode is both ethical and spiritual, acknowledging the value of relation between "self" and others, thereby acknowledging the ethical responsibility one has for helping another and for the betterment of the human conditions of others. The "principle of relatedness" gives way to "betweenness" that exists in and through the I-Thou mode of interacting or meeting other entities.

"Betweenness," by its nature and caring orientation, is a form of social discourse that connects one's self with others, the I-Thou, and enables the building and sustaining of communities that serve to transform the conditions in which individuals live and work. These communities are characterized by a caring relation that serves as a foundation for the social structure of the community, and by the nature of the social discourse that pervades and gives substance to the relatedness of individuals comprising the community. Caring, like "betweenness" is a form of social discourse. As such, it is concerned with "affirming and encouraging the best in others" (Noddings, 1992, p. 25) and involves an "orientation of deep concern that carries us out of ourselves and into the lives, despairs, struggles, and hopes of others" (Noddings, 1994, p. ix-x).

Other types of social discourse, such as discussion, debate, dialogue, and design discourse, may be critically examined and the nature or purpose of the discourse discerned by the type of relation that is present and supported. Where an ethic of care is applied as in "care as a moral orientation" then the social discourse can be said to be a moral discourse. Social discourse, such as dialogue or design conversation, when guided by caring helps to establish conditions in which participants seek to address issues of social justice, equity, and inequalities, and wherein participants recognize "deep and perhaps irremovable differences–differences which counsel against sweeping solutions that affect people's lives directly and preclude their effective use of self-chosen strategies" (Noddings, 1999, p. 19).

4. DESIGN CONVERSATION

Design conversation has been variously defined (Banathy, 1996, 2000; Jenlink, 2001; Jenlink & Carr, 1996) with a common ground being dialogue. Banathy (1996, 2000) distinguishes between two types of dialogue, strategic and generative, and suggests that it is the combining of these types of dialogue that form design conversation. Generative dialogue leads to a common ground and collective consciousness within the design community. Strategic dialogue focuses on addressing specific tasks and is applied in finding specific solutions in decision-oriented inquiry" (p. 389). Influenced by Banathy's work, Jenlink (2001) further explains that design conversation in the context of systems design:

> is viewed largely as a communicative action, providing a medium through which participants in the design process may engage in a multi-dimensional inquiry leading to the creation of a new system. Design conversation is not a singular type or form of social discourse, but rather a dynamic system comprised of different forms of discourse, each with a particular purpose and mediational importance. (p. 352)

4.1 Conversation as Medium of Design

Conversation as medium of design arises from an understanding that dialogue and design conversations give form to the reciprocal processes of systems design that make up the patterns of collective consciousness, of community. "The conversations serve as the medium for the reciprocal processes that enable [participants] to construct meanings toward a common purpose" (Lambert, 1995, p. 83). As a medium for design, the primary benefit of conversation "is in creating and maintaining the conditions for more conversation" (Burbules, 1993, p. 127). Relatedly, conversation serves as a medium through which participants connect to create a collective consciousness that enables a "community of mind" to form.

Conversation, as a medium, also contributes to the building of community. In this sense, community-building discourse is one of reciprocally engaging participants in " . . . mutual and dynamic interaction and exchange of ideas and concerns," seeking to create maturity in the community that usually "emerges from opportunities for meaning-making in sustainable communities over time" (Lambert, 1995, p. 95). Community building discourse provides a forum for exploring individual and collective concerns, examining common experiences, developing shared meaning, identifying core ideals, values, beliefs–what is *sacred*, constructing a change community language, and creating "community of mind" essential to a design community (Jenlink, 1996). Dialogue is essential to the collaborative process of community building. In a sense, design community is a discourse community, and the power of conversation is in its ability to weave "webs of relationships" among all stakeholders in an intersubjective fabric of cultural creativity.

4.2 Conversation as Communicative Action

Similarly, design conversation as a communicative method for social action, and medium for design inquiry, uses different forms of social discourse including generative, strategic and reflective dialogue, discussion, ethical discourse, critical discourse, post-formal discourse, aesthetical discourse, democratic discourse (see Banathy, 2000; Jenlink, 2001 for a more substantive examination). Importantly, these different forms of discourse comprise the communicative method and medium through and in which the design community situates itself and engages in design activities. Banathy's (2000) epistemological premises for conscious self-guided evolution are instructive in understanding evolutionary design conversation. Evolutionary knowledge, evolutionary thinking, and evolutionary consciousness are defining factors in the type of discourse that user-designers select as well as how the discourse is used as a communicative method for social action and conscious self-guided evolution.

Communication for social action that results in construction and reconstruction of evolutionary knowledge requires a generative form of discourse as well as considerations for ethical and aesthetical discourse. This is made apparent in the creation of an ideal image premised on evolution of a culture and its people. Communication for creating shared meaning, common ground, and collective (evolutionary) consciousness requires reflective and generative dialogue, but may likely be tempered by ethical considerations for diversity, equality, equity, caring. Communication for translating design solutions into reality may require strategic dialogue, but will likely draw on critical and democratic forms of discourse to ascertain "goodness of fit" as well as ensure inclusion and authentic participation. Each design activity, guided by evolutionary thinking, will seek out the form of discourse that enables the user-designers to realize the ideal social and societal systems.

Designing an ideal social system is guided by the conscious awareness of what the ideal is, and by the communicative actions essential to, realizing that ideal. Such consciousness and communicative actions characterize a meta-state of design discourse that may be called design conversation. Jenlink's work with design conversation helps shape the understanding of evolutionary design conversation as

> a meta-conversation, embodies each form of discourse throughout the design activities, enabling the stakeholders to engage in the objectivation and embodiment of the participant's subjectivity in the transformation of the ideal into reality. Key in this meta-conversation is the transition from one form of discourse to another, and knowing when each form is relevant to the design activities. (2001, p. 356)

Design conversation is essential to each phase of conscious, self-guided evolution, from the design of an inquiry process model that will be designed by the systems design community to guide the design of a social system, to the design of the evolutionary system (ES) that is a "*temporal, here-and-now manifestation*" of the new system to be designed (Banathy, 2000, p. 346, italics in original).

4.3 Design Conversation Characteristics

Jones (1984) offers the following in way of understanding design conversation when he explains the design process as

> unlearning what we know of what exists, of what we call the "status quo," to the point where we are able to lose our preconceptions sufficiently to understand the life, and the lives, for which we design, and where we are aware of the ways in which new things added to the world, can change the ways we see it. (p. 172)

Design conversation acknowledges the need for a "users language" which implies paying attention to actual users, that is people "in situ," people in the here and now, each with his or her own history, and his or her own peculiarities. Such peculiarities have to be transformed into resources for action. User languages will emphasize differences between actors, in terms of intended use" (de Zeeuw, 1993, p. 13). "The term "user" means the users of the enculturation learning systems, those who want to become competent in design and eventually will become designers in designing their own systems" (Banathy, 1996, pp. 245-246). Systems design language–the words, concepts, meanings informed by the philosophy, theory, and practical experiences of systems–provides the user with a language for communicating within and through the design conversation, and change process.

Returning to the conversational nature of design conversation, it is important to recognize that the systems design process requires multiple forms of conversation which enable the participants to access as many dimensions of the design process as is possible. This may include such types of conversation as dialogue, community building, sustaining, inquiring, and conversations with *self* and *context*.

Incorporating different types of conversation, design conversation is an ongoing discourse that moves through different phases of communicative interaction. Participants are "caught up" in the ongoing flow of interchange. At various times this flow and interchange can change its tone, its purpose, and its direction through choices of its participants. Recognizing how new and varied "moves" are "appropriate at some points, and not at others, is an important part of how one can remain within the spirit of dialogical relation while being flexible about just what that entails from moment to moment" (Burbules, 1993, p. 129). Key characteristics of design conversation include aesthetic, authentic, ethical, sustainable, and unique.

4.3.1 Aesthetic

Banathy (2000) has argued, related to the design experience, that there must be diverse and full participation of the all vested parties in the educational system. As a source of creativity in systems design, user-designers work within the *design space* to create new educational systems that reflect a symmetry of diverse perspectives; the experience of systems design shared by all members is enriched by the inclusive nature that defines both the experience and the design. The quality of aesthetics in design conversation, comes into play in two ways for user-designers:

> Through their participation, members of the design community can bring into the design individually and collectively their aesthetic values and ideas so that the design, once implemented, will be aesthetically pleasing. If this quality is explicitly sought and realized in the course of design, then the designers' involvement in the design inquiry will become in itself an aesthetic experience. (Banathy, 2000, p. 364)

Aesthetics in systems design is concerned with elegance and beauty in content and form, in the spatial and experiential dimensions of social interaction and design that is social creativity. Elegance in design conversation is concerned, in part, with qualities of discourse and practice that obtain a richness and dignity in interactions, relations, and participants. An elegant and beautiful design of a social system necessarily seeks to value the voice of all individuals and recognizes that the voices of disadvantaged and marginalized populations are critically important to designing a new system that is ethical and moral.

4.3.2 Authentic

The nature of design conversation as communicative and emancipatory action relies on discourses that are democratizing and authentically participative. Design conversation provides a public sphere, in which participants' voices are valued, listened to, and have a primary role in determining the conceptions and actions necessary for designing a new system. Design conversation is authentic if it is " . . . carried out by the stakeholders of the system. An authentic design has to build on the individual and collective values, aspirations, and ideas of those who serve the system and who are served and affected by it" (Banathy, 1996, p. 228). Authenticity as a defining characteristic of design conversation s concerned diversity of perspectives and recognizes the need for involvement of diverse stakeholder groups that reflect the makeup of the larger population that will benefit from the social system being designed.

The ethical consequences of systems design rests in large part with whether or not the design process, and therein design conversation explicitly, provided opportunity for authentic participation and was socially and culturally inclusive – active participation by all including marginalized populations and socially and politically disadvantaged publics. To authentically engage a citizenry in the design of its future will require multiple forms of social discourse to provide opportunity and space for the diverse citizenry to become engaged in the design conversation. This becomes even more apparent when considering the need to overcome or transcend.

4.3.3 Creative

Design conversation is purposed to create new systems of human activity in society. Systems design, Banathy explains (2000),

> is a creating activity that brings forth a potential-driven, intended novel sociocultural system in the . . . design space. In this space, alternative design ideas are proposed and tested for their viability and for their 'goodness of fit' with their enfolded potential and the environment, which becomes their life space. (p. 243)

Toward fulfilling this purpose of a creating activity, design conversation necessarily fosters an appreciation for creativity and imaginative possibility. To transcend levels of consciousness and thinking that have created existing systems, and the problems endemic in those systems, requires new levels of thinking hallmarked by the capacity for creative consciousness that recognizes, on one level, "that the formulation of a problem is more important than its solution" (Csikszentmihalyi, 1994, p. 138), and on another level, understands that creativity is response to social problems and issues too complex to be resolved by an individual or small group of individuals.

Creativity as an integral dimension of design conversation, functions within the system design process to illuminate and interrogate existing social issues and problems, creating knowledge in response to the ever- evolving problems humankind experiences as we advance in the new millennium. Creativity, in the sense of designing new social systems, is a process of change perhaps most uniquely human, and a such, "worthy human purpose and one most likely to produce knowledge that will provide leverage over the ever more challenging problems we most surely will confront as our species moves, however haltingly and poorly prepared, into a new millennium" (Csikszentmihalyi, 1994, p. 175).

4.3.4 Ethical

Ethical discourse is a governed by socio-cultural rules of right and wrong. Churchman (1982) makes an important point concerning ethics and conversation, "ethics is an eternal conversation…its conversation retains its aesthetic quality if human values are regarded as neither relative nor absolute" (p. 57). Banathy (1996) adds that ethical discourse is focused on "values, morals, and ethics…among the [user-designers]" (p. 181). User-designers must focus not only creating the ideal educational system, they must engage in explicit discourse aimed at finding common ground and developing collective consciousness. Bohm (1996) further explains the idea of common in his statement that communication should not be understood as the "attempts to *make common* certain ideas or items of information" but as the effort of two or more people to "make something in *common*, i.e., [create] something new together" (p. 2).

Ethical design conversation, in relation to creating and sustaining design conversation that embodies ethical responsibility, is a primary factor in the conceptions, decisions, and actions undertaken in the systems design processes, processes that are dialogical by nature. The design of social systems that is concerned with the creation of ethical systems of human activity brings to the foreground the importance of situating ethicality within social systems design. Banathy (1996) suggests that the design conversation can

> be termed to be ethical only if it enables the self-determination of the stakeholders and respects their autonomy and uniqueness. Design should be self-guided and self-directed by the users of the system . . . The ethical

> and liberating involvement . . . is based on the understanding that we have the right and responsibility for the design of our lives and for the design of the systems in which we live. (p. 230)

The ethical nature of conversation figures largely into both the path that conversation follows as well as whether the conversation takes the participants on a journey that conjoins them into common purpose and enables them toward the creation of something new. Design conversation that embraces an ethics of discourse, by its nature, is an *eternal conversation* that retains "its aesthetic quality if human values are regarded as neither relative nor absolute" (Banathy, 2000, p. 373).

4.3.5 Learning

Design conversation, as the medium for systems design, necessarily creates a social space for design learning, a space in which individual and collective capacity building and learning. Banathy (1996) views social systems design as a "Future creating disciplined inquiry" (p. 45), and therein design conversation creates a space through and for discourse and inquiry. As he further explains,

> Even if people fully develop their potential, they cannot give direction to their lives, they cannot forge their destiny, they cannot take charge of their future—unless they develop competence to take part directly and authentically in the design of systems in which they live and work, and reclaim their right to do so. That is what true empowerment is about. (p. vii)

By engaging in design, the stakeholders as user-designers—individually and collectively—have benefit of the design community as forum and context for collective involvement as a design community (Banathy, 1996; Isaacs, 1993). Design conversation not only creates a space for design activities, it also serves as the medium for dialogic exchange of ideas concerning the design of a new system, and in the process design conversation serves to animate design learning—learning that is focused on developing competencies and capacities necessary to authentic participation in systems design—and simultaneously enable the transcending of existing levels of consciousness and the evolution of consciousness to new levels of design cognition and understanding.

4.3.6 Sustainable

A design conversation is sustainable only if it " . . . is accomplished and put in place by the creative, collective, and unconstrained participation and contribution by all people in the system" (Banathy, 1996, p. 229). When stakeholders make these types of contributions, the participation enables people to understand their system as well as their role and relationship to the system. This in turn enables the creation of like-mindedness or a *community of mind* that generates genuine respect for each other as the process develops fellowship. Participation and contribution by stakeholders enables sustainability to takes place at a deeper level of commitment, giving way to a conscious or "deep" community, a consciousness essential to designing a new social system and which represents the stakeholders individual and collective values, ideas and decisions (Banathy, 1996).

4.3.7 Unique

Design conversation conveys a rich uniqueness that is sensitive to certain conditions. If systems designers aspire to create viable systems, the conversational process must understand the uniqueness of the human activity systems, which includes the systemic context, the nature of the system to be designed, the individual and collective readiness and capability of the people involved, the resources available, the design situation, the values and world views of the designing community, and time, space, and complexity factors (Banathy, 1996).

5. DESIGN CONVERSATION AS FUTURE CREATING

In the closing paragraph of *Order Out of Chaos*, Prigogine and Stengers (1984) noted that societies are immensely complex systems, highly sensitive to fluctuations, involving a potentially enormous number of bifurcations. This, the authors note, leads to both threat and hope. The threat lies in the realization that "in our universe the security of stable, permanent rules are gone forever. We are living in a dangerous and uncertain world that implies blind confidence . . . Our hope arises from the knowledge that even small fluctuations may grow and change the overall structure" (Prigogine & Stengers, 1984, p. 313).

The function of design conversation in designing social systems, in part, is to create alternative future possibilities; possible futures that transcend existing systems of human activity and offer hope for the evolution of humankind, imaginatively, socially and morally. It must be to prepare future citizens with capacity and capability to engage in creating a new and consciously renewed generative order that can guide the creative surge of conscious evolution. This requires an epistemological perspective of systems design that fosters social creativity in present and future generations of citizens, toward a goal of enabling the guided evolution of society. Laszlo (1987) explains that

> The evolution of our societies, and therewith the future of our species, is in our hands. If we make use of our capacity to think rationally and act

> purposefully, we can defuse immediate threats and promote the creation of more mature, more autonomous, more dynamically stable societies. (p. 102)

However, the evolution of society calls for the cultivation of capacities far beyond those that currently define the majority of humankind. Evolution, as Laszlo (1996) states, "is always possibility and never destiny. Its course is logical and comprehensible, but it is not predetermined and thus not predictable" (Laszlo, 1996, p. 23). Becoming conscious of creating alternative systems of human activity enables humankind to engage in conscious actions toward creating the future. Educating our selves, creating a consciousness of "educated hope" and "imaginative possibilities" works toward enabling humankind's conscious evolution. Taking responsibility for the future is not easy work; it is about advanced citizenship, about a concern for the welfare of our societies, a consideration for the future of humankind.

Increasingly, a widening range of social, ethic, racial, linguistic, cultural, and religious diversity defines society, both nationally and globally. Contributing to the growing complexities of our responsibilities in as systems practitioners, cultural and economic globalization has rendered the world smaller, and more interdependent than ever before. Simultaneously, we are confronted with the challenges inherent in social systems that are, in many respects, artifacts of an industrial era, symbolic of global capitalism, increasingly defined by threats of war. Ecological devastation amidst global terrorism has created a fragile socio-political ecology that threatens the potential of humankind's evolution. As Banathy (1996) notes, we are at a point in

> societal evolution where unprecedented fulfillment as well as loss of direction, despair, and destruction, are equally possible. However, we are not at the mercy of evolutionary forces but have the potential and the opportunity to give direction to societal evolution...provided we create an evolutionary vision for the future and develop the will and the competence to fulfill that vision in our own lives, in our families, in the systems in which we live, in our communities and societies, and in the global system of humanity. (p. 313)

Each generation of individuals in any society inherits the artifacts of cultural history, including technologies developed to support problem solving. The role of social systems, like the family, school, civic government, etc., in the evolution of society has been, historically, that of transmitting the artifacts of society, one generation to the next. As society grows more complex in its social structure, "what is most significant is the activities of the society itself. These may lead eventually to decay, more or less independently of the institutions, the will, and desire of the people who make up the society" (Bohm & Peat, 1987, p. 206).

6. DESIGN CONVERATION AS CONSCIOUSNESS EVOLVING

Critical to the design of social systems is the work to create and sustain public spaces in which democratic discourse and social action may be experienced as a part of the design process. Also critical in this view of systems design is self-renewal and evolution of consciousness as critical processes implicit and explicit in the design; contributing to the participant's ability to transform social structures and transcend existing systems as the ideal is created and realized through the actions of participants (Jenlink, 2001).

6.1 Consciousness and the Design Experience

David Bohm, in his book *Wholeness and the Implicate Order* (1995), is helpful in understanding consciousness as implicate order; an enfoldment of thought, perspective, worldview. An individual's consciousness is an enfoldment of many thoughts and perspectives over time, creating implicate patterns or relationships. Not dissimilarly, a collective or societal consciousness such as that represented in a particular culture or people may also be understood as being implicate order. When one individual's thought, perspective is enfolded on that of another person's, patterns of value, belief, assumption, become enfolded into the cultural fabric.

Language, as an artifact of a culture and its people, is an enfoldment of symbols and meanings that create an implicate order. Meaning enfolded in the words and structure of language creates implicate patterns of meaning through the use of language to generate thought and action. As individuals engage in communicative relationships, the meaning implicate in language is unfolded in the social or cultural groups through the discursive interactions. As meaning unfolds through communication, the implicate nature of meaning is made explicate, creating opportunity for the participants to generate common meaning through sharing. Such sharing moves from the individual consciousness level to a collective consciousness level. Thus, as Bohm and Peat (1987) note, "there is an internal relationship of human beings to each other, and to society as a whole" (p. 185). What is seen in society – the explicate order – on one level is seen as enfoldment inseparably within the consciousness of each individual member in the society. Therein, implicate order "is the *content* of the culture, which extends into the consciousness of each person" (p. 185).

6.2 Creating the In-between Space

Hannah Arendt (1958), in *The Human Condition*, wrote of the importance of diverse persons speaking to one another as "who" and not "what" they are and, in so doing, creating an "in-between" among themselves (1958, p. 182).

Such creating is a process of change, which is most uniquely human. It requires of us a evolving of our consciousness; it requires that we transcend our current level of thinking that has created the world as we know it, and to evolve our conscious level of cognition to new levels where we may see the complexity of problems and then set about to create solutions that do not perpetuate but rather mediate those problems. Design conversation affords us an opportunity to create an "in-between", to create the social space within which to collectively come together to design new systems. Creating such space is a social act, an act of working beyond existing social structures to create something that did not exist before. This "in-between" space offers a form of freedom that liberates and at the same time creates solidarity, providing for consciousness evolving. This need for freedom is deeply rooted in human beings; no deceptive claim can quite satisfy the longing people have to be self-determining, to be free.

Banathy (2000) is instructive here when considering consciousness evolving,

> In the course of this century we have developed evolutionary consciousness and more recently we grasped the potential of conscious evolution, the potential of giving direction to our own evolution and the evolution of the systems we inhabit, the evolution of our communities and our society by purposeful and deliberate design. (Banathy, 2000. p. 317)

To engage in the conscious evolution of our species will require individuals who have capacities and competencies to conjoin as design communities. Toward that goal, evolving our consciousness beyond the constraints of existing societal limitations will require forms of collective communication that enable us to create design spaces. Evolution in this sense, as Jantsch (1980) explains,

> is the result of self-transcendence at all levels . . . [It] is basically open. It determines its own dynamics and direction. . . . By way of this dynamic interconnectedness, evolution also determines its own *meaning*. (p. 14, italics in original)

Design conversation can play an enormous role in all our lives as we seek to create these evolutionary design communities. In the balance, between humankind's present and future, is the responsibility for evolutionary guidance. Ultimately, an educated citizenry will determine the evolutionary path chosen.

We have reached a time in our increasingly global society where we are charged with bringing into existence a space "in which things are saved from destruction of time" (Arendt, 1958, p. 57). The reality of such a space – an "in-between" – will rely

> On the simultaneous presence of innumerable perspectives and aspects in which the common world presents itself and for which no common

> measurement or denominator can ever be devised. For though the common world is the common meeting ground of all, those who are present have different locations in it. . . . Being seen and heard by others derive their significance from the fact that everybody sees and hears from a different position. . . . Only where things can be seen by many in a variety of aspects without changing their identity, so that those who are gathered around them know they see sameness in utter diversity, can worldly reality truly and reliably appear. (p. 57)

Creating this "in-between" space is the province of systems practitioners and users alike, and it requires conversations that lend to the design of systems that transcend existing systems; it requires new levels of consciousness that transcend those levels of consciousness that created the old systems. It is an epistemological undertaking guided by a social responsibility concerned with alternative possibilities of the future.

6.3 New Order of Creative Surge

What is needed, as Bohm and Peat (1987) note, "is not simply a new creative surge, but a new order of creative surge, one that extends into science, culture, social organization, and consciousness itself" (p. 210).

Banathy (2000) likens the work of those engaged in systems design to the work of the artist, "who—making a portrait—first attempts to capture the overall form (of the sitter) with an initial sketch. Then, within the overall form the image is made more detailed as the work progresses" (p. 147). The aesthetic imperative guides the design process, and, like the artist, those engaged in design are "always working from the generative source of the initial big idea," (p. 147) the center of social creativity that shapes the change in educational consciousness. The aesthetic imperative, "is the generative source that *enfolds* the idea. It is, then, the (enfolded) idea from which the picture is generated. It is by the process of painting that the work *unfolds* into ever more definite form" (Banathy, 2000, p. 147). Likewise, as the design community engages in designing a new system, the enfolded idea is unfolded as the old system is transcended and the new system comes into being.

The creative surge, a change in consciousness – consciousness evolving – expresses *and* affects changes elsewhere. Consequently, one avenue of design inquiry or design learning takes us into the formation of such consciousness. Another takes us into the way a system activity might be changed in order to both affect and respond to the social issues or problems in the intellectual and imaginative life of user-designers. Finally, it is the creative function of design conversation that is necessary to look to the life of society and its people, in the present and for the future. To identify when a human activity system is viable, and when it is not, to see what might be done to sustain viability. And at these points the study of individual and collective needs becomes the debate about what an individual or community shall do evolve society toward the improvement of humankind. As user-designers, each individual has to provide some justification for his or her beliefs and aspirations. At that moment, the analysis of consciousness is the analysis of his or her ideology and imaginative life as these abstractions live concretely in the times (Inglis, 1975, pp. 34-35).

6.4 Participatory Consciousness

Design conversation fosters a participatory consciousness, a "way of being in the world, that is characterized by what Schachtel (1959) calls "allocentric" knowing . . . a way of knowing that is concerned with both "*the totality of the act of interest*" and with the "participation of the total person" (of the knower) (p. 225). It requires an attitude of profound openness and receptivity. It involves, Schachtel (p. 181) states, a temporary eclipse of all the perceiver's egocentric thoughts and strivings, of all preoccupations with self, and self-esteem. One is turned toward other (human or nonhuman) "*without* being in need of it" or wanting to appropriate it to achieve something. The latter would point to preoccupations with self that antedate the experience of "I" as separate from the world and block full perception of other (p. 177).

PART TWO: ORGANIZING THEMES OF COMPENDIUM

7. FIVE ORGANIZING THEMES

The Compendium is organized by five themes. Section I: *Design Conversation* opens with *Chapter 2* by Bela H. Banathy, in which the author examines the history of the conversation movement, a 20-year history. Drawing from his experience of attending systems conferences, and as a pioneer in *Research Conversations*, Banathy guides the reader to an understanding of how certain essential features of the conversation movement evolved within the international systems science community. In *Chapter 3*, Harold G. Nelson explores design conversation, discussing the nature of design and systems. Importantly, Nelson examines service as an interrelationship that defines a design capable system. Contributing the philosophical exploration of dialogue, in *Chapter 4* Patrick M. Jenlink examines the power of dialogue in relation to social systems design, focusing on the meaning of dialogue and its role in overcoming paths of least resistance. Dialogue as social discourse is presented as a means to collectively transcend existing systems.

Section II: *Perspectives of Design Conversation* opens with *Chapter 5* by Bela H. Banathy, introducing the reader to the importance of searching together, examining the approaches, methods, and tools that animate dialogue and design conversation. Sabrina Brahms, in Chapter 6, explores uses of metaphor within the context of social systems design, focusing on the integral use of metaphor in conversation. Karen E. Norum, in *Chapter 7*, introduces the reader to appreciative inquiry as conversation. Norum explains appreciative inquiry as both a framework and process for creating ideal social systems, explicating the central principles of appreciative inquiry.

Gordon Dyer, in *Chapter 8*, examines the concept of rights and responsibilities as pertaining to conversation practice. Dyer is concerned a fundamental principle of systems design, that those affected by a system be involved in its design. To this end, Dyer emphasizes the need to ensure processes and methods of design conversation within systems design dynamics that give users of the system an opportunity to participate. Yoshihide Horiuchi in *Chapter 9* explores the need for transcultural orientations in conversations. Situating his discussion with the International Systems Institute conversations. Horiuchi discusses cultural communication and the need for mindfulness, focusing the reader on the importance of conveying respect and interaction. Larry A. Magliocca and Karen E. Sanders, in *Chapter 10*, conclude this section by examining the critical role of dialogue in emancipatory systems design. Magliocca and Sanders examine dialogue and the human burdens of engaging in collective conversation as well as the burden of solutions. The authors introduce the reader to the CogniScope™ System as a backdrop for their work and exploration of dialogue in systems design.

Section III: *Modalities of Design Conversation*, opens with *Chapter 11* by Patrick M. Jenlink and Bela H. Banathy, wherein the authors examine the relationship of dialogue and design conversation, specifically the importance of dialogue to transcending existing social systems and levels of consciousness. In *Chapter 12*, Alexander Laszlo and Kathia Castro Laszlo examine the notion of *learning conversations* and *design conversation* in social evolution. Concerned with creating new cultures, the authors delve into the importance of creating a vision of a sustainable and evolutionary future. Alexander N. Christakis and Kevin Dye, in *Chapter 13*, examine lessons learned from using the *CogniScope™* in *systems design*. In this chapter Christakis and Dye examine the state of the art in social systems design inquiry, explicating for the reader the basic underpinnings of collaborative design. In *Chapter 14*, Peggy Barnes Gill guides the reader in an exploration of narrative as discourse in systems design. Here Gill contrasts narrative inquiry to more traditional approaches to inquiry, arguing for narrative as key to systems design. Patrick M. Jenlink concludes this section with *Chapter 15*, exploring the mediational role of discourse in systems design, drawing from cultural-historical activity theory to provide a framework for understanding design conversation.

Section IV: *Practical Applications of Design Conversation*, opens with *Chapter 16* by Matthew Shapiro, providing a practical example of design conversation through an exploration of the *North End Agora*. Shapiro explores the powerful societal effect of sustained and constructive conversation within the North End neighborhood of Boise, Idaho. Douglas C. Walton and Halim Dunsky, in *Chapter 17*, argue for conversation as the communication method of choice, situating the reader in the resurged idea of civil society. The authors examine design conversation, providing a historical and theoretical examination. Then Walton and Dunsky provide a more practical exploration of design conversation, guiding the reader through the Asilomar Conversation connected to the International Systems Institute. In *Chapter 18*, Raymond A. Horn Jr. introduces the reader to post-formal conversation as an intervention strategy, incorporating the *Case of Texas Educational Model* as a backdrop. Horn examines the use of conversation in political contexts, drawing attention to the importance of understanding the historical and contextual nature of systems when attempting to affect change, as well as the focusing the reader on the critical importance of recognizing cultural and systemic patterns and the processes that both reify patterns within systems and are necessary to changing systems. *Chapter 19*, by Mario Cayer, concludes this section by providing insight to the experiences of long-term practitioners of dialogue. Cayer examines, in depth, Bohm's dialogue and provides a practical exploration of how dialogue is used to create change within social systems.

In Section V: *Reflections on Creating Future*, Patrick M. Jenlink offers reflective thoughts in *Chapter 20* on the learning journey presented in the Compendium. As well, Jenlink directs the reader to consider the work of creating our shared future through design conversation.

8. REFLECTIONS ON DESIGN CONVERSATION

We are, as a social species, communicative by nature. That we are also sentient beings places within our reach the capacity to be equally creative and destructive, whether through discourse or social action. When we consider, in relation to being a communicative species as well as sentient beings, that we are by nature extremely diverse, we are filled with great potentials, potentials waiting to be unfolded through genuine dialogue. Whether we choose, as sentient beings, to be creative or destructive, is a consequence of our choices in matters of communicative action within our cultures and societies.

The fact that we display great diversity – that there are differences – figures largely into the potential for either creative or destructive actions. Always present is the question of how we shape our actions through conversation, which in turn shape our future. In the end, the realization of our great potentials rests, in no small measure, with our collective capacity for cultural creativity and in our capacity to achieve new levels of consciousness.

REFERENCES

Arendt, H., 1958. *The Human Condition*. University of Chicago Press, Chicago, IL.

Banathy, B.H., 1996. *Designing Social Systems in a Changing World: A Journey Toward a Creating Society*. New York: Plenum Press.

Banathy, B. H., 2000. *Guided Evolution of Society: A Systems View*. New York, NY: Kluwer Academic/Plenum.

Berger, P.L., and Luckman, T., 1966. *The Social Construction of Reality: A Treatise in the Sociology of Knowledge*. Doubleday, New York.

Bohm, D., and Peat, F. D., 1987. *Science, Order, and Creativity*. Bantam Books, New York.

Bohm, D., 1998. On Creativity. (Ed. Lee Nichol) New York: Routledge.

Buber, M., 1970. *I and Thou* (W. Kaufman, trans.). Charles Scribner & Sons, New York.

Burbules, N.C., 1993. *Dialogue in Teaching*. Teacher College Press, New York.

Csikszentmihalyi, M., 1994. Memes versus genes: Notes from the culture wars. In D. H. Feldman, M. Csikszentmihalyi, H. Gardner, *Changing the world: A framework for the study of creativity*. Westport, CN: Praeger, pp. 159-172.

de Mare, P., Piper, R., and Thompson, S., 1991. *Koinonia: From Hate, Through dialogue, to Culture in the Large Group*. Karnac Books, New York.

de Zeeuw, G., 1993. The actor as a perfect citizen. In F. Stowell, D. West, and J. de Zeeuw (eds.), *Systems science*. Plenum, London, pp. 11-18.

Giroux, H.A., and McLaren, P., 1986. Teacher education and the politics of engagement: The case for democratic schooling. *Harvard Educational Review*, **56**: 213-238.

Hollingsworth, S., 1994. Sustained conversation: An alternative approach to the study and process of learning to teach. In S. Hollingsworth (ed.), *Teacher Research & Urban Literacy Education: Lessons & Conversations in a Feminist Key*. Teachers College Press, New York, pp. 3-16.

Inglis, F., 1975. *Ideology and the imagination*. New York: Cambridge University Press.

Isaacs, W.N., 1993. Dialogue, collective thinking, and organizational learning. *Organizational Dynamics*, **22**(2): 24-39.

Isaacs, W.N., 1996. The process and potential of dialogue in social change. *Educational Technology*, **35**(1): 20-30.

Jenlink, P. M., 2001. Activity theory and the design of educational systems: Examining the mediational importance of conversation. *Systems Research and Behavioral Science*, **18**(4): 345-359.

Jenlink, P.M., (ed.), 1995. *Systemic change: Touchstones for the Future School*. IRI/Skylight Training and Publishing, Inc., Palatine, IL.

Jenlink, P.M. (1996, February). *The power of dialogue in systemic change: Discovering our mental pathways*. Paper presented at the American Educational Communication and Technology National Convention, Indianapolis, Indiana.

Jenlink, P.M., 1997, August. *New Directions in Preparation: Creating and Sustaining Professional Learning Communities*. Paper presented at the 51st Annual Conference of the National Council for Professors if Educational Administration, Vail, CO.

Jenlink, P.M., and Carr, A.A., 1996. Conversation as a medium for change in education. *Educational Technology*, **36**(1): 31-38.

Jenlink, P.M., Reigeluth, C.M., Carr, A.A., and Nelson, L.M. (1996). An expedition for change: Facilitating the systemic change process in school districts. *Tech Trends,* **41** (1): 21-30.

Laszlo, E., 1987. *Evolution: A Grand Synthesis*. New Science Library, Boston, MA.

Laszlo, E., 1996. *Evolution: The General Theory*. Hampton Press, Cresskill, NJ.
Lambert, L., 1995. Leading the conversation. In L. Lambert, D. Walker, D.O. Zimmerman, J.E. Cooper, M.D. Lambert, M.E. Gardner, and P.J. Ford-Slack (eds.), *The Constructivist Leader*. Teachers College Press, New York, pp. 83-103.
Noddings, N., 1992. *The Challenge to Care in Schools: An Alternative Approach to Education*. Teachers College Press, New York.
Noddings, N., 1993. Caring: A feminist perspective. In K.A. Strike and P. L. Ternasky (eds.), *Ethics for Professionals in Education: Perspectives for Preparation and Practice*. Teachers College Press, New York, pp. 43-53.
Noddings, N., 1994. Foreword. In L.G. Beck (ed.), *Reclaiming educational Administration as a Caring Profession* Teachers College Press, New York, pp. ix-x.
Noddings, N., 1999. Care, justice, and equity. In M. S. Katz, N. Noddings, and K. A. Strike (eds.), *Justice and Caring: The Search for Common Ground in Education*. Teacher College Press, New York, pp. 7-20.
Prigogine, I., and Stengers, I., 1984. *Order Out of Chaos*. Bantam Books, New York.
Schachtel, E. G., 1959. *Metamorphosis: On the development of affect, perception, attention and memory*. New York: Basic Books.

SECTION I

DESIGN CONVERSATION

Chapter 2

THE CONVERSATION MOVEMENT

BELA H. BANATHY
Saybrook Graduate School and International Systems Institute

1. INTRODUCTION

In the course of the last decade an observer of the mode of discourse in our various societal groups and communities was able to observe the emergence of new approaches to intentional communication. This emergent mode has several designations, such as: Dialogue, consensus-seeking conversation, thinking together, and searching together. It seems to me that these modes of intentional social communication modes are, on the one hand, reactions to the ever degrading and hostile political discourse and, on the other, they represent an increasing yearning for civility, mutual respect, and dignity in our social discourse, and they are preferred modes of searching for common ground.

Engaging in these new modes of discourse, we are also discovering the surprising power of intentional collective communication, as we are searching together for solutions to often, intractable issues and problem situations. Most significantly, however, we find the use of these communication modes most rewarding as we aspire to take charge of our future, and collectively engage the design of systems, in which we live and work.

By now, the societal landscape is rich with a great variety of people, groups, intentional communities, community agencies, volunteer and professional associations learning to use, and using, these new modes of civil discourse. They create not only their own future, but create a wealth of social and intellectual capital. Collectively, I see these emerging groups, communities, organizations, constituting a new societal movement: The Conversation Movement. In this chapter, I introduce a modest but ambitious part of this movement, which started almost 20 years ago and created a number of conversation communities.

2. THE CONVERSATION MOVEMENT OF THE INTERNATIONAL SYSTEMS SCIENCE COMMUNITY

Having attended systems conferences for many years, by the late '70s some of us in the international systems movement began to question the usefulness of the traditional conference format as a forum for knowledge development and application exploration. Searching for a more meaningful and satisfying format, we initiated RESEARCH CONVERSATIONS as a method of choice of collective communication in systems inquiry. A group that engages in conversation is called a CONVERSATION COMMUNITY. Conversation as a mode of disciplined inquiry combines "generative dialogue" and "strategic dialogue." While "generative dialogue" aims at establishing "common ground" of values and beliefs and a "common frame of reference" in an intentional community, "strategic dialogue" involves the community in addressing an issue of shared interest and purpose. (This paper is linked with an earlier, called "Conversation Heuristics.")

Since our first conversation in 1982, we have asked: How can we use the insights gained from systems inquiry for the advancement of the human condition? By now, several conversation communities in several countries have explored this question. These communities jointly constitute the "The Conversation Movement" of the international systems community.

In Part One, I describe the work and organization of the Conversation Communities, the conversation method, phases of a conversation cycle, and the characteristics of a Conversation Community. In Part Two, I provide an example of how a team works, by summarizing the program of one of the Research Teams of the 96 Conversation.

PART ONE: THE WORK AND ORGANIZATION OF THE CONVERSATION COMMUNITIES

In this Part, I (1) provide a brief history of the various conversations of the Conversation Communities over the last two decades. (2) Characterize the conversation method. (3) Describe the phases of a conversation cycle. And (4) Introduce the characteristics of authentic conversation communities.

3. THE CONVERSATION MOVEMENT: A BRIEF HISTORY

The first conversation took place at the Fuschl Lake in Austria in April 1982. A group of systems scholars met in a small hotel at the Fuschl Lake, near Salzburg. Participants came from three continents, representing ten cultures. They were invited as leaders of various systems societies. The conversation was organized by the International Systems Institute. The group spent five days in two conversation teams, addressing the question: How can we apply the insights gained from systems thinking and systems practice to promote human betterment and to improve the human condition? By the end of the conversation, the teams defined eighty items to guide to the work of the various systems societies and become an agenda for the conversations that follow. Following the Fuschl Conversation, a group of us – officers of the International Federation of Systems research (IFSR) – attended the Board Meeting of the Federation, where the Board decided to provide funding for the Fuschl Conversations.

The various conversations that followed the first Fuschl event, have been organized and coordinated by the International Systems Institute, in cooperation with International Federation of Systems Research, and with several member organizations of the Federation. By now we have held thirty conversations; ten Conversations in Fuschl, Austria; eight regional conversations: two in Crete; one each in England, Finland, Greece, Hungary; and three in Spain. Since 1989, we have held twelve international Conversations at the Asilomar Conference Center in California and established the Asilomar Conversation Community (ACC) as a conversation community of the International Systems Institute.

The Asilomar Conversation Community (ACC) has extended the Conversation agenda by focusing on: (a) the creation of knowledge base for social systems design, (b) the development of resources and programs for systems and design learning, (c) conducting research on conversation itself, (d) the design of systems of learning and human development, (e) the design of authentic and healthy communities, (f) the development of design culture, (g) sociocultural evolution, (h) the design of evolutionary guidance systems, (i) the design of evolutionary learning communities, and (j) societal design.

4. THE CONVERSATION METHOD

In view of the fact that most of the conversations have focused on issues related to social systems design, I briefly review here the conversation approach to the design of social systems. A more detailed description of methods was offered under the chapter "Conversation Heuristics."

Design Conversation is a process that carries a stream of shared meaning by a free flow of discourse among members of a design community, which seeks to create a new system or recreate an existing system. In recent literature the notion of "***generative dialogue***" has gained prominence as the most viable form of collective social discourse. Generative dialogue – a precursor of collective design has the power to generate collective meaning, a common ground, and collective consciousness the attainment of which is critical in social systems design. Complementing generative dialogue, "***strategic dialogue***" is aimed at pursuing a specific task, such as the design of a new entity. CONVERSATION combines "generative" and "strategic" dialogues, as the most appropriate modes of social discourse in design inquiry. Based on two decades of experience with design conversation, we agree with Bohm (1987), that a free flow of ideas, beliefs, and meaning among members of groups "may well be the most effective way of investigating the crisis which faces society, and indeed the whole of human nature and consciousness today" (p. 240). Such a disciplined exchange of ideas is a fundamental means of "transforming culture and freeing it of destructive misinformation, so that creativity can be liberated." (p. 240).

In the following paragraphs, I present some of the key ideas of a research team on "Designing Conversations" (1995). The statement was developed at the Fifth Annual International Conversation on Social Systems Design. Operating in an open conversation zone, the group placed in the container of the conversation respect for each other and respect for the common experiences and meaning emerging within the group flow. They viewed conversation as collective learning and inquiry into the assumptions that structure common experiences. Since conversations happen in an open space of evolving relationships, group facilitators temporarily emerge, but none are anointed as leaders. The process moves the group from an unstructured to a structured, task-oriented conversation. (From generative to strategic dialogue.)

The group concluded that attributes of genuine conversation include the recognition of (a) the essential nature of authentic participation; (b) emotional commitment that becomes the group's source of energy towards change; (c) a stance of curiosity; and (d) conversation as a process "to be given into," and "surrendered to." The evolving functions allow the group to become self-renewing, as members pay attention to the unfolding process and engage authentically in it. The group seeks what is important and maintains compassion and respect for conversation partners. There is no attempt to "changing others," no right or wrong, just being witnesses to the ideas of others.

The process described above is the process of creating the container of the conversation. A container can be understood as a receptacle for assumptions, intentions, images, beliefs, and aspirations shared by members of a group. What is contained creates a collective atmosphere or climate (Isaacs, 1993). What is in the container is the "stuff" which is used in the course of the upcoming design process, for the collective creation of the desired future system. The reaching of this state (of readiness to proceed) is required in order to initiate a strategic dialogue, which becomes the communication method of choice in engaging in purposeful design.

5. PHASES OF A CONVERSATION CYCLE

The Conversation program of the International Systems Institute has evolved over the last eighteen years. The current state of practice of the process of conversations is described here. The long range Conversation Program can be seen as a never-ending process of successive yearly cycles. A specific cycle focuses on an evolutionary stage of a selected topic. Often, a conversation cycle builds upon the previous program cycle. Alternatively, a team of the

Conversation Community might start a new topic. A program cycle includes six phases: (a) the definition phase, the selection of a topic, (b) the preparation phase, which includes knowledge base building, (c) the conversation event of five to six days, which is an intensive learning phase, (d) the follow-up phase: a reflection on, and reporting of, the conversation event, (e) the application phase, which is the use of what has been learned in functional contexts and, based on it, (f) the redefinition phase, the extension, confirmation, re-vision, and redesign of the program and the approach. The description that follows is the time scheme of the 96-97 Asilomar conversation.

5.1 Phase One: Definition

The selection/definition of the research topics of the 96-97 program began in the course of the 95 Annual International Conversation on Social Systems Design, held at the Asilomar Conference Center. As an outcome, five research teams decided to continue work on their previous design topic, by extending their knowledge base, and exploring their topic more in depth. Research Team six decided to focus on Societal Design, by exploring the use of the Cogniscope system for societal design. Intensive E-mail conversation among members of the Research Teams between November 95 and March 96 led to a detailed, collectively agreed upon, definition of the various topics. .

5.2 Phase Two: Preparation

This phase started in March 96, as the programs of the teams were announced in the March Newsletter of the Asilomar Conversation Community. The preparation phase included seven tasks: (1) Respond to the March announcement and sign-up for a research topic. (2) Explore the selected topics and develop a "think paper" that: (a) specify the team members' interest and previous work on the topic, (b) state their key ideas about the topic, and (c) identify topic-relevant knowledge sources. (Team members had 3 to 4 months to develop their paper.) (c) Announce the names and addresses of team members, (d) mail think papers to members of the research team by the end of July and continue knowledge development for the next three months, (e) develop collectively a first draft of the "theme," based on a synthesis of "think papers," (f) mail the draft to team members by mid September, asking for comments and modifications, (g) return comments to the coordinator by mid October, and (h) develop second draft and bring it to the Conversation.

5.3 Phase Three: The Conversation Event

Participants were asked to take advantage of the half-year duration of the preparation phase, explore relevant knowledge bases, and come to the Conversation with a rich set of core ideas. (Special note: Over the years, participants indicated that the success of the Intensive Learning Phase depends upon the depth of preparation, on the generation of a substantive knowledge base.)

At the on-set of the Conversation, research teams review the summary of their preparation, develop triggering question(s) that guide their work, define the approach to their conversation, and assign conversation roles to members of the team. At the end of each day, the teams evaluate their progress. The evenings are set aside for self-organized conversations or joint meetings of the conversation groups. A special, inspirational, feature of the day is a silent self-reflective time of walking down to the sand dunes of the Asilomar beach and admiring the beautiful sunset.

The afternoon on the day before the last the teams prepare their summary report as well as their program plan for the next yearly cycle. The report and the plan are presented to the large group the last morning. The same time, each participant writes a "valuation" paper of the Conversation. The last afternoon, the entire group gathers for an idea session, which explores the future of the Conversation Community.

5.4 Phase Four: Follow-up

Three tasks are scheduled for the follow-up phase. (1) Research Teams develop the comprehensive report. This report is a collective product of the teams and will have a limited publication for distribution among participants. (2) Participants submit their final valuation/recommendations. (3) Participants submit their final individual papers, based on their work on their topic and their findings of the conversation. The papers were published in the Proceedings of the Conversation. Selected papers were slated as edited issues in systems journals.

5.5 Phase Five: Application

A basic premise of the Conversation program is that without application in functional, back-home contexts, learning is incomplete. In fact, learning gets its meaning from its application. Application is informed by what has been learned. Participants of the 96 Conversation were expected to apply what they have learned from their Conversation experience in functional context of importance to them. They gave an account of their application experience in the course of the reporting period as well as in their individual paper.

5.6 Phase Six: Re-Vision

In most cases, the Conversation program cycle of a specific year merges seamlessly into the next cycle. In this case, participants continue "virtual conversation" through E-mail exchanges, in the course of which they collectively confirm, re-confirm, modify, extend, re-vision, redesign, and eventually define the program of their next year cycle. In some cases, a new topic emerges, which initiates a new research cycle.

6. CHARACTERISTICS OF AN AUTHENTIC CONVERSATION COMMUNITY

This characterization of an authentic conversation community was developed collectively by the Asilomar Conversation Community (ACC):

> We created a Conversation Community, which has developed a deep commitment to a shared vision and purpose.
> We communicate with each other openly and honestly.
> We feel that we can make a difference in the world.
> We organize ourselves with a total absence of hierarchy.
> We are equal partners in stewardship and mutual assistance. As stewards:
> We take collective responsibility for the work and service of our community.
> We practice genuine and authentic participation.
> We make full use of our individual and collective knowledge, abilities, competence, and creativity. We take responsibility for the continuing development of these qualities.
> We apply what we learn in the systems in which we live and work and offer it in the service of the common good.
> We rely on, honor, trust, and support each other and live by a collectively defined code of ethics.

The characteristics described here have served and continue to serve our Conversation Communities as an ideal state we aspire to attain. There always is a "creative tension" space between where we are and where we want to be which becomes a design space in which we create our collective future and are always on the journey of becoming.

7. SUMMARY OF PART ONE

In Part One, entitled "The Work and the Organization of the Conversation Communities," I described: (a) the emergence and the history of the Conversation Movement, (b) reviewed briefly the conversation method, (c) and the various phases of a conversation cycle, and (d) set forth the desired characteristics of a Conversation Community. In developing this part my intent was not to provide a prescription to others follow, rather, I wish to offer an example of a journey of an intentional community which aims to serve the common good, and which is inspired by the idea of applying the insights gained from the systems, design, and evolutionary sciences to the improvement of the human condition.

PART TWO: AN EXAMPLE OF THE WORK OF A RESEARCH TEAM: THE PROGRAM OF THE SOCIETAL DESIGN TEAM OF THE 96 ASILOMAR CONVERSATION

In Part Two, I (a) describe the genesis and the definition of the program of the Societal Design Team (SDT), (b) specify the conditions of participation, (c) give an account of the preparation phase, and (d) present a brief report on the approach and outcome of the work of the SDT. I served the team as the coordinator of the Preparation Phase.

8. THE GENESIS AND DEFINITION OF THE LONG-RANGE PROGRAM

The long-range purpose of the program is to:

- Create an extensive knowledge base that enables us to engage in societal design.
- Formulate – based on explicitly stated shared core values and ideas, and ethical perspectives – an Image of an ***Ideal Society***; a society which will create a designing democracy in which people have the right and assume the responsibility to design their systems and take part in making decisions that affect their individual and collective future.
- Select/develop design approaches/methods/tools that are appropriate and useful to transform the Ideal Societal Image into a normative description, a systems model of the envisioned society.
- Select/develop design learning programs that enable people to engage in the design of their lives, their systems, their communities and their society; and take part directly in making decisions that affect their lives.
- Introduce pilot programs in a variety of contexts and describe these programs in case studies.
- Formulate and distribute a comprehensive statement about findings and disseminate it to serve as a resource for continuing research and development.

The 1996 conversation event was viewed as the first cycle of the long-range program and it aimed to address the first three tasks described above.

9. CONDITIONS OF PARTICIPATION

Three conditions must be met by participants:

Participation at an earlier Conversation and an understanding of what social systems design is as a disciplined inquiry, how it works, and why we need it.

Acceptance of a set of underlying assumptions, such as: (1) People in a society have the right and responsibility to shape their own future and take part in making decisions that affect their lives. (2) In social systems design people who live in the system are the experts. Nobody has the right to design social systems for others. (3) Once we have acquired evolutionary consciousness, became aware of the possibility of conscious evolution, and developed evolutionary competence; we realize that we can give direction to our evolution, the evolution of systems in which we live, the evolution of our communities and the larger society by purposeful design.

Commitment to engage in the following tasks: (1) Work with several proposed knowledge sources, e.g., a set of designated books, chapters in books and other relevant sources selected by each participant. (2) Develop an "input paper," based on the knowledge base that explores the topic of the conversation. Share the paper with other participants. (Time for this work was about five months.) (3) Based on insights gained from the paper of others and continuing work with knowledge sources, develop another research paper.

10. THE PREPARATION PHASE

The research program of the SDT was conceived as the first yearly cycle of several inquiry cycles that implement the long-range effort. The program proceeded by a Preparation Phase aimed at creating a knowledge base for the intensive learning and design phase. The preparation phase was guided by two sets of triggering questions:

> "What are Markers of an "Ideal Society," a "Good Society," an "Ethical Society," a "Democratic Culture," a "Cultural Democracy," and a "Self-creating Society?"
>
> "What Design Approaches/Models/Methodologies/Methods/Tools would be adequate to create a New Societal Way of Life?" (Note that moving from the design of a specific social system to the design of a society implies a quantum leap in design.)

Initiating the ***Preparation Phase***, four tasks were defined:

First, we must learn to become informed and enlightened about notions of an ideal society, a civic community, a creating and authentic community, and a democratic society. Once we have created an image of such a community/society, we share our findings with other members of the team, so that we can formulate a collective knowledge base, that help us to become informed participants of the research program.

Secondly, we have to learn about approaches and methods by which to engaging in social systems design. Then, we should share our understanding of the approaches and methods of systems design with others in the research team. Then, we can formulate collectively a design inquiry approach.

The ***third*** *task* seamlessly blends into the first two tasks. It is the creation phase that emerges from the learning phase of the two tasks described above. It emerges as we begin to create ideas that will become the "stuff" of design for the intensive phase. During the intensive (design) phase we shall merge our individually created "stuff" as we create collective solution images.

There is a ***fourth,*** most important task. We shall do our best to get to know each other, learn from each other, understand each other, respect each other, and become friends.

In the course of the preparation phase we had an intensive exchange of E-mail messages. These messages shared ideas among members, clarified tasks and procedures, as well as suggested resources to use in order to build an adequate knowledge base for us. An example of guiding the research team is a message, which I sent out early in the Preparation Phase:

> DEAR SDT FRIENDS: In response to an inquiry, I suggest that the operational base of our conversation and our design work is *not* an existing (operating) system but an emerging, collectively created knowledge system, a conceptual system. This system empowers us to become viable and informed stakeholders in designing an ideal image of a desired society. Such a system is to be based on a clearly articulated *public philosophy.* This philosophy will emerge as we engage in generative dialogue and build such philosophy as our common ground as a shared wisdom.
>
> To coordinate our knowledge building, I attach a reading list witch contains some books you have already worked with at earlier conversations and some new knowledge sources that are critical to our task. Working with these sources and some others – that you will select – you are asked to identify core ideas about an ideal society. We shall share these ideas with each other. By the end of August, we enter into the second phase of preparation and will set out for us collectively agreed upon tasks we want to complete before the November meeting.

11. THE INTENSIVE LEARNING AND DESIGN PHASE

This phase is described in the following paragraphs by (a) defining the methodology selected for the intensive phase, (b) describing the application of the methodology, and (c) reporting on the outcome of the conversation.

11.1 The Methodology of the Intensive Phase

Alexander Christakis, President and CEO of Christakis/White Associates and a member of the Board of Trustees of the International Systems Institute made an offer for the use of the CogniScope System. The CogniScope System is a groupware supported design approach involving stakeholders of an organization or a community, working in a specially arranged design facility.

11.2 The Application of the CogniScope System

The shared assumption of members of the SDT was that they represent a small group of stakeholders of the society. This assumption legitimized their involvement in the research project. The application of he CogniScope System enabled the SDT to engage in an open and focused conversation to (a) collaboratively generate a vision of an ideal society and (b) attain consensus in the design of a Design Inquiry Program for accomplishing the ideal vision.

In the course of the Conversation, the SDT generated – and clarified the meaning of – a large number of ideas about the ideal societal image and organized them into "idea pattern." These patterns integrated the diversity of perceptions on the ideal societal image and on the design of the design inquiry.

The next task involved the generation of the markers that represented the ideal image of a "Good Society." Once generated, the markers were organized into systems of markers by exploring relationships of affinity and influence patterns among them.

A similar process was followed that generated and clarified the markers of the design inquiry that empower the stakeholders to design the ideal society. The markers were organized into and option field for choice making by members of the SDT.

11.3 The Outcome: Creating the Markers

The SDT generated 97 markers of an Ideal Societal Image. The markers were organized into seven system/clusters. A continuing process of exploring "significance" reduced the number to 43 "high significance" markers. The third process explored the "enhancement" patterns by asking: How does the attainment of a particular marker enhance the possibility of achieving other markers in the set? The exploration of enhancement identified 18 markers for defining the ideal societal image. These are introduced next:

- Constant creation of a living space for creativity
- Necessity of society to be anchored in spirituality
- The ideal society should create mediating institutions, which are value generating and enable the authentic participation of its members in decisions that affect their lives filled with love

- Absence of material, intellectual, and affective poverty
- Adequate attitude and aptitude to transform contradictory oppositions into polar oppositions
- Equal opportunity and access to learning experiences that create balance in body/mind/spirit
- Understanding, tolerating, and valuing multiple ways of knowing, being, and doing in the world
- Increase quality and quantity of human communication: Personal: (reflective thinking and reflective feeling) and interpersonal
- Should be in continuous community making, sharing in self-governance, and deliberating with fellow citizens about the common good
- Necessity of spirituality anchored in morality
- Should have an ongoing conversation to articulate its moral standards and shared ethical systems
- Actions are taken in consideration of their impacts on many future generations
- Integrate cultural democracy and democratic culture based on clearly articulated public philosophy
- Develop socioeconomic arrangements in the service of the common good and recognize the significance and contributions of "social economy."
- Integrate work and love by educating for the love of work, and work for the realization of the beloved
- Foster co-evolution between the individual, culture, society, and the Ecosystem
- Promote science and technology in the service of a civil society and human betterment

What should be noted is that the 18 markers cannot show the richness of the original set of 87 seven markers, not even the 43 markers that were selected for significance. It is the full report, identified in the reference section that portrays both the process and the product of the work of the SDT. Furthermore, another SDT team would come up with another set of markers, possibly with significant overlap with the markers produced at this conversation event. The great importance of the kind of program described here is that it provides opportunities for all kinds of groups to approach the design of their systems and envision and design their future by the application of a method which is open and disciplined.

11.4 The Outcome: Creating Markers for Designing

During the last day of the conversation the remaining task was to create a set of markers for the design of a design inquiry, which has the quality and power to be applied to the design of an ideal societal image. Time constraint limited the completion of this task. We were able to complete only the generation of – what we called – the "universe" of markers. 87 markers were generated. There was no time left to explore select for "high significance," and for "enhancement." Therefore, I introduce here a set of markers from the large set that is somewhat arbitrarily selected and it is in random order:

- Build design readiness in all contexts
- Foster the use of creative conversation in all social contexts
- Honor the uniqueness of each design situation
- Dream impossible dreams
- Create systems design approaches, methods, and tools that people can use
- Develop vehicles for authentic participation by all
- Create authentic points of entry to the design process
- Generate user-friendly design language
- Create experiential contexts in which people can learn the specific psycho-social skills needed to participate in collective design
- Use conversation to articulate shared values to be brought life as bases in making design decisions
- Create mediating institutions that enable participative design
- Build rich and open knowledge base for design
- Create rich resource base and support for design
- Establish moral and ethical bases for the design inquiry
- Develop methods that enable user-designers to synthesize design ideas
- Integrate a sense of – and manifestation of – beauty in the design solution
- Establish methods for the "valuation" of design alternatives
- Engage the designing community in design learning
- (The whole set of 87 markers is described in the referenced report.)

11.5 Some Reflections

The 12 participants represented a great variety of background and perspectives, a variation in experiences with design inquiry. They also represented four different cultures. But they had a genuine belief of the significance of the project and maintained a strong commitment to it. In four days they have accomplished the dual task of the research effort in a remarkable fashion.

Attending to a task, such as designing an ideal societal image is continuously trying to cope with the complexity of the task and the complexity intrinsic to the design situation. The other "coping" task is for participants to come out from – and suspend – their "self-referentiality" and engage with others with an open mind in a genuine collective conversation. We have ample evidence that the members of the SDT were able to transcend their self-referentiality and embed themselves in a collective effort of learning from – and creating with – each other.

12. SUMMARY

My task has been to introduce the "***Conversation Movement*** " of the international systems community. Following the description of the "genesis" and the brief history of the movement, *i*n Part One, I described the work and organization of the various ***Conversation Communities***, the conversation method, phases of a conversation cycle, and the characteristics of a Conversation Community. In Part Two, I provided an example of the work of a specific conversation team, the program of one of the Research Teams of the 96 Asilomar Conversation. My hope is that this contribution to the compendium inspires groups and communities to engage in "searching together" on their journey toward their evolutionary future.

REFERENCES

Bohm, D. and D. Peat., 1987, *Science, Order and Creativity*. Bantam Books, NY.
Christakis, A., D. Conaway, D. Post., 1996, *Defining the Ideal Societal Image and Designing the Design Inquiry for Designing It*. CWA Project SDT-1198-1, CWA, Paoli, PA.

Chapter 3

DESIGN COMMUNICATION

Systems, Service, Conspiracy, and Leadership

HAROLD G. NELSON
Advanced Design Institute, Seattle, WA, USA

1. INTRODUCTION

Description and explanation do not prescribe what actions ought to be taken in any design situation, what solutions are best for any perceived design problem or what creative insight should be innovated. The most careful scientist using the most accurate instruments, calibrated to the closest tolerances cannot observe what, by definition, proceeds from human imagination as an outcome of intentionality and purpose (*telos*). The reasoning and logic of accurate description and explanation are not the same as the logic and reasoning used to determine what is desired to be in existence that is not already found in existence. The rules and principles of observation and description cannot transcend their own context and become an epistemological link to other frames of reference and designs of inquiry that may have their own rational structure or internal logic.

Prediction and control do not justify using any means towards any ends for the same reason. The deontic term *ought* is not equivalent to the instrumental term *can*. If something ought to be done in a certain way for a certain outcome it is not justifiable only because it can be done. For example, technology in the Western world often falls into the trap of assuming that something ought to be done because it can be done. The assumed link is typically lifted from a frame of reference of an economy where money as the measure of value and economic return on investment stands in for any deeper ontological aspiration. If not economic, the link is aesthetic; it is done because it is pleasing to make something that is "cool" in the vernacular of the high tech world.

2. DESIGN

The fruits of objectivity or rational discernment and technologic expertise, even communicated from a systems perspective, are necessary but insufficient for determining intentionality and purpose in a designed response to the human condition. Despite this the primary strategy for preparing for change and action is to engage in studies and research resulting in comprehensive reports. These studies and research programs, which consume an immense amount of time and money, cannot provide the necessary link to action, change or design. They cannot point to what should or ought to happen, thus the reports gather dust while requests for more research and studies are made in the hope that more information can somehow be transformed into wisdom.

The natural link to intentional action based on firm ontological understanding is design. Design is a comparatively underdeveloped tradition of inquiry and action but is possibly the first form of human inquiry. Humans have designed their cosmologies from the beginning. Design inquiry and action is not limited to the common contemporary definition of design as a process producing physical artifacts or structures with unique appearances and specific utility. Rather, design is an approach to human agency in a complex world and is inclusive of many fundamentals based on specific foundational ideas. One of the foundational ideas, underpinning design as a comprehensive human capacity, is that of systems thinking.

3. SYSTEMS

The term *system*, used both as a description of an embodied way of thinking (as in systems thinking, or the systems approach) and as a description of the thing which is being thought about, is like the term design; both a verb and a noun. As a noun, the Greek origin of the term system is *sustema*, meaning 'a composite whole', while the verb is a derivative of the compound term *sunistanai* which means 'to bring together' (*sun* – 'together' + *histanai* – 'to cause to stand'). Thus a systems thinking approach reflects a desire to know how things are *caused to stand together* as a composition or whole and how to be an agent in that process. Design is a *process of creative thinking and innovative action.* Systems thinking and design action is thus about how people and things are *caused to stand together through an intentional process of creativity and innovation.* The animator and medium of such a causal chain is *design communication.*

Systems scientists – those who describe and explain the interrelationships of things (concrete and abstract) and people in the world and the compound qualities which emerge as a consequence (i.e. systems and compositions) – need to be in a deliberate instrumental relationship with those who reside in social systems. The role of change agent is operationalized and justified by the facilitation of design communication with those who populate the social systems so carefully researched and documented by systems scientists. The fruits of objective and rational discernment, even from a systems perspective, are necessary but insufficient for determining intentionality and purpose in designed responses to the human condition. There is a need for collaborative design communication among those who work to change systems by design, systems designers, and those inhabitants of the systems that are being changed; the clients. Stakeholders, those affected by any change who are not clients, and stockholders, those invested in the outcome of any change but not directly served by the change, must be included as well. Taken together, these roles and relationships define a design capable system.

4. SERVICE

The interrelationship that binds, animates, and defines a design capable system is *service*. Service is a contractual relationship where purpose and intention is blended with instrumental skill and judgement (Nelson & Stolterman, 2000). Service is a word that has many meanings in different contexts. It has a sense of ennoblement at the same time that it has negative connotations for the reasons given by James Hillman (1995):

> Service offends deep strata of human dignity. We may all want service, but who wants to give it? For service still means menial service (not banking, brokering, telephoning, teaching, installing, diagnosing or writing). The first trouble lies in the word, which invites in it cousins—serf, servile, servant, servitude, servility, all descendants from the common Latin ancestor, *servus*, slave. Service, as it is defined in our culture, is hardly empowering, or empowering only to those persons who can command service and the system for which we slave. (pp. 66-67)

In education one studies the liberal arts, not the servile arts, a representative and enduring cleaving of mind from body and spirit from matter. It is representative as well of the aversion and fear of submissive relationships of control, in contrast to control over one's own self interests. Service is perceived as putting ones self in an inferior role at the beck and call of demands issuing from above or below depending on your station point.

These service relationships are without much appeal to anyone except martyrs or those who willingly enjoy sacrificing their own self-interest for the benefit of others. However, service can be seen in a more positive and more appealing light, as there are other systemic, service relationships that do not require self-sacrifice or martyrdom.

Service can be defined as a self-referential, systemic relationship as in *self-serving*. In the search for truth (scientific, artistic or religious) one serves ones own purposes i.e. artists express their own feelings and emotions while scientists follow their own curiosity and passion and *believers* search for a true god(s) and metaphysical invariance.

From a design perspective, service is defined as *other-serving*. Design service is the quality of empathy, embodied in design communication, which is mutual rather than unilateral. Service, from a design perspective, is very different from the kind of empathic relationship employed in *helping* or *fixing* as explained by Rachel Remen:

> Serving is different from helping. Helping is based on inequality; it is not a relationship between equals. . . . Service is a relationship between equals. . . . Helping incurs debt. When you help someone they owe you one. But serving, like healing is mutual. There is no debt. (1996, pp. 24-25)

Design service is defined by the contractual (formal or informal) relationships of mutual and diverse benefit. In a relationship where there is an exchange of value of equivalencies there is no inequity, inferiority, domination, obligation, or unilateral control. Design, as service, is dependent on the presence of an authentically empathic communication system.

5. DESIGN AS CONSPIRACY – *A BREATHING TOGETHER*

Design is best operationalized through the interaction of a rich compound of individuals, a social system, composed of diverse interests and roles that are closely interrelated through collaboration, motivated by design service so as to create a *conspiracy* (i.e. *a breathing together*) of intent. This compound is not a blended mix but is a mediated complex of differences and diversity. Design is often portrayed as an act of the creative individual. Or on occasion, design is seen as the actions of unique individuals – working together on a project that demands creativity and innovation (in the same way that jazz is an expression of individual musical skills of excellence and intellectual capacity within an ensemble of similarly diverse individuals). This conspiracy in design integrates the gift of individual creative insight with the collective capacity for service through collaboration and inclusion. Both individual creativity and collaborative strength are satisfied in the process.

Design conspiracy is dependent on motivation that arises from human desires (*desiderata*). Design however is too commonly defined as problem solving or, at best, creative problem solving. There is a problem with problems however. Problems that are easily defined are trivial while those of consequence are impossible to define or in the words of Horst Rittlel it is the difference between *tame and wicked problems* (Rittel 1972). Unfortunately nearly every action taken by any group is based on justifications spelled out in *problem statements* that assume wicked problems or tame problems are the *raison dêtre* for agency.

Design's potential is sub-optimized when the trigger for action comes from within a problem focus. Delineation of design problems are derived from the perceived needs of clients or from the imagined needs of customers, end-users or consumers by surrogates. Needs are responses to that which is missing, broken, hated, feared, hurtful, craved or frightening. When need and the concomitant negative impulses of problem solving replace desire, communication degrades from conspiracy to collusion.

Design activity is further weakened by the separation of clients, end users, stake holders, decision makers, customers, producers and designers into isolated categories. Communication is reduced to keeping one another informed, but at arms length, through formal or legalistic exchanges of information. However when design action is characterized as the activity of a conspiring social system, unified through intention, focused on service, animated by a process of creativity and innovation, a different conceptualization of design communication emerges. Design communication is, by necessity, multifaceted and supportive of imagination (creativity), empathy (service) and production (innovation).

6. DESIGN VOICE AND DESIGN LISTENING

The *design voice* is an essential element in design communication. Clients, surrogate clients, end users, or customers are served by enabling their *voiced* desires and needs to be expressed at whatever level of clarity they have the capacity for. Designers need to have a *voice* in order to provide access to the imagined realm of possibilities within their minds by others. Those who transform design concepts into reality need to have a voice that conveys what is being heard as *the world speaks back*, requesting modified actions in response to imposed intentions.

Design voice does not consist simply of spoken or written words, sentences and texts. Conversation and the more formal process of dialogue are essential to good design communication but when the desired possible is too large and complex for words there is a need for a means of communication that go beyond the imposed silence on words or *logos*. The design conspiracy is more than whisperings and shared confidences. It is an exchange of hopes, desires, and possibilities that flow among collaborating minds, hearts and souls; a *diathenic graphologue*.

Diathenic graphologue (*Grk; letting something be seen through its image*) is a means of communication that is transcendent of logos. It is not limited to drawings, pictures or graphics but is inclusive of any means appropriate for communicating the essence, nature or quality of something as a whole. A representation of a whole through process, flow, form, shape, space, compound, pattern or a composition, whether imagined or encountered in the experienced world. Diathenic graphologue utilizes every type of sound, sensation and visual signal accessible to human cognition. Images, symbols, signs, and other representation systems of the imagination, carry the greater share of the burden in design communication in distinction to the role words alone can play.

Having voice is of little consequence if nothing is being heard however. A great deal of attention is given to making sure everyone has a voice in participatory situations, as found ideally in design projects and other forms of democracy. Energy and time is put into *claiming one's voice* so that one is in *good voice* when it counts. The same attention has not been focused on behalf of *listening*; claiming responsibility for it and becoming good at it. Design listening involves the ability to hear what has not been said as much as it involves the ability to hear what has been clearly stated. It is about hearing what is pressing for expression as much as for what has been expressed.

Design listening is not just about hearing and registering utterances in the manner of a mere facilitator, but is inclusive of an embodied empathic understanding of that which has been encoded in the voices of those to be served. Design empathy transcends mere sympathy and other stand-ins for true empathy through its synthesis of voice and listening. A synthesis that is essential in design service.

7. DESIGN LEADERSHIP – COMMUNICATING DESIGN DIRECTION

Good leaders are designers. Leadership, as a designed outcome of design service, is a composition of interactions among human beings who are intentional in their approaches to change through design communication. This composition is a rich mix of systemic relationships in alignment with guiding purposes. This makes it distinct from the types of leadership that are vision driven, where communication is focused on transmitting vision to the non-visionary.

Leadership, as a relationship-based design attribute, is founded on both systems thinking and design action. Design leadership is distinguished as the emergent quality of people in effective design communication rather than in roles of authority. Where one stands relative to other people in a leadership relationship is important both abstractly and concretely in relationship to communication. When people stand or sit across from one another conceptually, as in positional seating arrangements of typical organisational meetings for instance, there is an impulse to communicate through confrontation rather than collaboration. This may be appropriate when the intended outcome of the meeting is a solution to a perceived problematic situation but it is not an appropriate strategy, abstractly or concretely, when the desired outcome is dependant on creativity and innovation as is the case in design.

Standing or sitting in front of someone blocks their view and yours as well. Individuals standing next to one another scanning the same field of vision for opportunity, looking in the same direction at a horizon of possibilities form design teams. Design communication takes on a different quality when the dialogue is with others looking into the same unknown rather than in a face-to-face passing of abstracted data and information.

Design leadership can be expressed through the animation of an arrow of time configured as a process of design communication made up of elements related to each other, in distinctive temporal phases (see Fig. 1). The outcome of the process is the revelation of a unique *parti*, a seminal, meaning-making image pulled from out of the immense *terra incognito* of human purpose that is thrust back into the *mysterium tremendum* of human experience as a small contribution to shared wisdom through created meaning. When being developed into a design concept, the parti is transformed along two dimensions of understanding. One is a vector whose aim is towards *telos* (human aspiration in the case of design) via purposes and ends. The other is a vector revealing the *heart and soul* of the design concept via image and vision. Vision, in this case, is entirely different from the way the term is commonly used in that it is the cause for leadership and not the consequence of leadership.

The elements brought into relationship by the dynamics of the design process, an arrow of time, include desiderata, motivation, intention, appreciation and alignment. Although *alignment* is the seminal emergent quality of design communication, the other elements are essential to the integrity of the design process as well.

Desiderata focus on desires rather than needs which is a way to reinforce the avoidance of design as a problem-focused concept. The ability to express desires as triggers for change are atrophied or under developed in most of our collective endeavours including government, business, volunteer civic and social organisations. *Motivation* focuses on making explicit the reasons that we engage in design from a positive impulse for change. *Intention* refers to the ultimate aim of the process that is an a priori agreement (designing in this case). *Discernment or appreciative judgement* is the context of any design project made explicit. It is a determination of what is to be treated as background and what is to be dealt with as foreground (Vickers 1995). It is a judgement of what is to remain unchanged in the process of designed change. It is also the crucible that contains or holds the creative energy of intense design activity and determines the nature of what is possible.

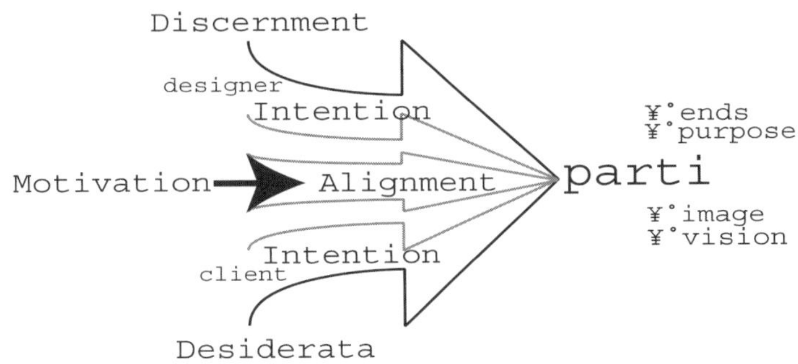

Figure 1. Design Process

8. DESIGN COMMUNICATION

As mentioned above, *alignment* is the emergent quality arising from the integration of design leadership with design communication. It is similar in purpose to the concept of cybernetics in systems science. Alignment is the result of the successful cyclic transitions of design communication through three different modalities of communication including; conversation, dialogue, and diathenic graphologue plus the closure-seeking transition activity of implementation (see Fig. 2).

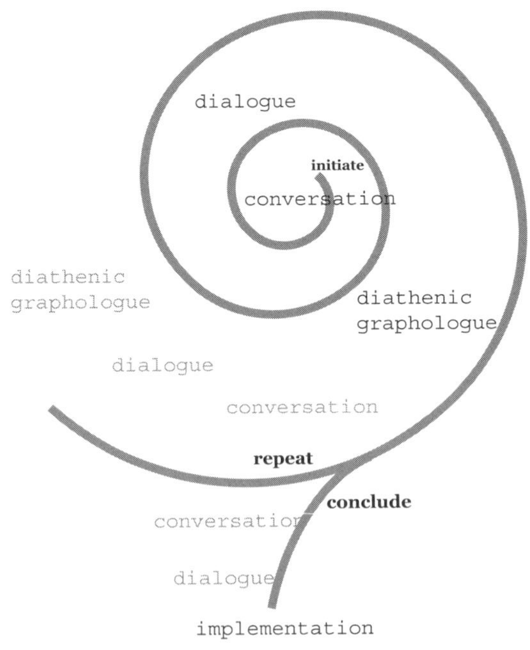

Figure 2. Phases of Design Communication

The three major phases of design communication, design conversation, design dialogue and diathenic graphologue, can be iterated as many times as considered adequate or brought to conclusion when deemed appropriate to initiate the innovation phase. Each of the three phases consists of an open beginning, ending in a commitment to particular outcomes. The design communication process and the attendant sub-processes are both cyclic and sequential emergent (see Fig. 3).

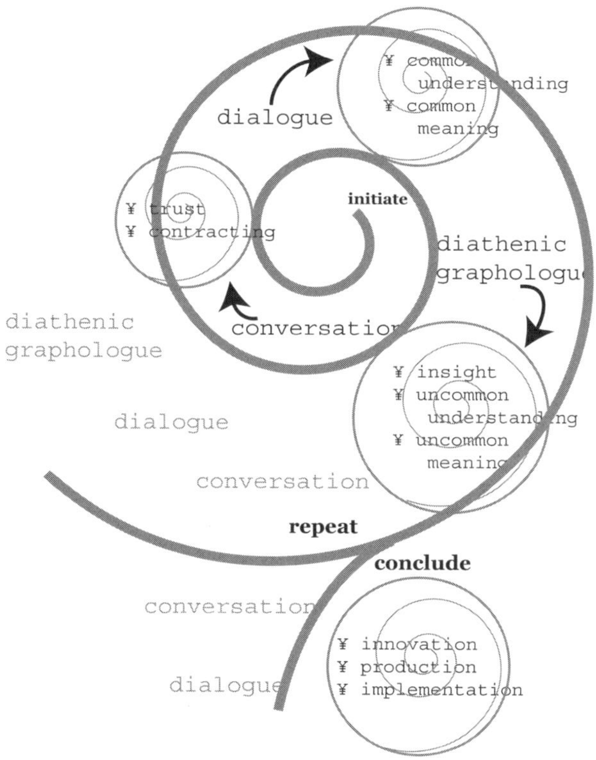

Figure 3. Design Communication as Cyclic and Sequential Emergent Processes

Design conversation (*a turning together*) serves the function of establishing trust as the necessary condition for the essential process of contracting (formally or informally) for design services. The design contract defines a common intention or aim for the subsequent communication stages. Formal dialogue (*meaning through words*) is a process that establishes common understanding for those participating in the design process (Senge 1990). With the establishment of common understanding or common ground, there is no expectation that it is necessarily a representation of a set of truths. It just means that there is agreement among everyone involved on the meaning of those things that have been the focus of their discussion. Common understanding gained through dialogue can come to the collective in successive measured steps or in a sudden bifurcating insight. In the latter case, each individual in the dialogue reaches the same shared understanding at the same time as everyone else.

Diathenic graphologue is the means by which *uncommon* understanding is reached among a collective and diverse group of individuals who have successfully established themselves upon some common ground sharing a common intention. The uncommon is represented through images presented in a rich variety of systems of communication. Cognitive art is the more common and well developed (Tufte 1990) but is representative of just a small part of what has been used traditionally by cultures and communities of people faced with new or novel possibilities in life. Communication in this mode is successful when the uncommon has been transformed into the common, through a renewed contracting process and a reengagement in dialogue at a different level of intention.

Innovation, as a creative form of implementation, is the act of bringing the imagined into the real world and making it a part of the larger experienced reality of others. It is the final stage in making the uncommon common when the design outcome is considered adequate and appropriate to the original contracted intention.

Even after innovation has been successfully carried out, design communication continues. The artifacts, the organizations, the tools the environments, all the things we create, begin to speak back and through their own voices affect more changes in our lives. They begin to redesign who we are, what we believe to be reality, what has meaning or value and what becomes the historical record of our lives.

9. CONCLUSION

Design communication is a complex multifaceted interchange among a diverse and complex collection of people in a variety of design roles who are aligned by a common intention, standing on common ground, reaching uncommon understandings together and making the uncommon a common reality through practical design action. Design communication enables possibilities to become realities in everyday life. It allows the collective to benefit from the creative gifts of individuals and it allows individuals to find commonality of purpose and intention with the collective. It allows the imagination to take form in real life. Good design communication not only enables the meeting of minds but the inclusion of hearts as well. It is the armature of a new form of leadership that is based on empathic service through creativity and innovation.

Design communication is part of the rich complexity of a tradition of creating that, which does not yet exist from out of the desires of people expressing their potential as humans more fully. Design is distinct from the other traditions of inquiry and action such as science, art or technology. Although it is common in everyday life it remains under developed professionally and academically. Although it is a very old tradition it is being newly rediscovered and made visible in contemporary life (Banathy 1996). Design capability and competence must become an integrated aspect of our institutions and organisations. Social systems that have the immense power and resources to bring dramatic change to our lives, but often without good reason and with undetermined value. Design, as an option, is concerned with evoking good reason and producing things of value.

REFERENCES

Banathy, B. H. (1996). *Designing Social Systems in a Changing World*, Plenum Press, New York, NY.

Hillman, J. (1995). *Kinds of Power; a Guide to its Intelligent Uses*, New Currency Doubleday, York, NY.

Nelson, H. G. and E. Stolterman (2000). Design as Being in Service. *Foundations for the Future; Doctoral Education in Design*. D. Durling and K. Friedman. Staffordshire University Press, Staffordshire, Great Britian, pp. 23-33.

Remen, R. N. (1996). "In the Service of Life." *Noetic Science Review*(summer).

Rittel, H. (1972). "On the Planning Crisis: Systems Analysis of the 'First and Second Generations'." *Bedrifts Okonomen* (Norway), **8**: 390-396.

Senge, P. M. (1990). *The Fifth Discipline*. New York, NY, Doubleday.

Tufte, E. R. (1990). *Envisioning Information*. Graphics Press, Cheshire, CN.

Vickers, S. G. (1995). *The Art of Judgment; A Study of Policy Making*. SAGE Publications, Thousand Oaks, CA.

Chapter 4

THE POWER OF DIALOGUE IN SOCIAL SYSTEMS[1]

PATRICK M. JENLINK
Stephen F. Austin State University and International Systems Institute

1. INTRODUCTION

The primary challenge to global society, as we pass the threshold of a new millennium, is to transform existing social systems, and therefore American society, in such ways as to embody a more inclusive, democratic, and open-ended communicative spirit. Necessary to this transformation is a social discourse that enables the power and potential of global citizens to be realized. Dialogue is such a discourse that enables citizens in unconcealing societal patterns and structures, which guide and direct individual and collective interactions within and across events and activities. Engaging in dialogue, as a social discourse of creative possibility, with others in our daily lives can contribute to change in our selves and society. This type of systems change requires that we unconceal the powerful, fragmented, and constraining forces that drive individuals and communities to continue using old systems of thought and beliefs. Unconcealing these systems of thought and belief that inform and otherwise reify existing social structures and cultures patterns is necessary to transforming society. Dialogue is transformative, and is understood as a form of conversation or discourse that creates a context in which individuals and collectives are

> . . . led beyond the intense and particular feelings accompanying our own deeply held values, and beyond the particular beliefs to which these feelings are attached, to a realization that the other—who feels intensely about that which I do not believe—is still one to be received . . . [Such] dialogue . . . is vital in every aspect of education. (Noddings, 1984, p. 186)

Rollo May (1969), in his work to understand human interaction, has furthered explained dialogue as a discourse that

> implies that [men and women] exists in relationship. The fact that dialogue is possible at all—that it is possible, in favorable circumstances, for us to understand each other, stand where the other is standing—is, in itself, a remarkable point. Communication presupposes community, which, in turn, means a communion between the consciousness of the persons in the community. This is a meaningful interchange, which is… a built-in aspect of the structure of human intercourse. (p. 155)

An important insight is presented in this understanding of dialogue. Communication, community, communion, and consciousness share a commonality in that each is based on relationships. Included in this commonality are relationships with self, with others, and with community and society. Human interaction, through conversation, is premised on the availability and capacity of individuals to engage in social discourse. Dialogue, as a form of social discourse, enables human interactions that contribute to building and sustaining communities as well as enables the creation of shared consciousness.

In contrast, Friere (1987/1985) notes that much of the present social discourse in society might be best characterized as "antidialogical," a discourse of power, authority, and control. Consequently, there are two challenges we must face in undertaking to change American society, and the complex social systems that make society. The first is to create communities of critical social discourse—conversations of critique and possibility—where citizens conjoin to transcend existing systems (Jenlink & Carr, 1996). These communities of social discourse must critically examine the social structures and cultural patterns that shape society, as well as past and present attempts to liberate individuals and enable them to change and transcend the existing social systems.

The second challenge, of no less importance, is that of creating a new public philosophy of American society, a new politic of social change which understands that the possibility of human betterment lies in connecting the *communities of difference*[2] that represent our fragmented and increasingly diverse society. In this matter, it is not the goal to divest these communities of their *difference*. Quite the contrary, what is important is to connect the communities in such a way that the power of their *difference* may create a sustainable democratic dialogue for change. Connecting these communities will require that existing assumptions driving our understanding of social systems and change in social systems be illuminated and then be suspended from the politically-charged arguments of the moment. In this suspension it is important to respect the cultural identities of individuals and communities, and the self-identification these individuals and communities have with their various points of view, and to accept the limits that individuals may have in identifying with the subjectivity of others. To achieve this form of connecting will require new forms of relations across and within difference.

At the heart of connecting these *communities of difference* is a new kind of *social relation*—a relation that draws its power and promise from the *spirit of dialogue*. Bohm and Peat (1991) explain the spirit of dialogue well.

> In dialogue… a person may prefer a certain position but does not hold to it non-negotiably. He or she is ready to listen to others with sufficient sympathy and interest to understand the meaning of the other's position properly and is also ready to change his or her own point of view if there is good reason to do so. Clearly, a spirit of good will or friendship is necessary for this to take place. . . . The *spirit* of dialogue is, in short, the ability to hold many points of view in suspension, along with a primary interest in the creation of common meaning. (p. 82)

Conversations of change must have a balance between critical social discourse which examines the dominant societal paradigms[3] and constraining ideologies, and the social discourse of possibility that seeks to suspend individual and collective mental models[4] and engage in a generative, creative future-oriented process of designing social systems. Transcending existing social systems will require that individuals as well as collectives and communities of individuals examine the mental pathways that have given rise to the dominant paradigms and ideologies, as well as to the mental models that govern our daily interactions within our personal and professional lives.

It is because of people's embeddedness in their mental models—in their active memories—and history, because of their fragmented and incoherent thoughts and actions, because they are so inextricably locked into individual and collective inferential processes that change is difficult. It is because of a confluence of these hidden mental forces on larger collective scales (organizations, communities, and social institutions) that change of society and of social change processes used in society cannot be conceived.

Becoming aware of how we and others with whom we work are enmeshed in our personal and professional lives, and how we are engaged with what surrounds, us as well as how we are subject to the demands of the moment, is important to transforming our thinking. But we must move beyond awareness and we must seek to free ourselves from this enmeshment if we are to begin to understand the influence of our mental models and how this influence governs our lives. As Merleau-Ponty explains:

> It is because we are through and through compounded of relationships with the world that for us the only way to become aware of the fact is to suspend the resultant activity...to put it out of play. Not because we reject the certainties of common sense and a natural attitude to things—they are, on the contrary, the consistent theme of philosophy—but because, being the presupposed basis of any thought, they are taken for granted and go unnoticed, and because in order to arouse them and bring them into view we have to suspend for a moment our recognition of them. (1962/1967, p. xiii)

What he is suggesting is exploring the possibilities of unconcealing what is ordinarily obscured by the familiar, what become the conscious as well as unconscious forces that guide our interactions with the work around us. The exploration of the mental models, which influence our lives, can only occur when we suspend ourselves from the phenomena of our experiences and engage in conversations given over to dialogic relations[5]. It is in unconcealing our uniqueness—our values, beliefs, and underlying assumptions—that we begin to explore the influence of our uniqueness, of the inferential processes that guide our actions.

In the following sections, I will begin by exploring the importance of overcoming paths of least resistance as related to dialogue. Following in the next section will be an examination of the origin and meaning of dialogue. Next, the process of dialogue will be presented, followed by an overview of important principles to consider when engaging in dialogue. In the next to last section, the power of dialogue in social systems change is examined. Finally, the last section presents concluding remarks.

2. OVERCOMING PATHS OF LEAST RESISTANCE TO DIALOGUE

Dialogue, as social discourse, begins with people. The challenge in dialogue as social discourse is to change old mental models of conversation, and abandon the *paths of least resistance*[6] that typify actions, such as engaging in argument and debate at the expense of sharing knowledge and learning from each other's experiences. Unfortunately, as Fullan (1993) notes, individuals have a better track record of maintaining the status quo than they have of changing themselves. "Changing the deep fundamental structures of a system begins with changing the internal mental structures of people, their mindsets, which govern and give rise to the external system structures" (Jenlink, 1995, p. 46).

Dialogue, as social discourse, is about people changing their thought and belief systems, it is about creating and sustaining new thought and belief systems "through development of new critical masses. Once a critical mass becomes a majority, we begin to see the system change" (Fullan, 1996, p. 423). Creating a critical mass means that individuals and groups must transcend old mental models, and generate new, commonly shared perspectives. Paths of least resistance are often forged through continuous engagement in old ways of thinking and acting within social systems. As social structures and cultural patterns become reified over time, individuals follow the paths of least resistance in their daily lives and work. An example would be the use of discussion or debate to argue, negotiate, or otherwise advocate a position. Over time, as individuals use discussion as their primary form of discourse, they follow this conversational path again and again, moving further away from building strong community relationships while at the same time reproducing existing social structures and cultural patterns. Dialogue offers an alternative to paths of least resistance in that this form of discourse will enable participants to identify and expose the existing politics, philosophy and ideologies that often control and otherwise dictate the actions of individuals as well as collectives, organizations, and communities. As social discourse, dialogue will enable stakeholders to prepare the social contexts necessary for undertaking creative activities that lead to liberation from existing systems and transcendence to new systems. In essence, dialogue creates a conversational space by exposing existing ideologies, thought and belief systems, and structures of authority and power, a space in which a new public philosophy and ideal of society can be created.

In order to create this new philosophy and ideal of society we must create a collectively shared vision, a new ideal of the social system. This vision will act as a self-generating *energy field*[7], an emanating energy source generated through collectively shared values and beliefs. Creating this *energy field* will require, among many important undertakings, that participants develop capacities in systems thinking and systems design, the respect for and understanding of the power inherent in diverse or multiple world views, conversational skills essential to critical and dialogical discourse, and the intellectual technology to bring these undertakings into a confluence of energies for change.

The power to create an *energy field* may be found by using dialogue as a conversational medium for change. Engaging individuals in generative social discourse, which frees individuals from the constraint of old social structures, cultural patterns, and mental models, is critical. Likewise, engaging people in a dialogue that is essentially inclusive, democratic, communicative, and creative is one of many important steps contributing to the journey along a new pathway leading to changing ourselves, our social systems, and our society.

3. ORIGIN AND MEANING OF DIALOGUE

The origin of dialogue dates back to the time of Socrates, who employed the use of dialogue in his teachings and is recognized as the father of creative social discourse (Banathy, 1996; Zeldin, 1994). Dialogue comes from the Greek *dialogos*. *Dia* is a preposition that means "through" or "between," and is akin to "dyo" or "two." *Logos* comes from *legein* and means "to speak." It may also mean thought as well as speech—conceived materially as spirit, *pneuma*. Therefore, dialogue is a speech, a conservation between, through two or more people (Capranzano, 1990).

Several authorities have contributed to the meaning of dialogue, and are instructive in understanding the use of dialogue in the context of social systems change. Seven sources are drawn on for purposes of exploring dialogue as conversation or social discourse, including Mikhail Bakhtin, David Bohm, Martin Buber, Nicolas Burbules, Maurice Friedman, William Isaacs, and Patrick de Mare. Dialogue, as a process and conversational medium for systems change, takes on a much different character than other more traditional forms of conversation, and draws heavily on the major dimensions reflected in the different forms dialogue assumes in the scholarly literature.

3.1 Mikhail Bakhtin

Mikhail Bakhtin's view of dialogue is one seen through the lens of language and speech, of the spoken and written word. According to Bakhtin, language is fundamentally dialogical. For Bakhtin, "Life is dialogical by its very nature. To live means to engage in dialogue, to question, to listen, to answer, to agree, etc." (Todorov, 1984, p. 97). Bakhtin's perspective provides an important and comprehensive understanding of the communicative process through which social knowledge is constructed. Communication is a collaborative act, and knowledge that is owned individually and collectively by individuals, is a product of that collaboration. It is the construction of *dialogic relations* through communication, through the meanings and understandings created in the words we use, that we come together to shape and reshape ourselves and those around us. When we consider the languages that each participant brings as his or her culturally acquired symbol system for communicating, and we become aware that there is a heteroglossia[8] of actual languages that come into play in the process of dialogue, we begin to understand that each language "tastes of the context and contexts in which it has lived its socially charged life" (Bakhtin, 1981, p. 294).

We live by communication, and many of our practices that define us as human, that give us our uniqueness and individual as well as collective identities, are outgrowths of the varying ways we communicate: our language, our thinking and thought systems, our ethical and moral character, and our social organization. The important dimension of dialogue presented by Bakhtin is that of the dialogic relation created through language and the understanding of how we construct meaning among ourselves in our patterns of communication.

3.2 David Bohm

Bohm (1996) takes a different direction with his theory of dialogue when he focuses on thought as a system. He defines dialogue as "a *stream of meaning* flowing among and through us and between us...out of which will emerge some new understanding" (p. 1). He distinguishes between dialogue and discussion. Discussion is often the form of conversation that we find in school and community efforts of change. Dialogue relates to "logos", or, "the word" which he defines as the "the meaning of the word."

Discussion, on the other hand, is an analytical process,

> where there may be many points of view, and where everybody is presenting a different one—analyzing and breaking up. That obviously

> has its values; but it is limited, and it will not get us very far beyond our various points of view. Discussion is almost like a ping-pong game, where people are batting the ideas back and forth and the object of the game is to win or to get points for yourself. (p. 1)

For Bohm (1985, 1994, 1996) dialogue, as a group process, discloses and brings into focus the systemic and reflexive nature of thought, eliciting self-images, assumptions and prejudices with their often attendant emotions—defensiveness, anger, fear, insecurity and many others. Dialogue uses a *cool inquiry*, a process of obtaining high energy and low friction or conflict in their interactions. Discussion, in contrast, is characterized as a hot inquiry, or one where judgment, advocacy, and argument charge the discourse with an energy that fragments and distances people from their collective potential. Implicit in this theory of dialogue is a focus on tacit infrastructures and tacit thought processes, which is largely a tacit ground, shared among people and that is mostly incoherent, creating social and personal fragmentation.

The importance of dialogue lies in bringing to the surface the fragmented thinking and incoherent patterns of thought that have constrained individual and collective growth and change. The virtue—critical dimension—of such an approach is that the group may be able to detect the *flow of meaning* passing amongst its members. This meaning may be the content of some particular subject; it may also be the quickened pulses that pass through the group as the result of conflict between two or more members. Typically there is not an agenda but rather we fall into dialogue as a form of social relation. Such dialogue holds out the possibility of direct insight into the *collective* movement of thought, rather than its expression in any particular individual.

3.3 Martin Buber

Martin Buber (1958, 1965, 1988) viewed the importance of dialogue as that of examining the evolution of relationships from an I-It to an I-Thou presence in matters of human relations. For Buber (1988), the essence was that

> . . . where dialogue is fulfilled, in its being, between partners who have turned to one another in truth, who express themselves without reserve and are free of the desire for semblance, there is brought into being a memorable common fruitfulness which is to be found nowhere else. At such times, men who have been seized in their depths and opened out by the dynamic of elemental togetherness. The interhuman opens out what otherwise remains unopened. (p. 76)

Buber's primary focus is on the *between* which was the nature of the relationship that one person has with another. The critical dimension found here exists in creating a dialogic space which requires moving from an object-subject relationship where one person is treated as an object to be manipulated rather than as a person, the *other* with whom we share our *self*. The direction of movement must be to one of wholeness in *being* where each person is recognized and honored as an *other* in the relationship with equal voice and contributions to offer to building the *between*, a higher level of consciousness in human relations.

3.4 Nicolas Burbles

Complementing this focus on *betweenness,* and elaborating on the importance of *dialogic relation*, Burbles (1993) indicates that engaging in dialogue requires the development of special relationships—relationships with ourselves, others, and the setting in which we conduct our personal and professional lives. As he notes

> . . . dialogue is not fundamentally a specific communicative form of question and response, but at heart a kind of *social relation* that engages its participants. A successful dialogue involves a willing partnership and cooperation in face of likely disagreements, confusion, failures, and misunderstandings. Persisting in this process requires a relation of mutual respect, trust, and concern—and part of dialogical interchange often must relate to the establishment and maintenance of these bonds. (pp. 19-20)

Important to laying the foundation for developing a dialogic *social relation* are considerations for emotional factors in dialogue such as concern, respect, trust, appreciation, affection, and hope. Additionally, there is a requirement for communicative virtues or general dispositions and practices "that help support successful communicative relations with a variety of people over time. They include such qualities as tolerance, patience, an openness to give and receive criticism, the inclination to admit that one might be mistaken, the desire to reinterpret or translate one's own concerns in a way that makes them comprehensible to others, the self-imposition of restraint in order that others may have a turn to speak, and—often neglected as a key element in dialogue—the willingness to re-examine our own presuppositions and to compare them with those of others; to become less dogmatic about the belief that the way the world appears to us is necessarily the way the world is (Burbles, 1993, p. 42). The important dimensions reflected here are, *communicative virtues* and the *social relation* premised on building bonds of trust through dialogue. As Stewart (1978) notes, the fundamental element of dialogue, as well as change in human systems, is an act of collaboration, an act that constitutes the "reciprocal bond" between collaborators which generates trust and from which their shared meaning can emerge.

3.5 Martin Friedman

Martin Friedman, a scholar of Martin Buber's ontology of dialogue, notes that creating reciprocal bonds of trust requires that participants in the dialogue unconceal their identities, their uniqueness—the dynamically intrinsic structures of values, beliefs, and assumptions. Engaging in genuine dialogue is to become vulnerable, unconcealing one's uniqueness, and yet feeling safe and confident in the context of trust. As Friedman notes, there is an

> inextricable relation between dialogue and the unique . . . The unique is only known through dialogue. Only in genuine dialogue do I relate and respond to the other for himself and for the sake of our relationship and not as a function of knowledge and use, comparison and contrast, reflection, and analysis, seduction and exploitation . . . The proper understanding of dialogue includes uniqueness; for it is only in uniqueness that there is real mutuality, presentness, and presence. Dialogue means a mutual sharing in reciprocal presentness of the unique. (1976, p. 152)

The important dimension of dialogue reflected in Friedman's work (1976, 1992) is the uniqueness that each person brings to the process and the role that dialogue plays in healing the relationship with "self" and between "self" and "others." In a therapeutic sense, moving to levels of trust where participants are able to unconceal[9] their unique identities, is about discovering the "self" as well as the "other" in the conversation. This discovery is essential in the *healing* of existing relations as well as the building and sustaining of new social relations important to change and growth of human systems.

Unconcealing is about uncovering and examining the hidden forces, which guide and influence our interactions in the world. Unconcealing the deep underlying assumptions, beliefs, mental models that contribute to our unique identities is to search for the intrinsic identity that makes us unique. And these identities are, in part, derived from the socio-cultural settings in which we conduct our lives. Unconcealing the uniqueness of each person is threatening and risky. Disclosing the inferential processes that one uses, opens a context that is uncertain, often filled with insecurity, fear, anger, and defensive routines.

3.6 William Isaacs

William Isaacs, in his examination of dialogue as a generative process integral to organizational learning and the building of learning organizations, defines dialogue as a "sustained collective inquiry into the processes, assumptions, and certainties that compose everyday experience" (1993, p. 25). As a process of collective creation, dialogue seeks to unveil the ways in which collective patterns of thinking and feeling unfold "as conditioned, mechanistic reflexes, and potentially as fluid, dynamically creative exchanges" (Isaacs,

1993, p. 26). In essence, the purpose of dialogue is deeper than discussion, in that it seeks to explore and unconceal fundamental, collectively held assumptions and collective pressures that mediate against creative interactions while simultaneously seeking to examine and alter the *ground*—the cultural background of values, beliefs, assumptions—out of which systems problems, and individual as well as collective predispositions to see certain circumstances as problems, arise (Isaacs, 1992).

Unconcealment of the hidden *ground* provides a similar benefit as the unconcealment of participant's uniqueness, only on a collective and cultural level. Moving into dialogue as medium for social systems change, like Isaacs's assertion of its power in organizational learning, must be premised on a theory of dialogue where

> . . . the premise that the tacit forces that guide the ways people think and act are fragmented and incoherent, and that this ground and its influence are largely invisible to human beings. Dialogue creates special environments in which people can begin to perceive, inquire into, and shift their underlying patterns of influence, and create entirely new kinds of individual and collective minds. (Isaacs, 1996, p. 21)

The important dimension found in Isaacs's work is how dialogue, as a process, enables organizational learning with a focus on examining the collective and cultural *ground* that acts as a hidden force in the actions of individuals and collectives. Dialogue requires the creating of a *container*—the formation of participants as a collective—that serves to hold the dialogue conversation. This dynamic container influences and is influenced by the nature of the conversation it holds. Suspension of assumptions contributes to the creation of the container. Also important is the suspension of judgment with respect to what participants' share as content for the discourse.

Where social change is reliant on changing the thinking and belief systems inherent within human systems, understanding the collective *tacit ground* in which existing conceptual frames and old routines or patterns of action persist is essential. Creating a space, an emptiness, a container for conversations of change requires creating a new collective *field* with the energy of systems thinking and emergent values of wholeness and generative learning.

3.7 Patrick de Mare

Similar to Bohm's theory of dialogue as a *meaning flowing through* and premised on a *tacit ground*, de Mare (1991) uses a psychological lens to focus on large group therapy. Not unlike Bohm's focus of a tacit ground among people in large groups, de Mare includes the cultural *background* as the medium that controls and shapes collective human experience. In this perspective of dialogue, de Mare builds on the concept of *Koinonia*, a Greek work meaning impersonal fellowship. As he notes,

> It refers to the atmosphere of impersonal fellowship rather than personal friendship, of spiritual-cum-human participation in which people can speak, hear, see, and think freely, a form of togetherness and amity that brings a pooling of resources. (p. 2)

The large group, in de Mare's theory of dialogue, is viewed as an externalization of internal mental processes and thus gives the participants a way of addressing the sources of mass conflict or violence, and in healing distorting cultural patterns, such as the way marginalized populations may have been excluded from the activities of school, community, and the dominant cultures. In this perspective of dialogue, de Mare believes large groups enable people to engage in understanding and altering cultural assumptions present in society. The important dimension found in de Mare's theory of dialogue is that of cultural meaning—the cultural background from which individual and societal problems emerge. Unconcealing this cultural background, and creating an impersonal fellowship among participants, is critical to changing human systems.

3.8 Reflections on Dialogue

In reflecting on the important dimensions presented in each of these theories, and through assimilating these dimensions into a more coherent whole, dialogue is viewed as a dynamic relational process of social discourse. As social discourse, it is also viewed as a medium for effecting change within individuals and collectives as well as change within the cultural backgrounds that influence human systems and society. These dimensions also lend to understanding the process of dialogue and some important principles, which should be considered in using dialogue as a generative process in social systems change.

4. PROCESS OF DIALOGUE

Dialogue is a generative process that enables those who participate to come together in an authentic openness and preparedness to learn about oneself and others, and the contexts that define their actions and existence. Dialogue is a simple process, yet one that requires some introduction and understanding prior to its undertaking. As a form of communicative action, dialogue requires that new forms of social relations—dialogic relations—be created among participants as well as across the various *communities of difference* that constitute different stakeholders in a school, district, and community. It is process that requires the creation of a container—an emotional, psychological, social, and intellectually creative space—that is greater than the sum of the whole of participants represented by the collective.

Dialogue also requires the attendance to and development of communicative virtues, which bring into the foreground of the conversation the importance for social tolerance—a tolerance and respect for differences in identity as well as worldview. Also important to full and honest participation in dialogue are the virtues of openness, patience, potential for accepting responsibility for one's actions, the capacity to step back and re-examine one's own beliefs against the background of many worldviews, and the strength to suspend one's own beliefs in full view of all participants in the conversation. Relatedly, participants must simultaneously suspend judgments of other's beliefs as they are shared through the conversation, and made public.

The process of dialogue requires that multiple voices of difference are active in the conversation. It also requires that those engaged fully explore and understand the purpose for coming together in conversation, as well as seek to create themselves as a collective voice of exploration and creation while holding in highest regard the power present within such voices of difference.

Importantly, Gadamer (1976) draws attention to the nature of the dialogical relation within the process of dialogue. Specifically, Gadamer notes that the dialogical relation "carries away" its participants, and "catches them up" in a human interaction that takes on a force and direction of its own making, often leading in new directions and resulting in new insights.

> When one enters into a dialogue with another person and then is carried further by the dialogue, it is no longer the will of the individual person, holding itself back or exposing itself, that is determinative. Rather, the law of the subject matter is at issue in the dialogue and elicits statement and counterstatement and in the end plays them into each other…We say that we "conduct" a conversation, but the more fundamental a conversation is, the less its conduct lies within the will of either partner…. Rather, it is generally more correct to say that we fall into conversation, or even that we become involved in it. (p. 66)

The nature of the force Gadamer alludes to is comparable to the *energy field* that dialogical discourse is capable of generating as individuals come together through conversation, and form a consciousness through communion together. Also noted in Gadamer's examination of dialogue is a note that more often than not this form of social discourse is less driven by an agenda, taking its direction from the ideas that the discourse spark. It is important to consider in light of this description of dialogue, the rationale for dialogue in the larger scope of social systems, and whether or not dialogue can be forced on people or used as a mandated instrument of social agents.

When individuals set out to engage in dialogue as social discourse, the purpose must be identified. Likewise, consideration must be given to such factors as the context in which dialogue is to be situated, the time and space available for individuals to engage dialogue, the capacity for individuals to engage in dialogue, the ethicality versus instrumentality of dialogue, the pragmatics versus idealism of dialogue, and the understanding individuals bring to the dialogue process. Intimate to the dialogue process is a set of guiding principles that form the foundation of dialogue and provide guidance in understanding dialogue as social discourse.

5. PRINCIPLES OF DIALOGUE

Drawing from the different theories and building on their important dimensions, the following set of principles is offered as a guidance system for successful dialogue as a process integral to social systems change.

5.1 Generative Language

Dialogue requires a shared language, a language of constructed meaning and understanding. We all communicate using a language of words that are socially charged in the contexts of their origin. We rarely seek the meaning that one person has for a word. Typically we hear a word but assign our own socially charged meaning. When conversations are not open to sharing and collectively creating a language of socially constructed meaning, communications breakdown and incoherence takes over. Honoring the differences in language is a critical part of honoring and respecting each participant's unique identity as well as honoring the process of dialogue.

5.2 Personal Commitment

Dialogue requires investment of time and energy, both protected commodities. Commitment is a critical element in the building of new social relations. Personal commitment is about valuing the dialogue process, but more importantly it is about valuing others, ourselves and the contexts in which we act out our lives.

5.3 Symmetry of Power

Dialogue requires that power be symmetrical, that each participant to see the other as her/himself with respect to status, position, authority. Importantly, symmetry in power is about social justice, equity, and tolerance, as well as about allowing each participant's voice to be heard. Symmetry suggests that that there is an ethical, moral, aesthetic, and spiritual nature to dialogue that recognizes the diversities that shape the identity of each person.

5.4 Dialogic Relationships

Dialogue requires that we enter into relationships that focus on the *betweenness* that emerges when one person sees another as a person to share with rather than an object to control. Dialogical relations are the living medium of dialogue, a medium through which our collective selves are able to engage in a generative process of learning. Such relations are reciprocal in nature.

5.5 Reflective Consciousness

It is important for participants, individually and collectively, to engage in inquiry that seeks to examine experiences and events that influence the dialogue process as well as the larger process of changing human systems. Reflective practice enables the participants to ask questions of the inferential processes they use both consciously and unconsciously in their daily lives. It is vital to understanding the mental pathways that govern personal and professional experiences, and serves to inform the process of change.

5.6 Communities of Difference

Dialogue seeks to encourage the inclusion of participants invited from across the *communities of difference* that represents the larger society. Often there are feelings of anger, disenfranchisement, being marginalized, and resentment toward the formal authority figures as well as institutions. Dialogue seeks to provide a medium of exchange in which all voices are honored, and where cultural healing is possible. Dialogue seeks to build new social relations that connect the communities of discord into a community of creative action. Dialogue also seeks to focus this diversity, and the energy it carries, into a collective energy, sustaining members as a generative, creative force for change.

5.7 Unconcealment

In dialogue, authentic participation is a process of making visible the tacit hidden forces, which guide our actions and interactions with others and the world around us. Each of us has an identity that is developed over time through our experiences. Our identity derives its substance from the context and cultural background of our personal and professional lives.

Unconcealing this identity is risky but as vital to the dialogue process as it is to changing human systems. Unconcealment often brings with it the tendency to defend our personal identities, seeking to protect who we are and how we understand our relationships with self, others, and the world. Defensive routines often accompany unconcealment. Successful dialogue understands the unconcealment process and the need to mediate tensions that might develop.

5.8 Deeply Listening

Dialogue, as a sharing through conversation, requires each participant to be available to listen to the other. Listening is an art and is essential to communication as a central tenet of dialogue. Listening deeply is about hearing with your mind what others are saying, suspending the need to formulate a response before the other person finishes their thought. Listening deeply means that we seek meaning in what another says, listening not only to the words but also to the expression of emotion and the socially charged nature of their language. We must seek to understand the unique identity that each person brings to the dialogue, recognizing that listening is more than just the thoughts that others share, it is the very essence of who they are.

5.9 Ethic of Care

Dialogue is premised on dialogic relations as well as a consideration for the other with whom individuals in a dialogue conversation interact. An ethic of care conveys a focus on the relation that exists between the self and others in the discourse. Caring for, with, and about self and others is important to establishing the dialogic relation that forms the foundation of dialogue as social discourse. Caring is concern for how individuals meet and treat one another, their ability to respect and value the other in the relation.

Dialogue is premised on the ethic of care as well as the capacity for caring. An ethic of care and capacity for caring are essential to the dialogic relation that is at the heart of dialogue as social discourse. Caring for others and with others, as well as for self, separates dialogue as social discourse from other forms of conversation.

5.10 Creating Community

Dialogue encourages individuals to collectively share their assumptions, ideas, and beliefs in an authentic and open manner. When a group of people self-organize in dialogue they bring their individual uniqueness, their identities together in an effort to understand each other and collectively engage in meaning making. When individual identities are collected together and people disclose the inner self or who they are, they form a community in which the dialogue takes form and is shaped by the nature of the community. Likewise, as the conversation is shaped over time, the nature of the dialogue begins to reshape the community through collective thinking and generative learning.

5.11 Respect for Self and Others

Dialogue seeks to create a respect for and honoring of the uniqueness of each person. At the heart of dialogic relations is a respect for oneself, for others, and for the process that brings us together. Respect is a sensitivity to the individual identities of all persons as well a mindfulness of the process in motion. In this sense respect is about finding our centeredness with others, the environment, and ourselves.

5.12 Suspending Assumptions

Dialogue encourages participants to unconceal their unique identities, to suspend their assumptions and opinions so as to move outside of their field of influence. By suspending the intrinsic elements of our identities, we contribute to the strength of the dialogic relations and the container. The capacity for suspension must be balanced by our understanding of the tendencies toward defensive routines and the ability to set these aside.

5.13 Suspending Judgment

Dialogue requires that participants set aside judgement of what others say, keeping an open mind to differences in opinion and perspective. Closely associated with suspending assumptions and making ourselves vulnerable to each other is the capacity to suspend our need to judge others' identities and what they share. Judgment is a form of defensive routine that often serves to protect our identity and challenge the identity of others.

5.14 Systems Thinking

Dialogue seeks to create wholeness by bringing people together in systemic and reflexive conversations. The power of systemic thinking lies in the understanding that each person derives his or her sense of identity from the whole, and as participants in the dialogue their different perspectives enable us to see more clearly the nature of the whole.

6. THE POWER OF DIALOGUE IN SOCIAL SYSTEMS CHANGE

Fundamental societal problems today have less to do with efficiency or economic productivity and growth. Rather, the problem confronting society is one of creating and sustaining a viable democracy capable of critiquing its own actions and capable of developing its own capacity to change. A viable democracy is one given to communities in which inclusive, democratic, and open-ended dialogues are a part of the daily lives of all citizens. This will require two things up front.

The first requirement is a new public philosophy that presents an ideal of society more clearly aligned with learner needs, present and future. The second requirement is a new politic of change that is essentially democratic and informed by a balance of critical social discourse and a dialogical discourse of opportunity. Neither the new public philosophy nor new politic of change are discretionary acts when we consider the magnitude of the problems society faces. In effect, at the heart of creating this new public philosophy and politic of change is the process of dialogue. Dialogue serves as a creative process of generative action that can lead to a new educational system, and a renewal of the public spirit of democracy. The power of dialogue, as the process has been outlined in previous sections of this text, resides in understanding that change is basically an individual and collective effort that transmutes into the larger societal and cultural background.

What will bind a community of people together in dialogue, and their efforts toward social systems change, is creating a shared knowledge and collective action, a core set of shared values that is continuously subjected to revision in what Booth (1974) calls the "court of communal exchange" (p.148). Creating the conditions that encourage and sustain a context for dialogue will likewise encourage and sustain efforts of change in complex social systems. This undertaking will require that certain basic principles of dialogue be accepted as crucial to bringing into existence the *spirit of dialogue* and negotiating dialogic relationships between individuals and across the various *communities of difference*. Individuals from within and across these communities will engage in dialogic relationships, honoring their differences while simultaneously creating a new public philosophy of society, a new ideal of society as an essentially inclusive, communicative, and democratic system.

7. CONCLUSION

Undertaking the design and change of social systems will require the creation of a new public philosophy of discourse as well as a new politic of participative democracy. This new politic must be guided by critical and generative discourse, the spirit of dialogue and a deep commitment to the democratic involvement of people at the grassroots level. We will need to necessarily overcome the fragmentation that has swept us into politically charged *communities of difference*. But more importantly, we must overcome the fragmentary thinking processes that have kept these communities apart, thus draining the powerful energy that difference offers to and is essential for creating systems change in society. At the heart of this effort to design and recreate our social system, and our society, is the "proposition that we must do better in involving and enabling people as participants in the ongoing dialogue that is a democratic society..." (Burbules, 1993, p. 18). The process of dialogue as a conversation for change presented in this chapter suggests a way to address this proposition. Changing social systems, and society, is not an easy challenge to face. We are cautioned to remember, though, that social systems change is like the dialogue. As Bakhtin explains,

> there is not first or last discourse, and dialogical context knows no limits (it disappears into an unlimited past and in our unlimited future). Even past meanings...can never be stable (completed once and for all, finished), they will always change...in the course of the dialogue's subsequent development...At every moment of the dialogue, there are immense and unlimited masses of forgotten meanings, but, in some subsequent moments, as the dialogue moves forward, they will return to memory and live in renewed form (in a new context). Nothing is absolutely dead: every meaning will celebrate its rebirth. The problem of the great temporality. (cited in Todorov, 1984, p. 110)

NOTES

[1] The idea for this chapter, as well as sections of text, were adapted from a paper presented by the author to the Systems Thinking SIG at the 2000 Annual Meeting of the American Educational Research Association in New Orleans, April (See Jenlink, 2000).

[2] *Communities of difference* as used in this article, reflect considerations for the diversity within and across sociological groups and, of equal importance, diversity in constructed worldviews of individuals and collectives. With respect to social systems change, it is important to acknowledge the diverse professional communities which constitute the essence of society and contribute to the working of society's myriad social institutions and human system. In full recognition of the meaning of *difference*, it is acknowledged that *difference* is not simply a matter of sociological group membership, but also of the constructed worldview and subjectivity of the persons who enter a social, communicative, or dialogical relation. Therefore, *difference* cannot always be inferred nor witnessed from the outside, but rather requires a full participation of persons in relations in such a way that the inside *difference* is unconcealed. Understanding and acknowledging the power and possibility inherent within *communities of difference*, when examined dialogical discourse, is at the heart of a self-renewing social system. When viewed through the more traditional non-systems lens, the lack of interdependent relationships connecting these *communities of difference* contributes to a lack of cohesion in thought and action, and an overall fragmentation society's various institutions and systems. See Bauman (1988-1989), Burbles & Rice (1991), Derrida (1982), Ford-Slack (1995), Young (1990) for further discussion on difference.

[3] A societal or cultural paradigm is civilization's fundamental view of the nature of things. It is an abstraction, which organizes our understanding of the actual world, or the world as it is. Kuhn (1970) and Schwartz and Ogilvy (1979) have examined the emergent and changing nature of paradigms, suggesting that on the societal level, paradigms influence patterns of social thought and belief that, in turn, influence individual and organizational patterns of thought and belief.

[4] Mental model is a concept that shares common meaning with terms like *mindset, worldview, and frames of reference*, and is often treated synonymously. Mental models may be held by an individual or shared collectively within collaboratives, teams, organizations, and communities. Mental models comprise what is considered as *active memory;* that is, an individual interprets his or her environment by applying certain conceptual frames that provide theoretical

understanding for the world. This is often referred to as *know why* or conceptual knowledge. Associated with these conceptual frames are learned routines or strategies—procedures—which enable the individual to interrelate to a situation or to effect action during an event. This is often referred to as *know how* or procedural knowledge. Mental models are influenced by our thinking and thought processes as well as our language and learning processes. And they are also influenced by the larger societal paradigms of the day.

[5] Dialogic relations refers to the relation that two persons, or collectives, fall into as a natural outflowing of the communicative process embedded within a dialogue conversation. Dialogic relations have a specific nature, they can be reduced to neither the purely logical nor the purely linguistic. It is a relation between and among people that exists as a shared energy created when people are drawn into a particular dynamic of speaking and listening with/to one another. The relation created through dialogue is such that when one enters into a dialogue with another, each person is carried along by the nature of their sharing, each sharing creating a context in which the person is drawn deeper into their own self-examination, able to examine their own beliefs while providing a similar opportunity for their partners self-examination. Dialogic relations are shaped through individuals participating in the conversation. Reflexively, the conversational discourse is shaped by the evolution of dialogic relations. Dialogic relations are not a tool or something we use instrumentally, but rather something we fall into or become involved in as a natural consequence of the conversation.

[6] The idea of *Paths of Least Resistance* originates with Robert Fritz (1989) and his work in understanding the creative force in our lives. Each of us has his/her own paths, created over time and guided by internal structures. The structures that have the most influence on our lives are composed of our desires, beliefs, assumptions, and objective reality itself. For further exploration of paths of least resistance as well as learning how to change our lives through creative forces, see this author's work.

[7] The concept of *field*, as used in this paper, refers to an energy of shared values, beliefs, and assumptions, a type of energy that emanates from collectively created and mental models. In part, this energy is generated from the collective tacit ground and thought as a system which is self-renewing and self-organizing. It is a *system of thought* that is created through a shared language system, collective thinking and the *spirit of dialogue*. Field in this sense derives its meaning from the collective works of Bohm (1983), who discussed quantum field theory in his exploration of wholeness and the implicate order in systems, Lewin (1951), who introduced the concept of field in social sciences and his pioneering work in social dynamics and human understanding, Sheldrake (1981), whose understanding of morphogenic resonance presented a biological field theory, Isaacs (1996), who introduced the concept of self-organized charged fields of meaning created through dialogue, Saussure (1959), who presented the argument for language as a system, a unified field of composed of parts not reducible to the sum of its parts, and Sheldrake and Bohm (1982), who collaborated on morphogenic fields and the implicate order.

[8] Bakhtin uses the concept of heteroglossia to refer to the many different languages reflected in the different voices associated with a social phenomenon. Each voice of a different participant in the phenomenon develops his or her own meaning for concepts, words used, their own language for communicating about and within the social phenomenon. Bakhtin also uses heteroglossia to explain the interweaving of these languages into complex patterns of emerging significance.

[9] Martin Heidegger (1971, p. 54) first used the term unconceal. Unconcealing refers to the conscious action, on the part of participants in the dialogue, to making visible that which they have concealed. Maxine Greene (1988) goes further to say that concealment does not simply mean hiding something; it means dissembling, presenting something as other than it is. To "unconceal" is to create clearings, spaces where a person or persons may interact by breaking through what is masked or half-hidden. The use of unconcealing in this text also carries with it meaning that a person or persons often intentionally hide or conceal aspects of their identities in attempts to protect or keep their identity secure as well as unknown. Identity in this sense refers to the intrinsic nature of a person, such as their perspectives, cognitive processes, and emotions. Identity also refers to the deep structures of values, beliefs, and assumptions, which influence our inner workings.

REFERENCES

Bakhtin, M., 1981. *The Dialogic Imagination: Four Essays by M.M. Bakhtin*/(Michael Holquist, Ed.). University of Texas Press, Austin, TX.
Bakhtin, M., 1986. *Speech Genres*. Dalton, New York.
Banathy, B.H., 1996. *Designing Social systems in a Changing World*. Plenum Press, New York.
Bauman, Z., 1988-1989. Strangers: The social construction of universality and particularity. *Telos*, **28**:23.
Bohm, D., 1983. *Wholeness and the Implicate Order*. ARK Paperbacks, New York.
Bohm, D., 1985. *Unfolding Meaning*. ARK Paperbacks, New York.
Bohm, D., 1996. *On Dialogue*. Routledge, New York.
Bohm, D., 1994. *Thought as a System*. Routledge, New York.
Bohm, D., and Edwards, M., 1990. *Changing Consciousness*. Harper, San Francisco, CA.
Bohm, D., and Peat, F.D., 1987. *Science, Order, and Creativity*. Bantam Books, New York.
Booth, W. B., 1974. *Dogma and The Rhetoric of Assent*. University of Chicago Press, Chicago, IL.
Buber, M., 1958. *I and Thou* (Second Edition). Macmillan Publishing Company, New York.
Buber, M., 1965. *Between Man and Man*. Macmillan Publishing Company, New York.
Buber, M., 1965/1988. *The Knowledge of Man: Selected Essays*. Humanities Press International, Inc., Atlantic Highlands, NJ.
Burbules, N.C., 1993. *Dialogue in Teaching*. Teacher College Press, New York.
Burbles, N.C., and Rice, S., 1991. Dialogue across differences: Continuing the conversation. *Harvard Educational Review*, **61**(4): 393-416.
Capranzano, V., 1990. On dialogue. In T. Maranhão (Ed.), *The Interpretation of Dialogue*. The University of Chicago Press, Chicago, IL, pp. 270-291.
de Mare, P., Piper, R., and Thompson, S., 1991. *Koinonia: From Hate, Through Dialogue, to Culture in the Large Group*. Karnac Books, New York.
Derrida, J., 1982. Difference, in *Margins of Philosophy*. University of Chicago Press, Chicago, IL, pp. 1-27.
Ford-Slack, P.J., 1995. Reflections on Community: Understanding the familiar in the heart of the stranger. In, L. Lambert, D. Walker, D.P. Zimmerman, J.E. Cooper, M.D. Lambert, M.E. Gardner, and P.J. Ford-Slack, *The Constructivist Leader*. Teachers College Press, New York, pp. 159-170.
Friedman, M., 1976."Dialogue of touchstones": An approach to communication and identity. *Communication*, **2**:143-157.
Friedman, M., 1992. *Dialogue and the Human Image: Beyond Humanistic Psychology*. Sage Publications, Newbury Park, CA.
Friere, P., 1970. *Pedagogy of the Oppressed*. Herder & Herder, New York.
Friere, P., 1985. *The Politics of Education: Culture, Power and Liberation*. Bergin & Garvey, South Hadley, MA.
Fritz, R., 1989. *The Path of Least Resistance: Learning to Become the Creative Force in Your Own Life*. Fawcett Columbine, New York.
Fullan, M.G., 1993. *Change Forces: Probing the Depths of Educational Reform*. New York: The Falmer Press.
Fullan, M.G., 1996. Turning systemic thinking on its head. *Phi Delta Kappan*, **77**(6): 420-423.
Heidegger, M., 1971. *Poetry, Language, and Thought*. (Trans. A. Hofstadter). Harper & Row, New York.
Isaacs, W. N., 1993. Dialogue, collective thinking, and organizational learning. *Organizational Dynamics*, **22**(2): 24-39.
Isaacs, W.N., 1996. The process and potential of dialogue in social change. *Educational Technology*, **35**(1): 20-30.
Jenlink, P.M., 1995. Educational change systems: A systems design process for systemic change. In, P.M. Jenlink (Ed.), *Systemic Change: Touchstones for the Future School*. IRI/Skylight Training and Publishing, Inc., Palatine, IL, pp. 41-67.

Jenlink, P. M., 1999, February. *Dialogue as Inquiry Method: Examining Issues of Relationships, Reflexivity and Self-consciousness*. Paper presented at the Advances in Qualitative Methods Conference, Edmonton, Alberta, Canada.
Jenlink, P.M., and Carr, A.A., 1996. Conversation as a medium for change in education. *Educational Technology*, **36**(1): 31-38.
Kuhn, T.S., 1970. *The structure of scientific revolutions*, Second Edition, Enlarged. The University of Chicago Press, Chicago, IL.
Lewin, K., 1951. *Field Theory in Social Science: Selected Theoretical Papers*. University of Chicago Press, Chicago, IL.
May, R., 1969. *Love and Will*. Norton, New York.
Merleau-Ponty, M., 1962/1967. *Phenomenology of Perception*. Humanities Press, New York.
Noddings, N., 1984. *Caring: A Feminine Approach To Ethics and Moral Education*. University of California Press, Berkeley, CA.
Sarason, S.B., 1990. *The Predictable Failure of Educational Reform: Can We Change Course Before It's Too Late*. Jossey-Bass Publishers, San Francisco, CA.
Saussure, F., 1959. *Course in General linguistics*. (Ed. Charles Bally and Albert Sechehaye, Trans. Wade Baskin). Philosophical Library, New York.
Schwartz, P., and Ogilvy, J., 1979. *The Emergent Paradigm: Changing Patterns of Thought and Belief*. Analytical Report of Values and Lifestyles Program. SRI International, Menlo Park, CA.
Senge, P.M., 1990. *The Fifth Discipline: The Art and Practice of the Learning Organization*. Doubleday/Currency, New York.
Senge. P. M., Kleiner, A., Roberts, C., Ross, R., and Smith, B., 1994. *The Fifth Discipline Fieldbook*. Doubleday/Currency, New York.
Sidorkin, A.M., 1999. *Beyond Discourse: Education, The Self, and Dialogue*. State University of New York Press, New York.
Stewart, J., 1978. Foundations of dialogic communication. *Quarterly Journal of Speech*, **64**:183-201.
Todorov, T., 1984. *Mikhail Bakhtin: The Dialogical Principle*. (Trans. by Wlad Godzich). University of Minnesota Press, Minneapolis, MN.
Young, I.M., 1990. The idea of community and the politics of difference. In, L.J. Nicholson (Ed.), *Feminism Postmodernism*. Routledge, New York, pp. 300-320.
Zeldin, T.,1994. *Intimate History of humanity*. Harper Collins, New York.

SECTION II

PERSPECTIVES ON DESIGN CONVERSATION

Chapter 5

SEARCHING TOGETHER

Approaches, Methods, and Tools

BELA H. BANATHY
Saybrook Graduate School and International Systems Institute

1. INTRODUCTION

In the course of the last decade, an observer of the mode of discourse in our various societal groups and communities has been able to see the emergence of new approaches to the heuristics of intentional communication, such as: Dialogue, consensus seeking, conversation, thinking together, and searching for common ground. It seems that these modes of intentional social communication are–on the one hand–reactions to the ever more degrading and hostile public, private, and political discourse and–on the other–they represent an increasing yearning for civility, mutual respect, and dignity in our social discourse.

Engaging in these new approaches of collective discourse, we are also discovering the surprising power of intentional collective communication as we are searching together for solutions to intractable issues and complex problem situations. Most significantly, however, we find the use of these communication modes most rewarding as we aspire to take charge of our future, and collectively design systems in which we wish to live and work. By now, the societal landscape is rich with a great variety of groups, intentional communities, community agencies, volunteer and professional associations, organizations in the private sector, learning to use–and using–these new modes of civil discourse. They create not only their own future, but create a wealth of social and intellectual capital.

In Part One, I introduce a general picture of definitions and approaches to "searching together." In Part Two, I explore various group methods and tools to purposeful collective communication.

PART ONE: DEFINITIONS AND APPROACHES TO SEARCHING TOGETHER

Part One addresses (1) the dialogue idea, (2) generative and strategic dialogue, (3) dialogue as social discourse, (4) dialogue as learning, (5) dialogue as culture creating, (6) the dimensions of dialogue, (7) dialogue is thinking together, (8) dialogue: transform conflict into cooperation, and (9) dialogue as evolutionary conversation.

2. THE DIALOGUE IDEA

Webster defines dialogue as "conversation between two or more persons." It defines the purpose of dialogue as "seeking mutual understanding and harmony." In the present paper, I modify the first definition as an intentional conversation between several persons–ideally seven plus/minus two.

In the course of the last several years we have seen a surge of attention devoted to a new mode of social discourse, called "dialogue." This mode is applied to generate a common frame of thinking, shared meaning in and a collective worldview of a group. The term dialogue is derived from the Greek "dia," meaning "through" and "logos," standing for the "meaning of a word." Dialogue is a free flow of meaning between people in a communication situation. In dialogue, people may prefer a certain position, but they are willing to suspend it, they are willing to listen to others in order to understand the meaning of their position. They are ready to change their point of view and blend it with others. In dialogue, people are able to face disagreement without confrontation and are willing to explore points of view to which they do not subscribe personally. "They will find that no fixed position is so important that it is worth holding at the expense to destroying the dialogue itself"(Bohm & Peat, p. 242). In what follows, I introduce various perspectives on the dialogue idea.

Zeldin (1994) traced back in history the invention of dialogue and suggested that Socrates was the first known conversationalist, who replaced the war of worlds by dialogue. He introduced the idea that "individuals could not be intelligent on their own, that they needed someone else to stimulate them." (p. 33). He proposed that if two unsure people engaged in discourse, they could achieve what they could not do separately. "By questioning each other and examining their prejudices, dividing each one of those into many parts, finding the flaws, never attacking or insulting, but always seeking what they can agree upon, moving in small steps from one agreement to another, they would gradually learn what the purpose of life was. He argued that it was inadequate to simply repeat what others said, or borrow ideas. One has to work them out for oneself" (p. 34).

Bohm and Peat (1987) make a sharp distinction between dialogue and ordinary discussion. In discussions, people hold relatively fixed positions and argue their views in trying to convince each other. At best, they say, this form of discourse may produce some agreement or compromise, "but it does not give rise to anything creative" (p. 241). The word discussion has the same root as concussion and percussion. It is a process of shaking apart and hitting. Bohm uses the metaphor of a ping gong game, in which we pass the ball back and forth with the sole purpose of winning the game. Furthermore, in the course of a discussion, when something of fundamental importance is involved, positions often tend to be non- negotiable and confrontational. This leads to either a situation where there is no solution or to a polite avoidance of the issue.

Yankelovich (1999) suggests that "To Buber we own the stunning insight that, apart from its obvious practical value (most problem solving demands mutual understanding), dialogue expresses an essential aspect of the human spirit. Buber knew that dialogue is a way of being. In Buber's philosophy, life is itself is a form of meeting and dialogue is the "ridge" on which we meet. In dialogue we penetrate behind the polite superficialities and defenses in which we habitually armor ourselves. We listen and respond to one another with an authenticity that forges as bond between us" (pp. 14-15). Yankelovich defines dialogue as a highly specialized form of communication that imposes rigorous discipline on its participants. "When dialogue is done skillfully, the results can be extraordinary. Longstanding stereotypes dissolved, mistrust is overcome, mutual understanding achieved, visions shaped and grounded in shared-purpose, people previously at odds with one another aligned on objectives and strategies, new common ground discovered, new perspectives and insights gained, new levels of creativity stimulated, and bond of community strengthened" (p. 16). The author notes, however, that until recently people assumed that no particular skill is required to engage in dialogue. Most people assume that dialogue is just another form of conversation, not requiring special discipline. In the course of the past decade, however, a growing literature and practice have demonstrated the uniqueness of dialogue when it is well done.

Martin (1999) defines dialogue as "the art of living creatively in the world. It is a way of being in relationship with things and people that enables a creative outcome–an insight, a shift of perspectives, a deepening of understanding–that enriches our participation in life's unfolding. Dialogue implies that what we call the **truth** is already present in some way, such that every question contains its own answer, the every relationship carries its own seeds of new possibilities, and that every painful encounter has its resolution hidden in its dark tension" (p. 5). Usual social discourse blocks truth and prevents it from emerging. The "dialogue way" leaves aside prejudices, allows us to be present without judgement. When it happens, something new erupts into existence: A different perspective, a new insight, shared meaning.

Isaacs (1999) defines dialogue as "a discipline for developing shared meaning among disparate groups of people." He also believes that at its core,

> dialogue accomplishes something else: It can produce a deep shift in our understanding of the nature of power. Dialogue is a leveler, transforming repressive forms of hierarchy as it frees each individual to acknowledge what they think and feel and express it. But dialogue, in its fullest expression, enables the emergence of genuine collective leadership, whose highest aim is ultimately to make contribution, to give, not take. The willingness to give one's ideas freely, without the sense of having to draw close boundaries to protect or preserve them, generates a common mind and common pool of meaning out of which much can be done. (pp. 394-395)

Yankelovich (1999) sees dialogue as filling an urgent needs, as we grow more isolated from each other, increasingly fragmented and pluralistic, we are likely to misunderstand each other more an more, social bounds fray and civility loses ground. He considers dialogue as a disciplined encounter, which penetrates the polite superficialities and defenses in which we armor ourselves. He suggests that the act of reaching beyond the self to relate to others in a dialogue is a profound human yearning. He then asks. If yearning for dialogue is universal, why is it so rare? He answers: Because it calls upon skills that impose a rigorous discipline on participants and most people don't have access to appropriate learning resources to acquire competence in dialogue.

3. GENERATIVE AND STRATEGIC DIALOGUE

In the design literature, it was Bohm and *Peat* (1987), who proposed "dialogue" as a method of attaining free flow of ideas, beliefs, and as a means of creating meaning among members of groups." Dialogue may well be the most effective way of investigating the crises that faces society, and indeed the whole of human nature and consciousness today. Such a form of free exchange of ideas and information is of fundamental relevance for transforming culture and freeing it of destructive misinformation, so that creativity can be liberated" (p. 240).

Recently, dialogue" has gained prominence as the most viable form of collective social discourse. It is defined in the social discourse scholarship as "*generative dialogue*." It is not aimed pursuing a specific task. It aims to "generate" a common frame of reference, a shared view of the world among the parties of the dialogue. As such, generative dialogue that aims at creating a common ground among members of a community is an important front part of collective inquiry. More recently, the notion of "*strategic dialogue*" has emerged as a mode of communication for collective decision-oriented disciplined inquiry

Beck (1994), contrasting Generative Dialogue with Strategic Dialogue, suggests that, while Generative Dialogue aims to create a common ground, Strategic Dialogue focuses on addressing specific tasks and is applied in finding specific solutions in decision oriented inquiry. The dialogue scholarship community has not yet addressed substantively "strategic dialogue." Still, Schein (1993) suggests that the test of the importance of generative dialogue will be "whether or not difficult, conflict ridden problem situations can be handled in a dialogue mode. Our experience with strategic dialogue in the Conversations of the International Systems Institute during the last two decades indicates that indeed it can. We designated *the combination of* generative and strategic dialogue *as* "***conversation***."

4. DIALOGUE AS SOCIAL DISCOURSE

In the course of the last several years we have seen a surge of attention devoted to "dialogue" or "generative dialogue." This mode is applied to generate a common frame of thinking in a group. Bohm (1996) considers dialogue to be a free flow of meaning between people in a communication situation. In dialogue, people may prefer a certain position, but they are willing to suspend it, willing to listen to others in order to understand the meaning of their position. They are ready to change their point of view and blend it with others. In dialogue, people are able to face disagreement without confrontation and are willing to explore points of view to which they do not subscribe personally. "They will find that no fixed position is so important that it is worth holding at the expense to destroying the dialogue itself" (p. 242).

5. DIALOGUE AS LEARNING

Dialogue provides opportunity for learning. We learn how thoughts and feelings weave together, both collectively and individually. Suspending our assumptions, we can generate shared consciousness. (The root meaning of consciousness is "knowing it all together.") In a dialogue the "knowing it all together of individuals and the groups" form a subtle higher unity and come together in a harmonious way. The "spirit of dialogue" emerges and guides the group on a journey toward shared perceptions and purposes (Martin, 1999). In the dialogue event people are able to be honest and straight with each other, they level with each other, they share ideas freely. They develop a common mind, a shared mind, and think and search together in a new and creative way. They awaken their collective intelligence and their feelings of genuine participation, mutual trust, fellowship, and friendship. They can think and talk together. Shared meaning and understanding flow freely in the group. However, they can do none of these if there is hierarchy or authority represented in the group.

6. DIALOGUE IS CULTURE CREATING

The organizing principle of dialogue implies a change of how the mind works. In true dialogue a new form of consensual mind emerges, generating a rich, creative order between the individual and the social as a more powerful force than is the individual mind alone. This creative order "arises from a spirit of friendship dedicated to clarity and the ultimate perception of what is true" (Bohm & Peat 1987, p. 247). People, who learn the potential power of dialogue, will be able to transfer the spirit of dialogue into their daily activities and social relationships and into the systems and communities in which they live. Dialogue, therefore, may create a new culture in a community of inquirers and, furthermore "members of the community can explore the possibility of extending the transformation of the mind into a broader socio-cultural context" (p. 247). Schein (1994) suggests that dialogue is a vehicle for understanding cultures and subcultures in organizations. And organizational learning depends upon such cultural understanding. It facilitates the development of a common language and collective mental models. Thus, the ability to engage in dialogue becomes one of the most fundamental and most needed human capabilities.

7. DIMENSIONS OF DIALOGUE

Isaacs (1993) proposes that dialogue has several dimensions: (1) it is collective learning and action. The level of learning that can take place in a collective setting cannot ever be mastered individually. In dialogue the group creates a pool of common meaning and new levels of coordinated action, (2) it is paradigm exploration. Dialogue enables people to step back from the context of specific problems, reflect upon what lies beneath them, and learn a new way of seeing and attaining a new kind of consciousness. (As Einstein said, "we cannot address a problem from the same consciousness that created it."), (3) dialogue enables us to create a bridge between diverse cultural differences. We can reach back into our shared cultural background, and create a common flow of meaning, and (4) dialogue fosters the power of collective creation. As we suspend our assumptions and begin to listen to each other in a deep way, new creative insights and new levels of wisdom emerges. We not only transform existing patterns of thought but also transmute them and create new levels of consciousness.

8. DIALOGUE: THINKING TOGETHER

In his recent book, *Dialogue: The Art of Thinking together*, Isaacs (1999) proposes that we can generate and sustain a new conversational spirit "that has the power to penetrate and dissolve some of our most intractable and difficult problems" (p. 6). Dialogue represents not only talking together but also thinking together. In dialogue, "human beings could contain their inner emotional, intellectual, and even spiritual selves, especially those that arise in conversation–and find ways turn them to creative uses" (p. 8). Isaacs offers guidelines to engaging in and conducting dialogue. In exploring his guidelines, I place those in the experiential context of working with the conversation movement of the systems community over the last twenty years. Isaac's guidelines include the following.

> * The *mode we enter* into a dialogue situation is of much importance. The mode of entry contains the seeds of success. Our experience in our International Systems Institute Conversation Community suggests that dialogue best unfolds by having mutual respect, willingness to suspend positions and listen unconditionally, honoring contributions, having equality, being open to ideas, and always seeking consensus.
>
> * Honor, respect, and take advantage of the *uniqueness* and the unique contribution of each and every member of the conversation community. Each member has unique background, a unique story to tell, unique perspectives, and unique ways to create meaning. This uniqueness offers a rich diversity of ideas that is so essential in dialogue.
>
> * A *container* for the dialogue must be created if we are looking for significant change. Isaacs (1999) offers a set of core practices that provide the framework for the conversation: (1) *Evoke the ideal.* "The promise of dialogue is that a small group of people might do something that impacts the world. Evoking this potential, supporting its articulation, and asking people to reflect on it can make an important difference to making progress" (p. 294). (2) *Support dreaming out loud.* "Dreaming about how things actually can be different, given honestly how things are now are, requires the critical ingredient of support. This means an unwavering sense of reinforcement that does not judge what is said and done" (p. 294). (3) *Deepen the listening.* "People must come to the point of realizing that they listen with their mind and heart, not through their ears…outcomes in a group require that people discover that their listening matters" (p. 294).

I close our conversation with a quote from Isaacs:

> Dialogue enables a 'free flow of meaning,' which has the potential of transforming the power relationship among the people concerned. As this free flow emerges it becomes quite apparent that no one person owns this

> flow and no one can legislate it People can learn to embody it, and in a sense serve it. This is perhaps the most significant shift possible in dialogue: That power is no longer the province of a person in a role, or even of any single individual, but at the level of alignment an individual or group has with the power of Life itself. We have a familiar way of understanding power of this kind, though it comes from an unusual source. Power that respects no one but includes everyone, that calls for the best in people, and that evokes great creativity, is Love. Dialogue can unleash the power of love, not in the sentimental or moralistic sense but in the genuine sense of creativity. It may be the case that few 'great groups' of powerful collaboration would publicly attribute their success to love, but I suspect it is often the invisible ingredient. (Isaacs, 1999, p. 395)

9. DIALOGUE: TRANSFORMING CONFLICT INTO COOPERATION

Yankelovich (1999)–in *The Magic of Dialogue*–defines dialogue as a highly specialized form of communication that imposes rigorous discipline on its participants. "When dialogue is done skillfully, the results can be extraordinary: Long-standing stereotypes dissolved, mistrust is overcome, mutual understanding achieved, visions shaped and grounded in shared-purpose, people previously at odds with one another aligned on objectives and strategies, new common ground discovered, new perspectives and insights gained, new levels of creativity stimulated, and bond of community strengthened" (p. 16). The author notes, however, that until recently people assumed that no particular skill is required to engage in dialogue. Most people assume that dialogue is just another form of conversation, not requiring special discipline. In the course of the past decade, however, a growing literature (and practice) has demonstrated the uniqueness of dialogue when it is well done. Yankelovich sets forth a set of considerations for "doing it well." Yankelovich identifies variations in the length of dialogue programs as critical. He suggests that it is appropriate to plan for:

* *a brief dialogue, when we aim at:* (a) examining an issue from various perspectives, (b) creating mutual understanding (by generative dialogue) prior to making decisions (strategic dialogue), (c) working with people with shared interest who trust each other, and (d) helping people understand each other on a specific issue.
* *a midrange dialogue, when we aim at:* (a) working through an emotional laden issue, (b) reaching across gender or sub-cultural issues, (c) preparing the ground for decision making on a sensitive issue (generative dialogue), and (d) seeking common ground among people who have opposing views on a controversial issue.

* *an extended dialogue, when we aim at:* (a) preparing the ground for negotiation between people who mistrust one another and are coming from different cultures, (b) achieving higher level teamwork among organizations with different subcultures, (c) working together to develop a new paradigm, and (d) making use of people's pooled experience in order to achieve higher level of thought, trust, and intimacy. (p. 124)

Yankelovich notes that not considering the time needed to accomplish the specific purpose of dialogues is a fatal mistake. We need to provide time for settling down and get beyond the inevitable showboating and plumage display to an honest expression of feelings, fears, assumptions, and convictions.

I wish to capture here some of Yankelovich's (1999) closing thoughts.

> Upon reflection, I have concluded that at this particular juncture in American history, I am not overstating the case for dialogue. My studies of the public reveal an immense pool of good will and good faith all over the country. Americans are hungry for an enhanced quality of life, for deeper community, for endowing communal life with spiritual significance. They are ready to accept truths over and above those of science and technical expertise without discarding their immense contributions. They are ready to meet on the edge of dialogue in order to endow their lives and those of others with a larger meaning. At the risk of overstating the case, I believe that greater mastery of dialogue will advance our civility–and our civilization–a giant step forward. Dialogue has the magic to help us to do it. (pp. 217-218)

10. DIALOGUE AS EVOLUTIONARY CONVERSATION

Evolutionary design is a process that carries a stream of shared meaning by a free flow of discourse among members of an evolutionary designing community, who seek to create an Evolutionary Guidance System (Banathy, 2000). It is proposed that the combination of generative dialogue and strategic dialogue comprises a comprehensive method of social discourse that is the most viable to use in an evolutionary designing community. We call this approach *evolutionary* (*design*) *conversation*. The root meaning of conversation is "to turn to one another." Members of an evolutionary design community turn to one another without reserve and in truth and openness, accepting and honoring each other. Before the design community engages in the substantive task of evolutionary design, it involves itself in generative dialogue. This involvement will lead to the creation of a "common ground" as the community focuses on the thoughts, values and world views of its members and creates a flow of shared meaning, shared perceptions, a shared world view, in a social milieu of friendship and fellowship. At this point the community is prepared to move on and engage in the strategic dialogue of evolutionary design inquiry.

PART TWO: SEARCHING TOGETHER: METHODS AND "POTHOLES"

In part two the most often used conversation methods are introduced as examples. One is called Nominal Group Technique, the other Idea Writing. In conclusion, I define the various roles of participating in conversation. Then, examine some potholes along road to dialogue.

11. NOMINAL GROUP TECHNIQUE (NGT)

Moore suggests (1987) that NGT structures small group conversations for developing ideas about issues of importance when faced with ill-structured situations of uncertainty and ambiguity, when–while we seek diversity–seek consensus among members of the group. The process of NGT includes: (1) the formulation of triggering questions for the issue to be addressed, (2) the silent generation of ideas in writing, (3) their recording and reporting to the group, (4) the discussion/clarification of ideas, and (5) the selection/designation of ideas. These five activities are now described.

* The Formulation of the Triggering Question (TQ) is a crucial task. The TQ drives the whole activity. The TQ should be simple and unambiguous. "It should elicit items at the desired level of specificity" (Moore, 1987, p. 25). Example: At the on-set of design it is too early to ask: What should be the outcome of our design? Rather ask the question: What values should guide our design?

* Members of the group should formulate and pilot test the TQ in order to determine if, in fact, it will produce the desired response. Additional preparation include: (1) a room in which the group can sit around a table, (2) newsprint that can be posted on the wall, and (3) markers and sheets of paper for members of the group to write their ideas. (The preferred size of the group is between five and nine.)

* The Silent Generation of Ideas can proceed in response to the triggering question which is recorded on sheets of paper distributed to group members. Instruct members to write phrases or short sentences and work independently in silence. Invite members to record as many items as they wish.

* A Round-Robin Reporting and Recording of Ideas maps the thinking of the group. Members report one idea at a time without any discussion, until all the ideas are reported and recorded and numbered on sheets of

newsprint. Members may "hitchhike" on the ideas of others and/or add new ideas.

* A Discussion/Clarification of Listed Items follows once all items are recorded. Invite comments and questions but avoid by all means expressions of judgement. Keep in mind that a most outlandish or even an impossible sounding item might become the most valuable later. We can now "pull" or edit items with the consent of the authors of the ideas. But resist combining items into larger categories.

* Prioritizing/Selecting Items. This activity proceeds in silence. Members of the group work with the large list and rank-order items in sequence of their preference. Prepare a tally sheet on the newsprint on which next to a numbered item we record the preference number. We can now list the items according to preference and might proceed with further clarification and discussion.

* If we have more than one group working, the next task is to assemble all groups and proceed with the integration/consolidation and final tallying of items. The product of this activity becomes input to the design program.

Over three decades of wide use of NGT suggests that the technique is easy to learn and apply and people using it enjoy it. It is usually very productive and does not take much time. The use of NGT is appropriate when collective idea generation is called for in the formulation of design issues. NGT is helpful in neutralizing dominant individuals. It is a starting place in the work of design groups and is usually followed by a technique of idea development/elaboration, described next.

12. IDEA DEVELOPMENT

Idea development focuses on the further elaboration of a specific item, which was an outcome of an NGT activity. Example: let us suppose that we produced a set of values that should guide our design. We then ask: What are the implications of a particular value in making decisions in the course of the design inquiry?

The ideas developed become a documented input to design. In general, this technique is helpful in exploring the meaning and implications of specific ideas to the content of design in the context of a specific design program. The activity involves several steps. These are: (1) preparation, (2) individuals respond in writing to the item or the triggering question, (3) members comment in writing to other members written responses, (4) members read others' comments to their response, (5) the group discusses the principal ideas that emerge from the written ideas and responses and the group develops and records on newsprint the findings of the activity. These five activities are now described.

* The Preparation is similar to the NGT activity. It includes the designation of a room, appropriate seating arrangement, means of recording and displaying the ideas, a careful formulation of the triggering question. A briefing on the procedure is also part of preparation.

* The Initial Activity involves each member's written comment on the item, formulated by a triggering question. Members should take five to ten minutes to formulate their ideas on a sheet of paper, and should note their name next to the triggering question. Once completed, members place their sheet in the middle of the table. After all members complete their writing and place their sheets in the middle of the table the response activity commences.

* The Response Activity. Now each person records his/her comments in response to the ideas of other members, using the idea writing sheets placed on the middle of the table. More presents a form that can be used for this activity.)

* Members read to the group their own idea and the responses of others to their idea.

* Following each reading, conversation develops that aims at clarification and further development of ideas.

* The last activity is the development of an integrated/consolidated statement that is recorded on newsprint, reported to the larger group and used as input to the design program.

The use of Idea Development is appropriate when the design group formulates design ideas, such as: (1) core values and ideas, (2) the image components of the future system, (3) guiding perspectives, (4) the purposes of the system, (5) systems specifications, (6) systems functions, etc. The ideas generated usually invite analysis and impact assessment.

13. ASSIGNING ROLES AND RESPONSIBILITIES

In order to ensure productive and focused experience to all members, I have found it useful to assign specific roles/responsibilities to each and every member of design groups when using the techniques described here. Such an arrangement is a way to share responsibility and leadership in accomplishing the task of the group. Suggested roles /responsibilities include the following:

Guardian and Guarantor of

Participation, in order to ensure that all members have equal time and opportunity to make contributions and no person shall dominate the group.

Keeping focus on the theme, on the specified issue, or on the triggering question.

Following the selected technique, so that orderly progress can be made toward the accomplishment of the task.

Documentation, ensuring that whatever is developed by the group is appropriately recorded and is made available for successive design work.

Honoring all contributions and guarding against criticizing or "belittling" ideas.

Living by the values that have been articulated individually and collectively as bases of making design choices and decisions.

"*Keeping the fire burning*." Keeping the team spirit at the highest possible level. (Ideas: an inspirational quote, a logo for the team, a theme song, a team ceremony, envisioning an ideal image of the design group, an empty chair representing future generations, and keeping the fire burning in the fireplace.)

Coordinating the work of the group, including an initial briefing on the task at hand, keeping time and ensuring that the group has adequate opportunity and resources to accomplish the task. Depending on the number of participants, a person might have more than one role to play. All the roles described above are open to all participants. Roles are assumed at the discretion of individuals and the group. Furthermore, the role designations can also change based on the experience of the group. Roles can be modified and new roles added as required.

I have introduced above suggestion that can guide a disciplined approach to conversation. What follows is a presentation of potential problems that would endanger the success of a dialogue. We should take the "how to do" and "how to avoid" as complementary as we engage in the disciplined inquiry of conversations.

14. POTHOLES ON THE DIALOGUE ROAD

Yankelovich (1999) suggests that there are deeply ingrained habits that undermine dialogue. Some are too self-important to respond to other people as equals. Others may not be to view life from any other point of view then their own. They cannot step out from their box. People with these habits will be unable to participate in a dialogue.

In the dialogue road there are potholes that make our travel difficult. It is wise to point out these as we embark on our conversation, even display them on a newsprint for all to see. Potholes include:

* *Holding back.* Some people are reluctant to participate, open up and commit themselves. This happens when we address some deeply felt issues. We should give adequate time to work through this.

* *Being locked in a box.* Yankelovich quotes a statement from an educator. Faced with a mired of problems, he said: "We are tireless in ransacking our own system for answers. We don't look outside the box for solutions. I am saying that it is something we never do" In searching for solutions many people restrict themselves to their own world.

* *Prematurely moving to action.* There is a strong tendency in many of us to rush into action. Faced with an issue, people are *quick* to ask: What should we do? This pothole is very dangerous. "A focus on swift action gets in the way of doing dialogue. It short-circuits the process of probing the depth of participants' thoughts, perceptions, feelings, and assumptions that can provide the foundations for informed decision making" (p. 134). (Earlier, we defined conversation as an integration of generative dialogue–that creates a common ground–and strategic dialogue–which moves us toward searching for e solution.)

* *Listening without hearing.* A most common problem in dialogue is a lack of willingness to make the extra effort to understand others. Emphatic listening requires the ability to tune in other people's feelings and the meaning of what they say." In our culture, not being heard is a conditioned response that is constantly reinforced" (p. 136). A useful technique to avoid this pothole is for participants paraphrasing what they think they heard. "Is this what you said?" "Would you please clarify this?"

* *Strongly felt preexisting points of view* requires a huge amount of time to work through and resolve thoughtfully and collectively, particularly if these issues are value-laden. Participants need to become sensitive to recognizing this "pothole" and allocate extra time to navigate through it.

* *Showboating* is an urge of self-expression and demonstrating self-worth. Often people can't resist to show how smart they are, how much they know, how much they can offer. These urges, however, conflict with doing dialogue. In dialogue "subordinating one's personality to a certain degree is needed to empathize fully with someone else's point of view" (p. 139). Often–given extra time–the show-boater settles down: "having displayed their feathers in their full glory" (p. 139). But show-boaters that must be on all the time are not cut out for dialogue.

* *Scoring debating points* is a temptation some people can't resist. "They listen attentively, but not for the purpose of understanding; Their impulse is to rebut" (p. 140). But dialogue requires setting aside our adversary impulses. Dialogue is not debate.

* *Contrarianism* is similar to debating. Some people can't help to automatically advance an argument that takes an opposite stance from a stated position. For them the 'firework of controversy' is more appealing

than the quiet pursuit of' mutual understanding. If we have ample time, contrarianism might expand the range of possible ideas and choices. But it can become obstructive if it takes the form of self-indulgence.

* *Having a pet preoccupation,* an obsession with some single idea or interest is another pothole on the road of dialogue. Such idea-fixes prevent people to from hearing/understanding what other people say. We should let the person to articulate their concern, indicate that we understand it and ask her/him to move on with us in continuing the dialogue.

* *Aria-Singing.* "Many dialogue has been sidetracked by the special pleading of a well-meaning participant" (p. 142). In an era of identity politics and extreme positions (how well meaning) people are "prone to singing heir arias at the slightest provocation–or even without any provocation" (p. 143). We can deal with this pothole by giving to it the same consideration as we did in the case of items 6 and 9.

15. SUMMARY

In Part One of this chapter, I developed a set of definitions and perspectives on dialogue that aimed enhancing our understanding of dialogue, what are its characteristics, and where it makes difference in the human experience. In Part Two, I proposed some methods and tools that are available to us to engage in dialogue as a disciplined approach to intentional, purposeful, collective communication. This section also addressed a description of suggested roles and responsibilities of the dialogue community as well as a list of problems that might arise in the course of the dialogue and ways that we can deal with them. It is hoped that the core ideas and the methods of dialogue introduced will be of help to the learner in initiating and engaging in dialogue.

REFERENCES

Banathy, B.H., 2000, *Guided Evolution of Society: A Systems View*. Kluwer Academic/Plenum Publishers, New York.

Beck, M., 1948, *The Concept of Dialogue.* Unpublished manuscript.

Bohm, D., and Peat, D., 1987, *Science, Order, Creativity.* Bantam Books, New York.

Bohm D. (1996). *On Dialogue*. New York: Routledge.

Isaacs, W., 1993. Dialogue, collective thinking, and organizational learning. *Organizational Dynamics*, **22**(2): 24-39.

Isaacs, W., 1999, *Dialogue and the Art of Thinking Together*. A Currency Book. New York.

Martin, D., 1999, *The Spirit of Dialogue*. International Communication for the Renewal on Earth. Cross River, N.Y.

Moore, C.M., 1987, *Group Techniques for Idea Building,* Sage, Newbury Park, CA.

Yankelovich, D., 1999, *The Magic of Dialogue*. Simon & Schuster, New York.

Zendin, T., 1994, *Intimate History of Humanity*. Harper Collins, New York.

Chapter 6

CONVERSATION: CREATING A LIVING METAPHOR

SABRINA BRAHMS, M.A.
Organizational Consultant, Madison, Wisconsin

We must engage in the very process we are studying.

– Donald Schön

1. INTRODUCTION

In this chapter, the author will discuss the many possible uses of metaphor within the context of social systems design and will particularly focus on the integral use of metaphor in conversation. First, an overview of the common understanding of metaphor will be provided and then, a brief historical perspective on the definitions and theories of metaphor within relevant philosophical and linguistic literature. The author will also introduce direct examples of metaphors that are used by different cultures to convey values as well as provide a narrative description of the author's experience with metaphor in the context of design conversation. In doing so, the text will move from the theoretical domain to the applied domain. This movement between theory and application will be evident throughout the chapter. The author is not only interested in the theoretical aspects of metaphor, but wishes to convey by illustration, something of its potential power.

2. DEFINING METAPHOR

The American Heritage Dictionary (1983) defines metaphor as follows: "a figure of speech in which a term that ordinarily designates an object or idea is used to designate a dissimilar object or idea in order to suggest comparison or analogy, as in the phrase*, evening of life*." We are taught in elementary school, at least in the United States, that metaphor is a literary tool, to be trotted out for use in poetic expression. This chapter will suggest that metaphor can have much broader application.

Aristotle (*Poetics*, K. McLeish, translation, 1999) was an early theorist concerning the purpose of metaphor. He defined it as an implicit comparison, but thought of it as a mere ornament in the process of communication. In this century, along similar lines, philosophers addressed metaphor as a "problem" in language (Wittgenstein, 1953). In their positivist paradigm, these theorists objected to the use of metaphor altogether because it can not directly state truths. Reality should be literally describable, from this viewpoint, and metaphor should be avoided. "To draw attention to a philosopher's metaphors is to belittle him—like praising a logician for his beautiful handwriting" (Black, 1962, p. 25).

As philosophers reflected on the notion of objective truth and as the notion of subjective truth was introduced, ideas about metaphor began to widen. No longer seen as a semantic problem, philosophers interested in language began to suggest new definitions (Pepper, 1948; Black, 1962; Schön, 1979; Kuhn, 1979, Lakoff and Johnson, 1980).

Pepper (1948) introduces the idea of a "root metaphor". He defines this term as an underlying metaphor, for which a person finds evidence in a context that he knows and which he then applies in order to understand new contexts. An example he gives is of someone who comes from a sea town and identifies greatly with water. This person may have a "root metaphor" that "all things are water". In order to understand new experiences, he applies his "root metaphor" to help him make new meaning. As we can see, this concept goes well beyond the idea of metaphor as a mere literary tool.

Black (1962) supposes that metaphors are not merely comparisons in language, but filters we use to understand and define our experiences. Again we find support from philosophers who consider metaphor to be intrinsically bound to the way we make meaning. For example, Black analyzes the metaphor "man is a wolf". We associate wolves with ferocity and dominance. When we view men in this light, we emphasize those characteristics and de-emphasize others. Black also suggests that our understanding of metaphor is not complete; we still don't understand why some metaphors work and others do not.

Schön (1979) positions metaphors as central to our ability to understand our world. He states that our metaphors influence how we view our societies and its problems. He further suggests that our ability to solve problems may be limited by the way we metaphorically describe them. As an example, Schön describes a company whose research and development team wants to find a way to improve the paintbrushes it manufactures. Only when a new metaphor for a paintbrush is conceived (a paintbrush is like a pump), the developers are successful in solving their technological problem.

Lakoff and Johnson (1980) extend the definition of a metaphor further by stating that nearly all our constructs of reality are metaphoric. They suggest that reality is constructed by creating shared meaning and our metaphoric language reflects these constructions. For example, when we say things as basic as, "that idea just won't sell" or "it's important how you package your ideas," we are implying that ideas are commodities. When we say, "that idea went out of style years ago," or "linguistics has become very chic," we refer to ideas as fashion. Lakoff and Johnson successfully illuminate the depth and breadth of our metaphoric use of language. They also provide a distinction between two types of metaphors, conventional metaphor, as described above, and novel metaphors that are capable of creating new meaning. Schön (1979) also identifies ways of using metaphors that create new possibilities. We will return to this idea later in the chapter because it is particularly relevant to social systems design.

Johnson's (1996) research reveals that the use of metaphor is a cognitive developmental task that manifests in all languages. Her research demonstrates that the developmental ability to interpret metaphors is evident even in a person's second language. Ivie (1996) suggests that metaphor can be used as a tool for critical thinking. He states that by analyzing our metaphors, we have the opportunity to understand our underlying assumptions.

3. DEFINING CONVERSATION AND SOCIAL SYSTEMS DESIGN

The author now turns to reviewing the concept of design conversation and the broader concept of social systems design. Defining these ideas thus creates a foundation for explaining how metaphor may be used in both. However, since the entire context of this book concerns itself with social systems design and conversation, only a very brief overview will be provided here for the purposes of presenting a cohesive chapter.

Social Systems Design (SSD) is both a creative and disciplined approach to transforming our social systems. Such social systems can range in size and scope to include families, schools and communities, as well as governments and societies. Social Systems Design is a different way of defining the problems inherent in our institutions by defining them in light of possible solutions. This form of inquiry is more interested in what should be than what is (Banathy, 1996). SSD is a values based inquiry, recognizing that at the foundation of every social system is the set of values of its creators.

Social systems design can be used as the basis for a theoretical analysis of a social system or it can be applied. In the application of social systems design, stakeholders gather to co-create a concept of new possibilities, a vision of a shared future. In order for this co-creation to be successful, communication among stakeholders (called co-designers) must be clear and productive. Herein lies the need for specific methods to facilitate successful co-design.

Conversation has emerged for precisely this purpose. The term conversation, defined more loosely in the general population, has a more specific meaning in this context. First, it draws on the principles of dialogue as defined by Bohm (1996). Dialogue is a collective method of exploring thought and honoring the individual's participation within the larger group context. Its purpose is to encourage greater consciousness and shared meaning. The structure of dialogue encourages a generative free-flow of ideas and this is its important contribution to the idea of conversation. However, the need for a more focused exchange has been identified (Banathy, 1996) as a necessary accompaniment to generative, non-goal oriented exchanges. This focused form of dialogue is called strategic and serves to address specific design purposes. The weaving of strategic and generative dialogue defines conversation within a design context.

4. DESCRIBING THE METAPHOR CONVERSATION PROCESS

The author of this chapter participates in a yearly conference, *The Asilomar Conversation* of the International Systems Institute. This conference has a different format than other conferences; it is different because participants do not sit as passive receivers of knowledge, listening to a few selected presenters. Instead, each conference participant chooses a research group based on her interest and then prepares an appropriate paper. The paper is distributed to the other group members before the conference begins and this is the official form of presentation. Then at the conference, the group spends five day, expanding individual preparatory work by conducting conversations with one another thus the conference's name. In this forum, there is no expert.

The background provided here is relevant to this chapter on metaphor because it was through participation in two such research groups that the author gained knowledge and experience about the effectiveness of using metaphor within the context of conversation. The first of two metaphor groups (Mapping the Use of Ethical Metaphor, 1998) resulted from the work of a preceding group on the role of social creativity (Report from the Social Creativity Group, 1997). The group members of the social creativity group, in the course of their week of conversation, realized that metaphor was worth its own line of inquiry, thus, the Metaphor Group of 1998 was born. It is this group's research that provided the foundation for the ideas expressed in this chapter.

The eleven participants spent the five days of conversation exploring the possible uses of metaphor within the design process. As is a landmark of every design conversation, there was no strict outcome goal and no linear process to be rigorously adhered to. Due to this stipulation, surprises in learning emerged. Many group members reported that their participation in the group was particularly rich and enlivening.

We began with what the author might define as an intellectual exploration of the topic. We grappled with the definition of metaphor, but didn't come to any conclusive answers in early phases of our conversation. We did decide that our use of the term metaphor would not only include metaphor as defined by the dictionary, but would include myths and stories as well. We included these to emphasize that myths, legends, stories, and parables perform the same functions as do simple linguistic metaphors, albeit in a longer format.

Our group explored the different cultural applications of metaphor and realized that many cultures use metaphor to perpetuate cultural values. Let's take for example, the cultural value of kindness. The African tale, *Mufaro's Beautiful Daughter's* (Steptoe, 1989), tells the story of two sisters who have the opportunity of becoming queen. The younger daughter, who is not only beautiful but also kind, passes the requisite tests due to her compassion toward animals and societal outcasts. A similar theme can be found in Norwegian folklore. Askelad, in *Askelad and the Silver Ducks* (Roll-Hansen, 1964) finds fortune, not only due to his strong work ethic, but due to his kindness as well. In *A Precious Life* (1989), a tale whose origin is associated with the Buddha, the great deer acts out of compassion when he saves the hunter who meant to kill him.

One group member shared the Greek Myth of Procrustes Bed (Bulfinch, 2000). In this myth, Procrustes, the inn owner had one bed size. If the guests were too long for the bed, their feet would be cut-off to fit. If the guests were too short for the bed, they were stretched it fit it. This myth, of course, serves as a metaphor for fitting to exacting standards. We noted that the metaphor itself could be used to promote or disapprove of exacting standards, depending on the context. The context in which a metaphor is used became an important group theme.

5. USING METAPHOR IN SOCIAL SYSTEMS DESIGN

We also explored possible social systems design situations that would benefit from the use of metaphors. The initial work conducted within the group has been expanded by the author and included as follows.

5.1 Defining Design Terms

As explained earlier, social systems design is a different way of looking at systems and since it represents a paradigm shift in thinking, the concepts require more explanation. Our design group agreed that metaphor would be an excellent explanatory tool for these concepts since metaphors have the unique ability to explain ideas that are not easily defined by more literal means. The best example that the authors know of the use of metaphor to explain design concepts is in Banathy's (1989) use of the "map" metaphor. Banathy defines the act of social systems design as a journey. The designer has a starting point and a desired destination. Banathy further explains that a map is useful when embarking upon any journey and points out that the map does not show the traveler what to do or where to go, but instead offers possibilities that the traveler, or in this case, the designer, must choose to act upon. Even the word journey is metaphorical and evokes the idea of unexpected outcomes and welcomed surprises.

5.2 Creating a Cohesive Design Group

Because SSD involves and honors all stakeholders of a system, a cornerstone of the design process is the identification of the shared values of these stakeholders. Discovering shared values may not be an easily accomplished task, especially in a complex social system where stakeholders have different priorities and points of view. As was stated previously, the group must become cohesive in order succeed at designing. Frantz (1992) writes about the need for group safety, that the design environment needs to be a place where people feel comfortable to share their values. She suggests that one way of creating this safety to be the sharing of stories that have metaphoric meaning for each participant. We can argue with people's ideas, but it is much more difficult to argue with their stories. Thorsheim and Roberts' (1989) research suggests a similar conclusion. In their work with women in the process of redesigning their lives, the researchers found that the sharing of life stories gave each individual woman a sense of connection with the group and contributed to a sense of group cohesiveness as well.

Our metaphor group also found this to be the case. Early in our group process, we each shared a personally meaningful metaphor. Some shared metaphors that they had heard and found interesting. Some shared metaphors that were personally constructed. One group member shared a dream that had strong metaphoric meaning. Through this experience, we came to know each other better, and were able to better honor individual contributions. The author believes that this sharing became an important foundation for the work that will be soon be described.

5.3 Honoring the Cultural Diversity of Group Members

One focus of our various group discussions was the question of whether metaphors are only applicable within their cultural context. We expressed concern about imposing one's own cultural metaphor onto someone from another culture. This led us to further discuss our ethical responsibilities in cross-cultural design. We came to a tentative conclusion that if one shares the context for a metaphor when using it, that the meaning of the metaphor could be transferred past cultural barriers. We also decided that it was most ethical to offer metaphors as opposed to imposing them. In addition, we realized it was equally necessary to listen for meaning instead of imposing our own interpretations. The authors found support for these ideas in the literature. Fitzgerald (1997) suggests that it is important not to make assumptions about another culture and its metaphors. Fitzgerald also suggests that while there are unique aspects to each culture, it is also important, when dealing in a cross-cultural framework, to recognize the qualities we all share as humans.

5.4 Facilitating Movement Passed Stagnation

Every design becomes stagnate at some point; this is the nature of the design process. But if dealt with properly, an impasse does not have to be lengthy or debilitating. We will look at an example from the literature that illustrates this point. Gregory and Noblit (1998) describe the cultural attitudes of a university in which they taught. According to the analysis of the researchers, this institution, which was located in Appalachia, operated with a pervasive metaphor of itself as a coal mine. The associated underlying assumptions included: a sense of unpredictability in work, and the idea that adversity can not be overcome, just endured. The premise of the article is that sometimes our metaphors can reflect the limitations in our thinking and these limitations can create institutional impasses. Only by examining the impasses to further progress is it possible for us to move past stagnation. This idea is also reflected in the work of Schön (1979) as well as Lakoff and Johnson (1980). Our group agreed that metaphor could be helpful in identifying stagnation and thus providing clarity on how to move beyond it. Not only did we theorize about this topic, but we experienced it as well, as will be described in the next section.

6. LIVING METAPHOR

We had already worked together for three of our five days when the group seemed unfocused; enthusiasm waned. We recognized ourselves at a point of stagnation and someone suggested we try an activity. The group member described that we could model a form of communication used by indigenous people in Australia called a “walk-about”. The idea presented was as follows: we would go outside with a certain design dilemma in mind. We would then listen to our surroundings for an answer to the chosen dilemma. When we “received” our answer, we would bring back a representation of the experience to share with the group.

To offer some background, ours was basically a rational group—a group of academics and consultants for social change—not people necessarily open to receiving messages from trees. The other background that’s important to provide is that the setting for the conference was a beautifully rugged conference center along the steel gray Pacific Ocean. This environment offered us the sand, Monterey Cypress trees, and the ever-present sound of the tides. We all agreed to the activity, perhaps only looking forward to being released from our little meeting room for an hour.

We wandered, we occasionally ran into other group members who were also wandering. When we reported back, we discussed that we each had spent much time wondering, wondering if we would indeed "hear" an answer, wondering whether we were crazy. Then slowly, each person presented his "walk-about" story. We witnessed the ceremonious presentation of each associated metaphoric object. Often, between stories, we shared silence. We were amazed that each person had returned to our meeting room with a new solution to their current design dilemma. The "walk-about", though different than a Western mode of solution seeking, had indeed worked.

After all the presentations were made and honored, we returned to our intellectual mode to evaluate the experience we had just shared. One group member suggested the power to be found in the silences, in the spaces. She said she'd call it "the space between". Thereafter, that metaphor came to represent our shared experience. The power resided not in the metaphorical phrase, but in the experience that the phrase represented. To this day, each of the authors of this chapter needs only to think of that phrase and conjure the entire experience. We were struck by how much more powerful is the metaphor when the representative situation is experienced, that we created the term "living metaphor" to distinguish this kind of metaphor from others.

7. CONCLUSIONS

As was stated at the beginning of this chapter, Black (1962) suggests that it is not yet clearly understood why some metaphors work and others don't. The author proposes that the effectiveness of metaphor may be associated with the strength of connection one feels to its meaning. As was experienced in our group, our metaphor was so effective not only because we felt a connection to its meaning, but because we had actually experienced the situation which the metaphor represented. We came to understand that metaphors are necessarily connected with the meaning given them by their user, that the context as well as the metaphor itself must be conveyed and honored. To understand a person's metaphor is to partially understand a person.

In addition, our metaphor research led us to understand that there is definitely an ethical way of using metaphors. This entails offering metaphors, but not imposing them upon others. We also learned the importance of listening to others as they present their metaphors, not assuming that we understanding without confirming. We recognized that we tend to underestimate the power of this kind of listening. As one group member said, "we are always searching for new systemic, disciplined design methods when we have a simple one (listening) always available to us." Ethics became a primary focus of our learning and a topic for further exploration.

In reference to social systems design and the process of conversation, we discovered that using metaphors can be helpful in many design and conversation contexts. It can be used to help the group become cohesive and productive. It can bridge cultural gaps. Metaphor can also be used to explain design terms and to move past stagnation in the conversation process.

Most importantly, we discovered the power of creating conditions for metaphors to emerge. Indeed it is helpful to apply metaphors, but it is also a very effective design conversation strategy to create the conditions for metaphors to present themselves. If we make space for metaphors to emerge from the conversational process, those metaphors will be particularly meaningful and useful to the participants. First, we create the conditions, then we must listen for the metaphors to speak.

ACKNOWLEDGEMENTS

The author would like to first thank Muriel Adcock for the early conversations that encouraged the writing of this chapter. In addition, heartfelt thanks are sent to each member of the 1998 Asilomar Conversation metaphor research team.

REFERENCES

A precious life: A jataka tale, 1989. Dharma Publishing, USA

American Heritage Dictionary, 1983. Dell Publishing, New York

Aristotle, 1999. *Poetics*. (K. McLeish, trans.). Theatre Communications Groups, New York.

Banathy, B.H., 1989. *Design: A Journey to Create the Future: A Map of the Journey*. Paper presented at the First Pacific Rim Fuschl Conversation. Asilomar, California.

Banathy, B.H., 1996. *Designing Social Systems in a changing World*. Plenum, New York.

Black, M., 1962. *Models and Metaphors*. Cornell University Press, Ithaca, New York.

Bohm, D., 1996. *On Dialogue*. (L. Nichols, ed.). Routledge, London.

Bulfinch, T., 2000. *Bulfinch's Greek and Roman Mythology: The Age of Fable*. Dover Publications, Mineola, New York.

Frantz, T.G., 1992. *Stories for developing systems designers*. A paper presented at the Fourth Annual Conference on Comprehensive Systems Design in Education. Pacific Grove, California.

Fitzgerald, T.K., 1997. Understanding diversity in the workplace: Cultural metaphors or metaphors of identity? *Business Horizons,* **40**: 66-70.

Gregory, R. & Noblit, G.W., 1998. The "coal miner" university: Explorations into metaphors on education organizations. *The High School Journal* **82**(1): 43-48.

Ivie, S. D., 1996. Metaphors: Tools for critical thinking. *McGill Journal of Education*, **31**: 57-68.

Johnson, J., 1996. Metaphor interpretations by second language learners: Children and adults. *Canadian Modern Language Review* **53**: 219-241.

Kuhn, T. S., 1979. Metaphor in science. In *Metaphor and Thought,* (A. Ortony ed). Cambridge University Press, Cambridge, pp. 409-419.

Lakoff, G., & Johnson, M., 1980. *Metaphors We Live By*. University of Chicago Press, Chicago

Mapping the use of ethical metaphor to the systems design/evolutionary guidance, 1998. In *Proceedings of the Tenth International Conversation on Comprehensive Design of Social Systems*. International Systems Institute, Asilomar, California, pp. 27-34.

Pepper, S.C., 1948. *World hypotheses: A study in evidence*. University of California Press, Berkeley, California.

Report from the social creativity group: Metaphorming of systems designers, 1997. In *Proceedings of the Ninth International Conversation on Comprehensive Design of Social Systems*. International Systems Institute, Asilomar, California, pp. 71-82.

Roll-Hansen, J., 1962. Askelad and the silver ducks. In *A Time for Trolls: Fairy Tales from Norway*. Johan Grundt Tanum Forlag Publishing, Oslo, pp. 11-18.
Schön, D.A., 1979. Generative metaphor: A perspective on problem-setting in social policy. In *Metaphor and Thought* (A. Ortony ed.). Cambridge University Press, Cambridge, pp. 254-283.
Steptoe, J., 1989. *Mufaro's Beautiful Daughters: An African Tale*. Scholastic, New York.
Thorsheim, H.I., & Roberts, B.R., 1989. Praxis through shared lifestoires: A design process for community empowerment through social elaboration of learning. In *Proceedings of the 33rd Annual Meeting of the International Society for the Systems Sciences*. (P.W.J. Ledington (ed.)). Available from ISSS, Institute of Safety and Systems Management, University of Southern California, Los Angeles, pp. 188-136.
Wittgenstein, L, 1953. *Philosophical Investigations*. Macmillan, New York.

Chapter 7

APPRECIATIVE INQUIRY AS CONVERSATION

KAREN E. NORUM
Gonzaga University

Have patience . . . try to love the questions themselves. Live the questions now. Perhaps then, someday far in the future, you will gradually, without even noticing it, live your way into the answer.

– Rainer Maria Rilke

1. INTRODUCTION

Banathy describes social systems design as a "process that carries a stream of shared meaning by a free flow of discourse among the stakeholders who seek to create a new system" (1996, p. 213). Later, he suggests that we "set forth an image of the system we design that is the best we can create, that reflects our highest aspirations and expectations" (p. 307). Appreciative Inquiry provides both a framework and a process for creating "ideal" social systems using a stream of positive conversation.

2. WHAT IS APPRECIATIVE INQUIRY?

Appreciative Inquiry (AI) was developed by David Cooperrider, professor of organizational behavior at Case Western Reserve University. As opposed to problem solving, AI begins with a search for the best of "what is" rather than looking for exactly what is wrong or what needs to be "fixed." According to Cooperrider, every system has good and bad in it. We are trained to look for the "bad" and "fix" it. We typically study "problems" such as alienation, stress, or conflict but we could just as easily choose to study joy, vitality, or cooperation (Ludema, Cooperrider, & Barrett, 2001). There are problems to study only if we agree there are problems (Gergen, 1999). AI gives us a structure for searching out the "goodness" in the system, allowing us to appreciate "what is" and use that as inspiration for what "could be." Assumptions about people, organizations and relationships are deliberately affirmative (Ludema, Cooperrider, & Barrett, 2001).

Six principles are central to the theory base of AI: the Constructionist Principle, the Principle of Simultaneity, the Poetic Principle, the Anticipatory Principle, the Positive Principle, and the Wholeness Principle (Cooperrider & Whitney, 2000; 2001). Each will be briefly described in the following section.

Four phases comprise an "appreciative" inquiry: discovery, dream, design, and destiny. Each phase can be considered a type of conversation. The phases and the kind of conversation associated with each will be described in a later section.

3. CENTRAL PRINCIPLES OF APPRECIATIVE INQUIRY

The six principles that are central to Appreciative Inquiry are based on generative-theoretical work. Propositions are associated with each principle. The principles and propositions contribute to an "affirmative basis of organizing" (Cooperrider, 2001, p. 48).

3.1 The Constructionist Principle

This Principle asserts that organizations are living, human constructions. They are constructed based on what we think we know, thus what we know and how we know it becomes fateful (Cooperrider & Whitney, 2000). "[T]he truth about an organization is what those involved agree the truth is" (Zemke, 1999, p. 30). This principle is strengthened by the proposition that stakeholders in the organization carry in their minds some sort of shared idea of what the organization is, how it should function, and what it might become (Cooperrider, 2001). Because these "guiding images are not detailed objectives but are paintings created with a larger brush stroke" (Cooperrider, 2001, p. 49), this principle calls us to unearth and examine the mental models (Senge, 1990) that we hold about an organization and consider how those mental models have effected the fate of the current system.

3.2 The Principle of Simultaneity

Change begins the minute we ask a question. The questions posed set the stage for what is found. What is found becomes the data we use to re-construct the future. "Even the most innocent question evokes change" (Cooperrider &

Whitney, 2000, p. 18). Thus, change is not something that happens after an analysis is conducted; change begins with the analysis. A corresponding proposition encourages us to create the conditions for organization-wide appreciation to "ensure the conscious evolution of a valued and positive future" (Cooperrider, 2001, p. 54). If as Cooperrider and Whitney state, "inquiry is intervention" (2000, p. 18), we can consciously create the conditions for organization-wide appreciation by being attentive to the language we are using and the questions we are asking. Our language generates change.

3.3 The Poetic Principle

"There is no such thing as an inevitable organization" (Cooperrider, 2001, p. 49). According to the Constructionist Principle, organizations are social constructions. That means they can be re-constructed. Just as a poem can be interpreted and re-interpreted as we bring new meaning to every reading of it, so can organizations be re-interpreted as the system they are embedded in changes. This is "The Poetic Principle": "organizations are genetically constituted socially in and through the images born in transaction among all participants" (Cooperrider, 2001, p. 49). As the stories of the people in and attached to the organization change, the organization changes. "Past, present and future are endless sources of learning, inspiration and interpretation" (Ludema, Cooperrider, Barrett, 2001, p. 189). This principle reminds us that we can choose what to study in an organization: the good or the bad, the joy or the alienation, the creativity or mediocrity (Cooperrider & Whitney, 2000; Ludema, Cooperrider, Barrett, 2001). A related proposition tells us that no matter what the previous history, every system can be altered and re-invented (Cooperrider, 2001).

3.4 The Anticipatory Principle

This Principle asserts that the image of the future guides the current behavior and actions of the system (Cooperrider & Whitney, 2000). Positive images of the future lead to positive actions; negative images lead to negative actions. A system's image exists "deep within the internal dialogue of the organization" (Cooperrider, 2001, p. 50) and becomes the "referential core" of the system, determining its essential characteristics (Capra, 1996; Wheatley, 1999). It is possible for this image to be incoherent or unclear or even for it to be pathetic. Many organizations are better at articulating what they do not want than at being clear about what it is they DO want. An image that is based on what we do not want is likely to engender negative behavior and actions. Malaise, mediocrity, angst, and dysfunction are likely to be present in such an organization. This principle is supported by the proposition that systems are limited only by their imaginations and capacity for creating truly shared vision (Cooperrider, 2001; Senge, 1990). Paradoxically, even the best future images can hold the system back if those positive images become so cherished, they cannot be given up for even better images (Cooperrider, 2001). This proposition reminds us of the Constructionist Principle: our positive images are constructions and at some point, need to be left behind for even better ones.

3.5 The Positive Principle

The more positive the question asked is, the more longer lasting and successful the change effort will be (Cooperrider & Whitney, 2000). Inquiring into the good, the better, and the possible promotes enthusiasm, creativity, and new insights. Problem solving is a null-sum game, directing the focus to what is wrong (Zemke, 1999). Building and sustaining momentum for change requires large doses of hope, inspiration, caring, excitement and commitment. Seeking out positive experiences and past successes and using those to build the future engenders positive affect and social bonding. The "heliotropic" proposition tells us that systems, like plants, move in the direction of light or positive imagery (Cooperrider, 2001). Thus, organizations move in the direction of what they study. To move in a positive direction, the system has to be studying the positive, not the negative. A related proposition asserts that the more an organization experiments with conscious evolution of positive imagery, the better it will become as its heliotropic and affirmative competencies strengthen (Cooperrider, 2001). This heliotropic tendency needs to be appreciated and understood—this proposition directs us to appreciate rather than fix our organizations (Cooperrider, 2001; Zemke, 1999).

3.6 The Wholeness Principle

This is an emerging Principle, based on the idea that when the wholeness of a system is deeply felt, it brings out the very best of the system (Cooperrider & Whitney, 2001). "We seek for connection and restore the world to wholeness" (Wheatley & Kellner-Rogers, 1996, p. 88). Systems move towards wholeness, seeking coherence. Empowering human potential and giving wings to imaginations helps the system to connect with itself (Barros & Cooperrider, 2001); with its referential core. Living systems appear to have a fundamental unity, a "sacred wholeness" (Gergen, 2001a) that is the essence of their integrity (Senge, 1990). Whole systems are healthy systems because their integrity is honored. As they connect more and more with their referential core, they become more "themselves." At the same time, there is recognition of a connection outside the system to something greater (Wheatley, 2002). The system's place in the wholeness of life becomes clearer. The Anticipatory Principle is related to the Wholeness Principle: the image of the future not only guides the current behavior and actions of the system, it determines its health as well.

4. DISCOVERY, DREAM, DESIGN, DESTINY: THE 4 D'S OF APPRECIATIVE INQUIRY

Cooperrider and Whitney describe Appreciative Inquiry as "the cooperative search for the best in people, their organizations, and the world around them" (1999, p. 10). This involves a search into what gives life to and sustains life in the system. AI employs a 4-phase model to conduct this search: Discovery, Dream, Design, Destiny (4-D's, Figure 1).

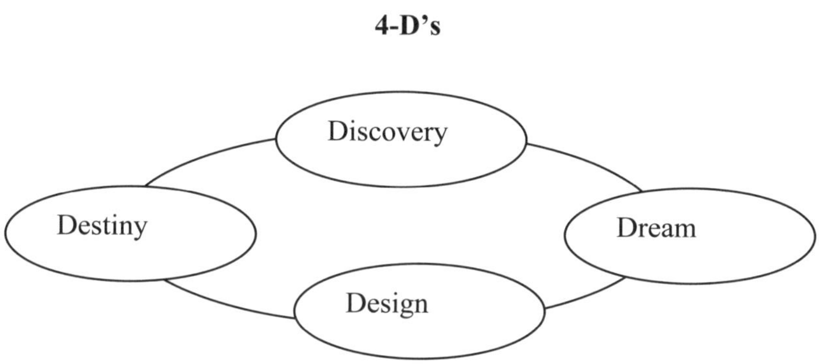

Figure 1. 4-phase Model

The cycle begins by crafting positive questions that are designed to uncover the life-giving and life-sustaining forces of the system. The inquiry begins through an interviewing process. During this process, the task is to discover the best of what already is—to appreciate the good things about the system. The "best of what is" is used to inspire "what might be": the Dream phase. Possible ideal futures are envisioned. The next step is to create the infrastructure that is needed to support "what should be." This is the Design phase. The new system is co-constructed. As the new system is implemented, the question turns to how to sustain and maintain this new system. The focus of the Destiny phase is how to continue to learn, improvise and adjust so that the system can continuously strengthen its affirmative capacity. This often leads back to the first phase in the cycle: Discovery. A new inquiry begins into the "best of" what has just been re-created.

A hallmark of AI is the kinds of questions asked during the inquiry: questions that are unconditionally positive (Ludema, Cooperrider, & Barrett, 2001), designed to strengthen the positive potential in the organization. The inquiry is based on the assumption that people carry many stories around with them and many of those stories are about value, wonderment, and joy (Gergen, 1999). The inquiry also assumes that what is "working" can be amplified and fanned throughout the system (Bushe, 2000b). According to the heliotropic hypothesis, systems grow towards what they persistently, actively, and collectively ask questions about. This premise is at the heart of the Appreciative Inquiry process. The importance of the language we use and its power to construct the systems we live in will be discussed in more detail later.

Each phase and the conversation associated with it will be described next.

4.1 Discovery

The purpose of the Discovery phase is to identify the life-giving and life-sustaining factors of the system as it currently exists. It is a search for the best of what is and an act of appreciation. The assumption is that there is "goodness" in the system that can be sought out. The task is to discover, disclose, and understand the positive capacity of the system (Cooperrider & Whitney, 2000) as well as or better than we discover, disclose and understand its problems.

In this phase, the conversation is anchored by carefully crafted positive questions. "Understanding depends upon how we look at facts, upon what questions we have asked in arriving at them" (Zohar, 1997, p. 65). Because the questions asked determine what is found and the data gathered determines what is designed, a fair amount of time is devoted to crafting "good" questions. "Good" questions are affirmative, posed as an invitation, and evoke storytelling about peak experiences. These questions are meant to engender wonder, surprise, and imagination for both the interviewer and the interviewee (Cooperrider, 2000).

As people share their stories about highlights of their experiences, ferreting out the life-giving and life-sustaining core factors that already exist in the system, hope grows and community expands. A positive, creative energy is created as the conversation turns from negativity, criticism, and "spiraling diagnosis" to that of "discovery, dream, and design" (Cooperrider, 2000, p. 123). Identifying the best of "what is" promotes dreaming about what might be. The outline of a feasible ideal system begins to emerge (Banathy, 1996).

4.2 Dream

While the conversation in the Discovery phase focuses on "what is," the conversation in the Dream phase turns to "what might be." Using data from the stories shared in the Discovery Phase, "people listen together for moments when the organization was 'alive'" (Cooperrider & Whitney, 1999, p. 14). What "might be" begins to take shape as people begin to expand the realm of the possible (Cooperrider & Srivastva, 2000). Data from the Discovery phase is mined; curious inspiration allows people to dialogue about their dreams and concerns in affirming ways (Cooperrider & Whitney, 2000). The power of hope is released as a valued and vital future is prefigured, becoming a powerful catalyst for change and transformation (Ludema, 2001). A shared vision of the future begins to take shape. A collective confidence is born as people dialogue about how to make their dream a reality.

4.3 Design

The dream becomes institutionalized in the Design Phase. The conversation in this phase revolves around what it would take to bridge "what is" with "what might be." What kind of infrastructure is needed to support the image evoked by the Dream Phase? Because the image of the future is based on actual examples from the organization's positive past (Cooperrider & Whitney, 2000), the "ideal" is still within the realm of possibility. Seven principles have been identified to guide the conversation in this phase (Mohr, 2001).

The Inclusion Principle asks that the whole system be involved in the conversation. This makes for a rich conversation and increases the possibility of true innovative design. The Aspirations Principle reminds us that when we seek our highest aspirations, constraints are removed and we move from better to breakthrough imaging of innovative structures and systems. Using positive past history to create a positive future is the Continuity Principle. By designing based on "what is," the new ideal is feasible. Sometimes truly innovative ideas take us beyond our comfort zone and require bold leadership for their institutionalization in the system. This is the Innovation Principle. The Chaordic Principle is a call to trust in the self-organizing capacity of the system. The glue of shared vision allows for freedom and autonomy as well as collaboration and cooperation as the system determines how to re-arrange itself. The fact that no design is perfect and regular cycles of inquiry are needed to deepen the understanding of what is working and why is the Re-Inquiry Principle. A continuous quest to discover the best needs to be embraced. The last principle, the Homegrown Principle, prompts us to remember that the more the stakeholders are involved in the design framework and process, the better its chance of successful implementation.

Information obtained in the Discovery Phase and mined in the Dream Phase provides the foundation for the Design of the new system. As a deliberately inclusive and supportive environment is created for conversation, interaction, and dialogue (Ludema, Cooperrider, & Barrett, 2001), new forms, new practices, new structures, new directions materialize. The dream becomes reality as the infrastructure to support it appears through conversation. The next challenge becomes how to nurture and sustain the new design.

4.4 Destiny

As its members live into the system they have designed, they learn to improvise and develop the capacity to tinker. Sustaining, maintaining, and nurturing the newly designed system comprises the conversation in the Destiny Phase. The talk is also about identifying the signs that indicate a new cycle of Discovery, Dream, Design, and Destiny needs to begin. The tendency to plan and coax the new Design into solid being needs to be abandoned. Improvisation and tinkering is called for instead: "Tinkering opens us to what's possible in the moment. Analytic plans drive us only toward what we think we already know" (Wheatley & Kellner-Rogers, 1996). Developing the capacity to improvise involves exploring and experimenting without knowing where it will lead (Barrett, 1998). Conversation centers on how to continue to learn, improvise and adjust so that the system can continuously strengthen its affirmative capacity. A "tinkering plan," one for sustaining, maintaining, and improvising the design, is developed in this phase.

4.5 A Stream of Positive Conversation

A free flow of discourse, characterized by a stream of positive conversation, pours through the four phases. Beginning with the Discovery Phase, the conversation starts with crafting and asking positive questions about positive factors already existing in the system. If this is "what is," then dreaming about what "might be" furthers the stream of conversation. The conversation spills into strategizing what it would take to support the "dream." The Design Phase turns the conversation to identifying the infrastructure necessary to support new forms, practices, structures and directions. With the dream becoming a reality, a conversation focused on nurturing, sustaining, and maintaining the new system takes shape. The conversation often turns deeper into itself as new discoveries are made, inciting members of the system to more deeply appreciate rather than fix the system. A conversation that begins with an inquiry into the best of what already exists in the system leads to the shaping of an ideal future, limited only by the imaginations and courage of the system's members.

5. CONVERSATION AS DESTINY

There is great power in how we talk about things. "The words we use and the way we use them are powerful indicators of how we see, of our particular vision of reality" (Daloz, 1986, p. 233). Language becomes central to the way we carry on our lives together (Gergen, 1999). We are constantly telling each other stories that form and inform our lives (Widdershoven, 1993). As we share our stories, we also share what we believe about the way the world works and begin to fashion our futures. The stories we tell have the potential to expand our imaginations and enlarge our vision of what could be (Feige, 1999). When we tell stories about our organization in the break room, over the water cooler, in the hallway, we socially construct what we want our organization to be (Abma, 1999; Bushe, 2000b; Gergen, 1999). These stories then end up guiding the organization's practices and policies (Abma, 1999; Cooperrider, 2000; Cooperrider & Whitney, 1999). Wheatley suggests, "...simple conversations that originate deep in our caring give birth to powerful actions that change lives and restore hope to the future" (2002, p. 23). This is the Constructionist Principle in action: a system can be reconstructed by changing the stories that are being told (Norum, 2000). Bushe reminds us, " . . . what is critical to creating change is not the generation of new images/theories but the telling and retelling of stories that create new and more efficacious meanings that support organizational evolution" (2000b, p. 105)

Gergen asserts the "meaning of our world is generated through the way we use words together" (1999, p. 35). We socially construct the system that we have by how we talk about it, the words we choose, the language we speak, the stories we tell. Gergen suggests four working assumptions that guide social constructionism.

5.1 Working Assumption One

According to Gergen, the "terms by which we understand our self and our work are neither required nor demanded by 'what there is'" (1999, p. 47). There are several ways to describe any one situation. However, once we have offered a description or interpretation of a situation or event, we have also constrained it by defining what it is and is not. We then make decisions based on what we think we now can and cannot accomplish, based on how we have chosen to describe the situation. Ackoff (1999) suggests most of the constraints we labor with or against are self-imposed and thus can be disposed if we so choose.

5.2 Working Assumption Two

"Our modes of description, explanation, and/or representation are derived from relationship" (Gergen, 1999, p. 48). Words do not mean anything on their own. As we put words together in sentences, paragraphs, and pages they become meaningful because a context has been created. "Meanings are born of coordinations among persons—agreements, negotiations, affirmations" (Gergen, 1999, p. 48). Wheatley tells us, "We live in a universe where relationships are primary" (1999, p. 69). We do not know what something is until it is seen in relationship to something else, when it takes on a context.

5.3 Working Assumption Three

We fashion our future through what we describe and explain (Gergen, 1999). Language does not simply describe actions, it is a form of action (Gergen, 2001b). We use lenses of our own making and choosing to filter and select what we will pay attention to (Wheatley, 1999). Gergen calls for generative discourse: "ways of talking and writing ... that simultaneously challenge existing traditions of understanding, and offer new possibilities for action" (1999, p. 49). He suggests that we can become poetic activists through carefully choosing the language we use to create our futures (1999).

5.4 Working Assumption Four

Making time to reflect on our forms of understanding is critical to our future well-being (Gergen, 1999). Based on what we think we know and understand, we make choices about what to save, what to destroy, what to resist, what to create. Unearthing and examining the mental models behind our forms of understanding is integral to helping us see other alternatives (Gergen, 1999; Senge, 1990). Gergen goes as far as stating, "If we are to build together toward a more viable future then we must be prepared to doubt everything we have accepted as real, true, right, necessary, or essential" (1999, p. 50). As assumptions are examined and suspended, dialogue that can lead to a new "ideal" shared vision is promoted.

5.5 Change the Conversation, Change Destiny

By how we talk about it, "we largely create the world we later discover" (Cooperrider & Srivastva, 2000, p. 92). Our current social vocabulary is largely deficit based, making it a challenge to consciously move to hopeful and appreciative language (Ludema, 2000). How we think and talk about things largely influences our destiny. If we think and talk like something is impossible to accomplish, it probably will be. By how we think and talk about it, we create the world we later discover.

If it were true that we socially construct the systems that we have by how we talk about them, then it would follow that if we change the way we talk about the system, the system will change. The language we use matters dearly because there are consequences to it. The words we use carry meaning; how that meaning is interpreted leads to assumptions and actions. Gergen asserts, ". . . we live [our] discourses" (1999, p. 62).

In Appreciative Inquiry, the questions asked determine what is found; what is found determines what is designed; what is designed becomes destiny. Thus, the questions asked become powerfully consequential. If we can choose to inquire into either the good or bad of a system, why not choose to focus the inquiry on the "goodness" of it?

6. APPRECIATIVE DESIGN CONVERSATION

"Design brings forth novelty, something that does not yet exist, something that has to be envisioned and then created" (Banathy, 1996, p. 196). It is a journey toward the ideal. Appreciative Inquiry provides a framework for identifying and creating that ideal. It engenders conversation that is positive, hopeful, creative, energetic, generative, and transformative.

If our goal is to design a system that is the best we can create, one that reflects our highest aspirations and expectations (Banathy, 1996), the kind of conversation promoted through Appreciative Inquiry is the kind of design conversation we seek. The determination to seek out the goodness of the system, to discover what is life-giving and life-sustaining about it so those factors can be amplified and fanned throughout the system, is AI's trademark. Such design conversation engages the mind, heart, and spirit and moves us toward the realm of wonder and learning (Cooperrider, 2000; Kahane, 1999). It allows for what Banathy describes as "heartstorming": freedom from the requirement to fit into the normal modes of scientific inquiry (1996, p. 220).

Such design conversation also asks us to begin with the questions, paying careful attention to what we are asking and how. The poet Rilke implores us to love the questions themselves, for we live our questions. Using AI as a framework, we can craft unconditionally positive questions (Ludema, Cooperrider, Barrett, 2001), and in the process of exploring those questions, heartstorm our way into an ideal design.

REFERENCES

Abma, T. A., 1999. Powerful stories: The role of stories in sustaining and transforming professional practice within a mental hospital. In R. Josselson and A. Lieblich (Eds.), *Making Meaning of Narratives: Vol. 6* (pp. 169-193). Thousand Oaks, CA: Sage Publications, Inc.

Ackoff, R. L., 1999. *Re-creating the Corporation*. New York: Oxford University Press.

Banathy, B. H., 1996. *Designing social systems in a changing world*. New York: Plenum Press.

Barrett, F. J., 1998. Coda: Creativity and Improvisation in Jazz and Organizations: Implications for Organizational Learning. *Organization Science*, **9**(5): 605-622.

Barros, I. O., and Cooperrider, D. L, 2001. *Appreciative Inquiry Fostering Wholeness in Organizations: A Story of Nutrimental in Brazil: How Wholeness, Appreciation, and Inquiry*

Bring the Best in Human Organization. Retrieved 6/6/02: http://appreciativeinquiry. cwru.edu/ research/bibPublishedDetail.cfm?coid=191

Bushe, G. R., 2000a. Advances in Appreciative Inquiry as an Organizational Development Intervention. In D. L. Cooperrider, P. F. Sorenson, Jr., D. Whitney, and T. F. Yaeger (Eds.), *Appreciative Inquiry: Rethinking Human Organization Toward a Positive Theory* of Change. Stipes Publishing, Champaign, IL, pp. 113-121.

Bushe, G. R., 2000b. Five theories of change embedded in appreciative inquiry. In D. L. Cooperrider, P. F. Sorenson, Jr., D. Whitney, and T. F. Yaeger (Eds.), *Appreciative inquiry: Rethinking human organization toward a positive theory of change*. Stipes Publishing, Champaign, IL, pp. 99-109.

Capra, F., 1996. *The web of life: A new scientific understanding of living systems*. Anchor Books, Doubleday, New York.

Cooperrider, D. L., 2000. The "child" as agent of inquiry. In D. L. Cooperrider, P. F. Sorensen, Jr., D. Whitney, and T. F. Yaeger (Eds.), *Appreciative Inquiry: Rethinking Human Organization Toward a Positive Theory of Change*. Stipes Publishing, Champaign, IL, pp. 123-129.

Cooperrider, D. L., 2001. Positive image, positive action: The affirmative basis of organizing. In D. L. Cooperrider, P. F. Sorenson, Jr., D. Whitney, and T. F. Yaeger (Eds.), *Appreciative Inquiry: An Emerging Direction for Organization Development* Stipes Publishing, Champaign, IL, pp. 91-125.

Cooperrider, D. L., and Srivastva, S., 2000. Appreciative inquiry in organizational life. In D. L. Cooperrider, P. F. Sorenson, Jr., D. Whitney, and T. F. Yaeger (Eds.), *Appreciative inquiry: Rethinking human organization toward a positive theory of change*. Stipes Publishing, Champaign, IL, pp. 55-97.

Cooperrider, D. L., & and Whitney, D., 1999. *Appreciative Inquiry*. San Francisco: Berrett-Koehler Communications, Inc.

Cooperrider, D. L., and Whitney, D., 2000. A positive revolution in change: Appreciative inquiry. In D. L. Cooperrider, P. F. Sorensen, Jr., D. Whitney, and T. F. Yaeger (Eds.), *Appreciative Inquiry: Rethinking Human Organization Toward a Positive Theory of Change*. Stipes Publishing, Champaign, IL, pp. 3-27.

Cooperrider, D. L., and Whitney, D., 2001. Discovery. Plenary Session The First International Conference on Appreciative Inquiry. Baltimore, MD. Sept. 30-Oct. 3, 2001.

Daloz, L. A., 1986. *Effective teaching and mentoring: Realizing the transformational power of adult learning experiences*. Jossey-Bass Inc., Publishers, San Francisco.

Feige, D. M., 1999. The legacy of Gregory Bateson: Envisioning aesthetic epistemologies and praxis. In J. Kane (Ed.), *Education, Information, and Transformation: Essays on Learning and Thinking*. Prentice Hall, Inc, Upper Saddle River, NJ, pp. 77-109.

Gergen, K., 2001a. *Fireside Chat*. The First International Conference on Appreciative Inquiry. Baltimore, MD. Sept. 30-Oct. 3, 2001.

Gergen, K. J., 2001b. *Social Construction in Context*. Sage Publications, London.

Gergen, K. J., 1999. *An Invitation to Social Constructionism*. Sage Publications, London.

Kahane, A., 1999. Scenarios for changing the world. In P. Senge, A. Kleiner, C. Roberts, R. Ross, G. Roth, and B. Smith (Eds.), *The Dance of Change*. Currency/Doubleday, New York, pp. 511-518.

Ludema, J. D., 2000. From deficit discourse to vocabularies of hope:

The power of appreciation. In D. L. Cooperrider, P. F. Sorensen, D. Whitney, and T. F. Yaeger (Eds.), *Appreciative Inquiry: Rethinking Human Organization Toward a Positive Theory of Change*. Stipes Publishing L.L.C., Champaign, IL, pp. 265-287.

Ludema, J. D., Cooperrider, D. L., and Barrett, F. J., 2001. Appreciative inquiry: The power of the unconditional positive question. In P. Reason and H. Bradbury (Eds.), *Handbook of Action Research*. Sage Publications, London, pp. 189-199.

Mohr, B.J., 2001. *Design: Plenary Session*. The First International Conference on Appreciative Inquiry. Baltimore, MD. Sept. 30-Oct. 3, 2001.

Norum, K.E., 2000. *Storying Change: The Power of the Tales We Tell*. Proceedings of the ALARPM/PAR World Congress, University of Ballarat, Victoria, Australia. September 10-13. Retrieved 7/6/02: http://www.ballarat.edu.au/alarpm/docs/Norum,_K_-_Paper.doc
Rilke, R. M. *Letters to a young poet*. Retrieved 7/6/02: http://www.sfgoth.com/~immanis/rilke/letter4.html
Senge, P. M., 1990. *The Fifth Discipline: The Art and Practice of the Learning Organization*. Doubleday/Currency, New York.
Wheatley, M. J., 1999. *Leadership and the New Science: Learning About Organization from an Orderly Universe (2nd ed.)*. Berrett-Koehler, San Francisco.
Wheatley, M.J., 2002. *Turning to One Another*. Berrett-Koehler, San Francisco.
Wheatley, M. J., and Kellner-Rogers, M., 1996. *A Simpler Way*. San Francisco: Berrett-Koehler Publishers.
Widdershoven, G. A. M., 1993. The story of life: Hermeneutic perspectives on the relationship between narrative and life history. In R. Josselson and A. Lieblich (Eds.), *The Narrative Study of Lives: Vol. 1* (pp. 1-20). Sage Publications, Inc., Thousand Oaks, CA.
Zemke, R., 1999, June. Don't fix that company. *Training*, **36**(6): 26-33.
Zohar, D., 1997. *Rewiring the Corporate Brain: Using the New Science to Rethink How We Structure and Lead Organizations*. Berrett-Koehler, San Francisco.

Chapter 8

RIGHTS AND RESPONSIBILITIES IN CONVERSATION PRACTICE

GORDON C. DYER
Open University in the East of England, Hills Road, Cambridge, UK

1. INTRODUCTION

This Chapter links to previous descriptions of conversation in this book as a collective disciplined inquiry, for exploring issues of social/societal significance, or as a process to aid the design of future social systems. This Chapter focuses on the concept *of rights and responsibilities* as a basis of techniques to support the growing field of conversation practice and purpose. Guidance for a group to sustain their conversation during the designing stage and to continue to develop their conversation and design during stages of implementation and review, are offered.

Thus *rights and responsibilities* are examined in two separate contexts. The first is the aim to ensure processes and methods of conversation within the group that are likely to lead to continuous positive creative synergy. This is explored through the metaphor of enthalpy, a concept from thermochemistry. In short we will exploit the notion of the "chemistry" of conversation.

Then, an approach for a group to develop a *Bill of Rights and Responsibilities* is introduced. This can serve as a symbol and affirmation of the group's collective ideals and intents for their social system, and also provide an instrument to allow progress towards their ideals to be reviewed and discussed.

It is a fundamental principle of systems design that those affected by a system should have the opportunity to participate in its design, and this includes young people. Some results from action research are described including those from an adaptation with 9-10 year old children in the UK primary school system, in which they derive their own bill of rights and responsibilities for their learning situation.

2. INITIATING AND SUSTAINING CONVERSATION

2.1 Pre-requisites for Effective Conversation

An effective conversation is one where the interaction between participants is such that their potential for creative synergy is maximised. We have all experienced some conversations when this did not happen. The development of a conversation can be uncertain; it can flicker and then die. Another might develop with considerable excitement and then fade. There will be those where the excitement and exchange will be sustained until the planned end of the meeting, which in the case of ISI conversations may be for 4 or 5 days. Two domains influence the success or failure of conversations: one is linked to the enthusiasm and thus energy within the group for the topic, the other to personal freedoms and psychological conditions e.g. mutual respect within the group which encourages their continuing participation. These domains are now examined in turn.

2.2 Energy Domain

Metaphors have been introduced elsewhere in the book as a powerful aid for systems designers. A simple yet powerful metaphor for conversation is that of lighting a fire. Gordon Rowland (Rowland, 1996) suggested that the activity of lighting a fire in a wood stove in terms of its preparation, ignition and feeding, could be explored as a metaphor for the development of a conversation. The key points arising from the Rowland metaphor are shown in Table 1 below.

A fire is a restricted type of chemical reaction, in that it is not reversible. The general case of chemical reactions, which may be reversible and may also involve what are termed catalysts, can offer further insights into conversations and we will now examine these through the enthalpy metaphor.

2.2 Enthalpy Metaphor

Enthalpy change is a term used in thermodynamics to reflect the energy changes during chemical reactions. In earlier papers the author (Dyer 1996, 1, 2) has demonstrated that enthalpy is a useful metaphor to provide insights as to how and why some conversations ignite and develop, and others do not. The comparison of the characteristics of chemical reactions and group interactions then leads to help with planning for a conversation and in sustaining the process.

The key feature of the metaphor is that the bonding changes within a chemical reaction can release energy (the technical phrase for this is negative enthalpy change) into the chemical system and sustain further reaction. This can be related to the interactions within a conversation group that can give rise to inspiring creative synergy. The basic metaphor is widened to include the concept of *activation energy* that is required to initiate chemical reaction and group interactions, and the role of *catalysts* in reducing the activation energy.

Table 1. Key Points of Rowland Metaphor

Preparation	Share input papers	⇒	Gather seasoned wood
	Offer ideas	⇒	Split wood or get kindling
	Offer shreds of ideas	⇒	Crumble newspapers
	Layer ideas - arrange loosely	⇒	Stack kindling over paper with space for oxygen
Ignition	Find trigger questions	⇒	Light a match
	Ask the question of ideas	⇒	Hold match near paper in several places
	Remain open	⇒	Ensure air available to fan flames
Feed	Introduce main issues	⇒	Add logs when kindling ablaze
	Develop conversation	⇒	When logs lit shut stove
	Seek patterns	⇒	Set damper for efficiency
	Allow conversation to take its course	⇒	Let fire burn; add wood periodically

The inherent stability of chemical compounds arises from the fact that the atoms from which they are made are chemically bonded together. The bonding itself involves energy, so the first step in any reaction is therefore an input of energy to break the bond that already exists in the reacting substances. The second step is to make new bonds that will exist in the products. The amount of energy required to start a reaction is called the activation energy – it is simply the height of the "energy barrier" or "energy hill" to overcome the bonding of the first molecules to enable them to react. Assuming the energy released from first interactions between molecules is greater than the activation energy then more molecules can react. As the reaction develops, eventually the energy release is sufficient not only to maintain combustion but also to release excess heat to the environment.

This provides a partial description of what happens within a situation involving human interaction. The participants will arrive with *external bonds* still partly intact. Hence the equivalent of *activation energy will be required to be input to the group to break these bonds*. Some form of spark must either be generated within the group or be brought in from the environment. The "wheelspin", which is a common experience in conversations can be seen as sparks which had inadequate energy to overcome the energy hill the group faced. But once the right spark has been found, and the group r*eforms bonds and works collectively, then energy is released* – through negative enthalpy – to enable them to interact further and "perform". As the group increases its bonding something akin to the chemical chain reaction is taking place. But something more dramatic will happen in the case of human interaction in that as the starting energies of the participants varies from day to day, not only are the activation energies different but so will the enthalpy changes. Thus *any chain reaction which results appears to provide at least a partial explanation to what is usually described as the synergy of human activity systems* i.e., the non-repeatability of group interaction and the capacity to produce unexpected results, which are sometimes very creative and positive, and sometimes the opposite.

Now to the notion of catalyst. A catalyst is a substance that alters the rate of a chemical reaction but may itself be unchanged at the end of a reaction. They work in various ways but their purpose is always the same – to reduce the activation energy. Using a catalyst will allow reaction to proceed with lower energy input. There is another advantage to catalysts, as conditions arise where several reactions, which can lead to different end products, are possible. Catalysts can sometimes be found that are specific to particular reactions taking place. So using the right catalyst, the reaction that produces the desired product can be enhanced at the expense of other possible reactions. We can use these ideas to help us in conversation design. A number of factors can operate individually and collectively as catalysts:

- external environment, which can be a source of inspiration. The ISI tries to choose venues e.g., Fuschl, Asilomar, Crete which potentially can have this catalytic effect.
- internal environment, e.g., warmth, comfort and the seating arrangements. With the latter aspect, care must also be taken to ensure cultural needs are taken into account.
- circulation of input papers
- opportunities to meet others informally in pairs, or small groups before the conversation,
- introductions and welcomes at the first plenary
- group exercises
- tabling of ideas from input papers
- previously shared or recently emerging ideas from the systems field
- imposition of time pressures, or deadlines.

The notion of the *reverse* chemical reaction, which is associated with using energy to break bonds that have just been formed, alerts us to the dangers within groups when relationships break down. Energy is then expended to break bonds at the expense of that available for joint creativity. These conditions are most likely to happen if any member of a conversation group feels that they are not being given adequate opportunity to contribute, or when their freedoms of expression, action, or participation are being impaired. Thus the challenge in conversation planning, stewardship or participation is to ensure that behaviours and actions is such that the energy of activation and interaction is maintained. This may be related to rights and responsibilities that are offered as the first part of the guidelines for conversationalists (see Table 2).

Table 2. Domain of Energy in Conversation

	Rights to:	**Responsibilities to:**
Energy of Activation	Expect that: -others will be prepared and will offer catalysts for the discussion -others will seek out opportunities for bonding and energy release	Offer catalysts for the group to respond to Respond to the needs of the whole group as well as those of self, i.e., respond if possible to catalysts offered by others
Energy of Continuing Group Interaction	Expect that: -others are aware of temperature and energy level within group -others will look for opportunities for bonding thereby fostering energy release	Prepare to participate fully Prepare to offer alternative catalysts as conversation changes direction and new activation energies are needed to get to grips with new topics Remain alert to energy level and temperature of the group: -particularly when group is operating synergistically and positively -and, alternatively, to any danger signs that bonds are breaking down, particularly because individuals rights under the various dimensions of freedom (see Table 3) are being ignored

2.3 Domain of Personal Freedoms

The second domain of rights and responsibilities (see Table 3) are those that any member of a small social unit (including a conversation group) should expect to have in participating in its on-going activity and design. This domain incorporates consideration of seven individual freedom areas, which have been adapted from previous work (Pinchot 1993) and (Dyer, 1995):

- freedom of expression
- freedom of learning
- freedom of action
- freedom to act as a team
- freedom of belonging to a community of differences
- freedom of own networks
- freedom of participative democracy

Combined together the rights and responsibilities for the domains of energy and individual freedoms in Tables 2 and 3 provide guidelines for the planning, development of process and facilitation of effective conversation. The list is intended to apply to a cover the case of a relatively large conversation group (12+), and is comprehensive. Depending on the context and size of the group, particularly if it is small, some domains and rights and responsibilities may be more important than others.

One of Banathy's key propositions of social systems design (Banathy, 1996) is that it is unethical to design social systems for someone else. So, what is offered above as possible rights and responsibilities within these areas of freedom would need to be considered by any conversation group at the start of their meeting. They would decide the extent to which they would accept the list, delete from it or otherwise modify it.

Activity #1

(1) Reflect on the rights and responsibilities for conversation in Tables 2 and 3, against your experience of conversation in your school or college. How far does your experience match the ideals in the lists? How far do you think you might be able to go in a conversation in trying to adopt these ideals in either school or college? Consider the additions, deletions or amendments you would like to propose. Write down your thoughts in your workbook.

It is recognised that what is offered above is much influenced by the author's own value system which is western-orientated and linked to a set of beliefs about fundamental human rights and equality of opportunity. The list might have to be considerably modified for use a non-western conversation context. Elsewhere in this book Yoshi Horiuchi has described how as an alternative, *rules* for conversation were adapted for multicultural conversations. However, what is crucial is that conversations must proceed on the basis of an agreed set of understandings, be they framed either as *rights and responsibilities*, or *rules* of conversations. What is also essential is that there is sufficient cultural commonality within the group for the conversation process to begin.

Table 3. Domain of Personal Freedoms in Conversation

	Rights to:	**Responsibilities to:**
Freedom of Expression	Freedom to communicate one's views without fear Equal opportunity to express one's views	Not to dominate conversation time Appreciate and bring out the many sides to every issue See the good in others' ideas and express it; try to build on those ideas Tell the truth
Freedom of learning	Freedom of inquiry Develop one's knowledge and competence	Be curious, persistent and aware Learn from past failures and successes Keep learning and growing Help others in group to learn and grow
Freedom of action	Offer choices of topic Equal opportunity of action Take individual decisions Limits to burdens: physical, emotional, or caused by the decisions of others	Commit to something worthwhile Achieve goals Recognise the possible consequences of individual decisions and face up to them if required
Freedom to act as member of a team	Freedom of team decisions	Recognise the possible consequences of team decisions and face up to them Care for team members Build the capabilities of every member
Freedom of belonging to a community of differences	Full membership of the group A group that cares for your welfare An ethical group	Neither show nor tolerate bias or prejudice Balance self-interest against the common good Work toward worthwhile common vision and values Find value in diversity
Freedom of own networks	Freedom of association Choice of friends Freedom to make and honour commitments	Make commitments wisely Deliver on one's commitments Use others' time wisely
Freedom of participative democracy	Equal opportunity to participate	Listen to others and support their rights Stand for what one believes in Use incentives, not mandates, whenever possible Not manipulate or coerce Reward service to the whole

Activity #2 (Group Activity)

Seek the co-operation of a colleague group, e.g., your class or a club or association to which you belong, and invite them to discuss and agree a set of rights and responsibilities for conversation. Have the discussion and then evaluate the outcome. How does this compare with your list of rights and responsibilities and your reflections under Activity # 1? Record your findings in your notebook.

3. BILL OF RIGHTS AND RESPONSIBILITIES

Gifford and Elizabeth Pinchot proposed the idea of a Bill of Rights and Responsibilities in the context of the workplace and in what they call the *intelligent organization* (Pinchot, 1993). Their scenario was the need to deal with issues of change and conflict in workplace rights e.g. between employee and employer, or between individuals or teams, which might occur in the workplace. It is a statement of ideal behaviours for maintaining an appropriate balance between rights and responsibilities. It shows how an "intelligent organization – one which develops and engages the intelligence, business judgements and wide-system responsibility of all its members", would operate. The implication is that the Bill would be published and everyone in the organization would in future aspire to hold to the statements contained within it.

The author adapted this idea (Dyer, 1995) for use in the context of systems design within any small social unit. There are two major differences arising in the adaptation. The first arises from the key proposition of *authenticity* that underpins systems design as defined by Banathy. He argues "The design of a social system is authentic only if it is built on the individual and collective values, aspirations, and ideas of those who serve the system and who are served and affected by it" (Banathy, 2001). Thus a Bill of Rights and Responsibilities must be viewed as a dynamic document that will continue to evolve and be discussed, reviewed and probably modified as new members join the community. We must also find ways of encouraging and educating the young to take part in the design process.

The second difference in the adaptation from Pinchot is that a Bill of Right and Responsibilities was proposed as part of a four stage methodological framework intended to help a small social unit design its own future. That framework is described fully elsewhere (Dyer 1996, 1), and as we are concentrating on rights and responsibilities, a short summary is sufficient here. An introduction stage involves the group sharing and (hopefully) reaching agreement over values and ideals. It culminates in a discussion on whether a symbol of affirmation of the group's collective ideals and intents for their social system will be made, e.g. a written statement. The group itself will set any norms, agree collective responsibilities, and any adjustments in relationships will be mutually agreed through their conversation process. The aim is to reconcile autonomy with the interdependence of the group; it is not take away independence. In the second stage the group may develop a draft declaration reflecting their values and ideals (see for example the online Draft Declaration of Interdependence of the International Systems Institute (ISI) at http://myweb.clark.net/pub/nhp/isi/principl/index.html.

The third stage is intended to allow all members to explore and recognise others' vulnerability to their actions. If there is a previous group history, individual accounts of when being in the group worked, did not work, and where alternative behaviour might have avoided a problem, can be very useful. At this point the group may want to draft a Bill of Rights and Responsibilities. It is where meaning is added to the ideals within any declaration that they might have drafted at Stage 2, and in effect, how they define interdependence. If a draft bill of rights and responsibilities is agreed, this can now serve as a *social contract*. The group is now in a position *to be* and *to behave* as the social system they have defined. The bill can also serve as the instrument to allow progress towards their ideals to be reviewed and discussed, which is the fourth stage in the framework. The review allows the group to consider how far their social system is meeting their statement of rights and responsibilities, and whether either behaviour, or their social contract, needs to be changed.

4. ACTION RESEARCH

There are two basic ways for a group to proceed to develop their own Bill of Rights and Responsibilities. The first is to use an existing document as a starting point and then adapt or modify it as the particular community wishes. For example, the list at Table 3, which was drafted to as a list of rights and responsibilities to guide a *process* of conversation, could simply be retitled as a draft Bill of Rights and Responsibilities, and used as a starting draft by the group as their collective wish for a social contract. The second method is to develop a document from scratch.

The way chosen will clearly depend on the context or context differences. The group may wish to consider questions of the following kind:

- do they share a vision that there should be a Bill of Rights and Responsibilities for the group?
- what vision of a Bill do they have?
- would they feel a personal commitment to making an effort to achieve that vision of a social contract?
- does the draft capture for them the key issues in terms of areas of commitment which ought to be addressed?
- could they sign up to it? as it stands (unlikely)? or with some modification and clarification (more likely)?
- what statements need amending? deleting? or need clarification?
- are there major omissions?
- is the language appropriate?

Activity #3 (Group Activity)

Part (1) Return to your colleague group and invite them to consider using a Bill of Rights and Responsibilities as the basis for a social contract for the social system that you are members of. Generate the Bill together using either of the two approaches mentioned above. Agree an appropriate time period for review, e.g., 3 months or term. Make a note of the agreements that the group reached.
Part (2) At the review time; invite the group to share their experiences, and views on whether behaviours and/or the social contract need to be changed. Record the outcome in your workbook. The author will be pleased to have a report on your action research.

The author has worked with a number of groups to help them develop a Bill. Others have also attempted to apply the approach. The language of the Bill in Table 3 is quite sophisticated and it was not surprising that in some contexts, such as in planning for a conversation in Japan, or when used with young people a simpler approach and simpler language will be necessary. Involving young people in design is particularly important and challenging, and so the Chapter will conclude with a description of successful efforts by a co-worker to develop bills of rights and responsibilities with 9-10 year old children in London schools, and with newly qualified teachers.

4.1 Westminster Behaviour Support Team

The approach has been adapted (Wye, 2000), within a primary school setting in Westminster City Borough in London, England. Wye, a member of the Westminster Behaviour Support Team (WBST), was responding to recent UK Government interest in rights and responsibilities for curriculum development in citizenship. He set out to explore the model as a vehicle to progress this. The WBST is established to help support teachers in Inner London schools in dealing with possible issues arising in multiethnic schools. The project has so far covered working with 3 classes of Year (Grade) 5 children, aged between nine and ten years old. The aim was to see whether children of this age could engage with the concept of a bill of rights and responsibilities, and whether they would see value in trying to develop one. The classes were around 30 in size, of mixed ability groups, and typical of inner city school populations, had a very high ethnic mix. The class whose work is shown below was half Afro-Caribbean, and included four children of oriental origin.

For children of this age a very simple approach to the process of conversation is needed, and Wye set the discussion up as for school *circle time*

(Mosley, 1993, 1996) Following his introduction, the children ran the session by themselves using a talking-stick. The ideas that emerged from one of the classes are shown in Table 4.

This outcome is very interesting and encouraging given the age of the participants. It demonstrates greater insights than the author would have imagined. The children themselves recognised that their rights were privileges. Wye pointed out that the items they identified link to Maslow's hierarchy of needs, viz. (1) and (2) relate to Maslow's "safety" need; (3) to "belonging" need; (4) to "self-esteem" need; and (5) and (6) to "self-actualization".

The exercise was clearly meaningful for the children as they had some very clear ideas on how they wanted to progress their work, i.e.:

- one class planned to use their bill to introduce themselves to their new teacher at the start of the next school year
- another class to use to help individuals to frame a letter of application to be a class monitor
- another wanted to have their work included as school rules.

Table 4. A Basic Social Contract developed by Children aged 9-10 Years

Privileges and Rights	**Responsibilities**
To feel safe	To look after yourself and others by being aware of how people feel
To feel that your property is safe	To look after property by keeping things safe and tidy
To have friends	To be loyal
To be listened to seriously	To think before speaking, be fair, polite, and take turn
To play	To play fair
To be given responsibility	To do your best

Wye is working with the management of the school with the aim of getting the children's' ideas incorporated into a revamped school policy. This will take time to achieve.

Wye also reported on his work using the model as the theme for induction workshops with newly qualified teachers in the Borough. This covered a number of investigations: one group covered their rights and responsibilities as teachers, other groups role-played as children of different age groups to get some feel for the language and statements that might would be appropriate. There was a good match between this outcome and that from 9-10 year olds. Wye comments: "The ideas that the teachers came up with when in role as children showed the excellent empathy they had with their pupils. Children do come up with these quite strict and onerous responsibilities and, in my experience, are quite realistic about their rights" (Wye 2000)

The teachers extended the workshop activity to consider how teaching activity and practice might develop the appropriate attitudes towards the responsibilities that had been identified. Table 5 gives a summary of the group's ideas. Readers will note that this is based on a slightly different set of responsibilities to those identified above by children, but this is not important here. The objective is to indicate the further potential of the rights and responsibilities model, in this case to indicate possibilities for curriculum activity.

Table 5. Application to Curriculum Activity

Rights to:	**Responsibilities to:**	**By the provision of:**
Work in peace	Leave others alone	Setting clear personal boundaries
Be stimulated	Listen, share ideas, teach each other	Circle time; appoint peer mentors
Access to resources	Take care of resources	Share pupils own resources; monitors,
Feel secure	Care for each other	Role play; drama; medical assistants; friendly bench; circle of friends
Make mistakes	Be willing to try	Scaffolding; not blaming; "golden time"; modelling positives; praise
Communicate with peers	Recognise suitable timing and voice control	Whisper sessions; level indicator,
Feel valued	Respect and value others	Teacher model; circle time; praise poster listing positive attributes
Be heard	Listen to others and let them speak	Circle time chair; hot seat; teaching listening and questioning skills
Access to the whole curriculum	Be open minded and receptive	Reward self motivation

The teachers found their experience using the model very worthwhile and have decided to continue their investigations with it through a voluntary conversation group.

NB. Many of the entries in the right hand column of Table 5, e.g., *golden time* and *hot seat* are associated with the *circle time* methods of Jenny Mosley (1993, 1996). The interested reader may follow this up through online publications and description shown at: http://www.circle-time.co.uk.

5. CONCLUSION

This Chapter has focussed on a conceptual model of rights and responsibilities for conversation *practice* and *purpose* in social system design. It emerges as a powerful tool in both contexts. The metaphor of enthalpy change in chemical reactions, and linked ideas of activation energy, specific catalyst, and reverse reaction, allowed us to focus on their equivalents in human interactions, and thus to how conversations can be initiated and sustained. We found a partial explanation for the variation of the synergy in human interaction. Along with consideration of some key personal freedoms we developed a list of rights and responsibilities for planning, and facilitating effective conversations, i.e. those with the maximum chance of sustained creative synergy. The list of rights and responsibilities provided (Tables 2 and 3) derive from Western value system. Consistent with the ethics of systems design, the list would be modified by any particular conversation group as they see fit.

A list of rights and responsibilities for conversation might also be regarded as the *social contract* for the implementation phase of a social system design, and thus the basis of its review. In this case the list of rights and responsibilities might be titled as a Bill of Rights and Responsibilities.

Action research has revealed that 9-10 year olds relate well to the concept of rights and responsibilities, and can be imaginative in drawing up a bill using appropriate language, and in ways to progress the ideas. Related work with teachers has indicated the development potential of the model, for example, towards curriculum design. More action research, including through the activities suggested in this Chapter, is needed to explore the model further. The author will be pleased to hear from anyone who completes the group activities, or undertakes related projects.

REFERENCES

Banathy, B.H., (1996), *Designing Social Systems in a Changing World.* Plenum, New York.

Banathy, B.H., (2001), *Guided Evolution of Society: A Systems View.* Plenum, New York.

Dyer G.C., (1995), Developing a Family Declaration of Interdependence: a Methodology for Systems Design within a Small Social Unit. *Systems Research.* **12(3):** 201-208.

Dyer, G.C., (1996, 1), Enthalpy: A Metaphor for a Design Guide for Conversation. *Educational Technology.* **36(1):** 46-53.

Dyer, G.C., (1996, 2), Enthalpy: A Metaphor for the Chemistry of Conversation. *Systems Research.* **13(2):** 145-157.

International Systems Institute (1996), *Declaration of Interdependence* [online], International Systems Institute. Available from: http://myweb.clark.net/pub/nhp/isi/principl/index.html [Verified 17 January 2002]

Mosley, J., (1993), *Turn Your School Around.* Cambridge: LDA.

Mosley, J., (1996), *Quality Circle Time in the Primary Classroom.* Cambridge: LDA. See also Jenny Mosley Consultancies [online]. Available from: http://www.circle-time.co.uk [Verified 17 January 2002]

Pinchot, G., and Pinchot, E., (1993), *The End of Bureaucracy and the Rise of the Intelligent Organization.* Berrett-Koehler, San Francisco.

Rowland, G., (1996), Lighting the Fire of Design Conversation. *Educational Technology* **36(1):** 42-45.

Wye, B., (2000) Correspondence with author covering various reports February – August 2000 on work with children and newly qualified teachers in London schools.

Chapter 9

TRANSCULTURAL COMMUNICATIONS

Theory and Practice

YOSHIHIDE HORIUCHI
University of Shizuoka, Shizuoka, Japan

1. INTRODUCTION

The goal of this chapter is to explain: why we need transcultural orientations as part of International Systems Institute (ISI) conversations; what we have achieved in past conversations, and, what we hope to achieve in future. Appropriate theory will be used to support the explanation where relevant.

First, by referring to theories on transcultural communications by Stella Ting-Toomey (1999) and Doris T. Allen (1990), we will review the ISI Transcultural Council (TCC) activities at ISI conversations. We will also describe the planned Japan Conversation, which is to be the first multi-cultural and bilingual ISI conversation. The Implicit Rules for ISI conversation and its relations with transcultural communications are then discussed.

The TCC has been encouraging efforts for transcultural communications at ISI conversations for some 8 years. At the 2000 Asilomar Conversation, we had participants from a larger number of nations than previously which enabled us to continue with these efforts. We can do more and better to make ISI conversations more transcultural, and intend to do so as described in this chapter and elsewhere.

2. CULTURAL COMMUNICATIONS

Stella Ting-Toomey (1999) has provided some useful definitions of culture and cultural norms. According to Ting-Toomey, a function of culture is to create "us" and "them," and that this causes anxiety for those individuals who are dissimilar to the majority group:

Culture: "(A) complex frame of reference that consists of patterns of traditions, beliefs, values, norms, symbols, and meanings that are shared to varying degrees by interacting members of a community" (Ting-Toomey, 1999, p. 10).

Cultural norms: "(R)efer to the collective expectations of what constitute proper or improper behavior in a given situation" (Ting-Toomey, 1999, p. 11).

Functions of Culture: "(W)ith people from a dissimilar membership group, we constantly have to perform guessing games. We tend to 'stand out,' and we experience awkwardness during interaction. The feeling of exclusion or differentiation leads to interaction anxiety and uncertainty.... The group inclusion need also creates boundaries between 'us' and 'them'" (Ting-Toomey, 1999, p. 13).

"Cultural communication provides us with a set of ideals of how social interaction can be accomplished smoothly among people within our community. It binds people together via their shared linguistic codes, norms, and scripts" (Ting-Toomey 1999, p. 15).

Transcultural Communications: Shimada et al., (1991) defines transcultural communications as "various forms of communications between people with diversified cultural backgrounds." (Shimada 1991, p. 45. Translation by the author.) Ting-Toomey (1999) states the transcultural communications competence as follows: "An incremental learning journey whereby intercultural communicators learn to mutually adapt to each other's behaviors appropriately and flexibly.... By signaling to the other party that we are willing to adapt our behaviors in a culture-sensitive manner, we convey our respect for the other's cultural frame of reference" (p. 263).

The author proposes the goal of transcultural communications of ISI as follows: "Stimulate communications among its participants with various cultural backgrounds by respecting diverse cultures and providing provisions for the participants to feel free to express opinions in their own culturally appropriate forms." Ideally, ISI prepares the conversation in such a way that the participants do not necessarily have to adapt the western-style expressive communication patterns, but they can choose their own preferred forms of communications. We could design ISI conversations with such an ideal state in mind as our long-term goal.

A short-term major task of the Transcultural Council (TCC) of ISI is to make all the conversation participants feel welcomed and respected so that they have the shared sense of the "conversation culture." We will discuss the implicit rules of ISI conversation in the latter part of this chapter.

Another task of TCC is to expand the range of the ISI conversation participants to a wider range of nationalities, cultures, age groups, etc. At the Asilomar 2000 TCC meeting, we were delighted to have participants from a larger number of nations than previous Asilomar Conversations, including Cyprus, Japan, Mexico, United Kingdom, and United States.

At the Asilomar 1999 TCC meeting, we paid special attention to intergenerational issues. As Alexander Laszlo, co-chair of the 1999 TCC meeting with the author, summarizes in the 1999 TCC report: "The fact that cultural differences are not only an issue of space and place but also of age and epoch brought forth an important aspect of TCC interests: how to create effective strategies for the facilitation of communication among and between generations" (1999 TCC Report, Section 2.1).

"We heard from a group of three youth (13, 15, 16 years of age) about intergenerational issues of cultural communication. They helped us understand how transcultural issues also occur in a generational context - both within and between/generations. To them, an important message for their generation is to 'be thankful for what you've got, even if it's not a lot'" (1999 TCC Report, Section 2.2).

3. MINDFULNESS

According to Ting-Toomey (1999), "*Mindfulness* means being aware of our own and others' behavior in the situation, and paying focused attention to the *process* of communication taking place between us and dissimilar others.... (I)f intercultural communicators continue to ignore group-based *and* person-based factors that have an impact on their encounters, their misinterpretations may spiral into major escalatory conflicts. Alternatively, individuals may stay in a very superficial relationship..." "Intercultural communication is defined as the symbolic exchange process whereby individuals from two (or more) different cultural communities negotiate shared meanings in an interactive situation" (Ting-Toomey, 1999, pp. 16-17, italics in original).

"With clarity of understanding, we can mindfully choose words and behaviors that make dissimilar others feel included and affirmed" (Ting-Toomey, 1999, p. 22).

One of the TCC goals is to "encourage cultural diversity and the development of culturally sensitive and culturally appropriate modes of conversation in ISI type events around the world (noting that the full adoption of all ISI ideals is not required)" (TCC 1999 Report, Section 2). "The idea... was to encourage conversation participants to appreciate the cultural difference among themselves" (TCC 1999 Report, Section 1.5).

The TCC has been making efforts to make ISI conversations open to various cultures of the participants at recent Asilomar and Fuschl Conversations. Such efforts were intended to make all the participants in a certain conversation group feel comfortable in expressing and sharing their ideas with others.

4. CONVEYING RESPECT AND INTERACTION SATISFACTION

Ting-Toomey (1999) notes the importance of conveying respect as follows: "By signaling to the other party that we are willing to adapt our behaviors in a culture-sensitive manner, we convey our respect for the other's cultural frame of reference.... Effectiveness refers to the degree to which communicators achieve mutual shared meaning and desired goal-oriented outcomes" (Ting-Toomey, 1999, p. 263).

A major task of the TCC is: "how to balance multiple perspectives. . . . (T)his concern includes attention to the difference between the ways in which people in one culture may formulate/assess the active contribution of ideas (through moderated or facilitated group processes) versus the way they may formulate/assess the passive contribution of ideas (through voluntary and self-generated presentations). Such differences beg for the creation of conditions that heighten sensitivities to issues of what might otherwise be considered 'implicit understandings' or notions of 'proper place'" (1999 TCC Report, Section 1.0).

Ting-Toomey (1999) also writes about the importance of interaction satisfaction in a transcultural setting: "(T)o the extent that the important identities (e.g., cultural or gender) of the intercultural communicators have been positively addressed and sensitively dealt with, they will experience interaction satisfaction." (Ting-Toomey, 1999, p. 265)

"(W)e can identify seven clear attributes associated with an effective transcultural communicator: tolerance for ambiguity, open-mindedness, cognitive flexibility, respectfulness, situational adaptability, verbal and nonverbal sensitivity, and creative thinking" (Ting-Toomey, 1999, p. 271).

5. IMPLICIT RULES FOR ISI CONVERSATIONS

The TCC Chairperson Gordon Dyer reports below on the Implicit Rules, and, later, their application to a Japan Conversation. While Rule Set 1 is universal, the Rule Set 2 could be culturally dependent; hence, it is up to any particular conversation coordinator(s) to decide which rules to adopt, in harmony with the cultures of the host country. These ideas are amplified in the following extract from the report on the Japan Conversation Design, which appeared in the *Journal of Administration and Informatics* in 1996. (Hereafter JCV Plan)

> Before we could formulate a plan for the conversation, we needed to have a clearer understanding of the norms of "rules" to which participants of ISI conversation generally subscribe. It might then be possible to see how far they might be applicable to Japanese culture. As far as our Group was aware, such rules do not exit in an explicit form, they are implied by the underpinning core values of the ISI which relate to equality and openness.
>
> We noted one recent attempt (Dyer, 1996) to express a set of ideals for conversationalist in terms of "rights" and "responsibilities," but the Group felt it would be useful to develop their own set of rules. Following discussion, the following set of rules emerged. They are structured into

two sets–the order of the sets relating to an ascending order of difficulty of application in Japan:

Set 1:

1. Display tolerance, patience and consideration to others.
2. Honor and respect each other.
3. When a colleague participant is speaking, one should listen, attempt to understand the point of view being expressed, reflect and respond.
4. Not to dominate.
5. Not to offend.
6. One must never lose control.
7. All ideas are viewed as contributions to the group for consideration, accepting that all ideas are not used.

Set 2:

1. Free exchange of ideas public ownership of ideas.
2. Equal opportunity to participate.
3. Stand for what one believes in.
4. Equal opportunity of action and decisions; but take responsibilities for actions and decisions.

6. COMMONALTIES AS FELLOW HUMANKIND

For an ISI conversation to be truly interactive among and participative of all its participants, it is vital that we recognize and respect various cultures of the participants. Yet, it is equally important to see the other side of the coin, as Greg LaPointe succinctly emphasized at the 2000 TCC meeting at Asilomar 2000. His point was that if we focus too much on the cultural differences among the participants, we could overlook the fundamental commonalties we have as fellow humankind.

Shimada et al., (1991) made a similar observation: "Books on transcultural communications tend to overemphasize cultural differences among people. . . . For those who wish to make effective transcultural communications, it is imperative that they recognize the similarities among people and respect them as ties among people with different cultural backgrounds. To understand similarities among us humankind is not a philosophical matter, but a pragmatic task" (p. 285, translations by the author.),

Also, Ting-Toomey (1999) writes, "In practicing mindful intercultural communication, we need to develop an understanding of the valuable differences that exist between identity groups; yet at the same time, we need to continuously recognize the commonalties that exist on a panhuman identity level," (p. 22) and, "With clarity of understanding, we can mindfully choose words and behaviors that make dissimilar others feel included and affirmed" (p. 22).

6.1. Transcultural Orientation at C.I.S.V. and ISI International Singing Night

At the age of eleven, the author had unique experience in the Children's International Summer Villages (C.I.S.V.) – an UNESCO-registered international nonpolitical movement for transcultural understanding. American psychologist Dr. Doris T. Allen who established C.I.S.V. in 1950 explains its goals:

> "C.I.S.V. was founded with the knowledge that transnational education programmes are most effective before substantial cultural perspectives and prejudices are formed. Accordingly, CISV programmes begin with pre-adolescents (age 11). Continued participation is possible through other activities for your people (age 12-18) and adults" (Allen 1990, p. 3).
>
> "The Village programme is a four week international 'summer' camp unique to CISV. Delegations of two boys and two girls (age 11) and adult delegates from 12 nations plus staff and four to six junior counselors (age 16-17) from three to five nations participate in a multi-language camp featuring typical camp activities and emphasizing international friendship, cross-cultural communication and cooperative living. Day by day the children learn, in a natural way, that despite national or cultural differences they have much more in common as members of the human family in an increasingly interdependent world" (Allen 1990, p. 4).
>
> "One of the most interesting research findings has been that communication seems to have struck a deeper level in situations where there has not been a common language. Warm human feelings seem to take over in face of a language area deficiency" (Allen 1990, p. 5).

The author's C.I.S.V. experience is that children could discover commonalties especially well in singing, dancing and playing games and sports. At Fuschl and Asilomar Conversations, we hold an International Singing night for the participants to discover their commonalties, and also to appreciate the various cultures of the participants. International Singing Night also helps families of the conversation participants to be a part of the conversation, rather than be shadows of the participants and be left forgotten during the six days of the conversation.

7. ETHICAL RELATIVISM AND DESIGN OF JAPAN CONVERSATION

Ting-Toomey (1999) contrasts what she terms ethical relativism against the ethical-absolute approach. At ISI meetings held in 1994 and 1995, some TCC members designed a Japan Conversation as the first multi-cultural ISI conversation with such cultural relativism in mind. As Ting-Toomey (1999) notes,

> The ethical-absolute approach often results in marginalizing or muting the voices of non-dominant groups in domestic and international arenas. . . . Most of the current 'universal ethics' approaches, unfortunately, are 'imposed ethics' that rely heavily on Eurocentric moral philosophies and principles to the exclusion of other cultural groups' voices and standpoints. . . . A more realistic alternative for guiding our actions in contemporary society may be that of *contextual relativism*, that emphasizes the importance of understanding the problematic practice from a contextual perspective. (Ting-Toomey 1999, pp. 273-74)

At the Asilomar 1994 and 1995 Conversations, the Transnational Conversation Design Group worked on designing a conversation to be held in Japan, with Japanese culture as "official culture," and Japanese and English as two official languages. The following extracts of the design group's report highlights the challenges and the steps proposed to overcome them.

> Formulating a Transnational Design Inquiring System Using Japanese Cultural Values in a Japanese Setting: Report of the Transcultural Design Group, 1995 Asilomar Conversation
>
> This paper summarized the efforts in the Transcultural design group at Asilomar 95, which was largely directed at planning the designing for the first Transcultural Conversation of the ISI....The report covers: the challenges identified in holding a system design conversation in an Asian culture setting the rational for a design which involves a cross-cultural sensitization phase and then a sequence of conversations which explore the extent to which the ISI model can be introduced in this environment. In accordance with the key principles of systems design the extent to which the conversation style moves towards the ISI ideals, will be decided by the Asian participants. It is hoped that the form of phased conversation design described here will provide a useful first step in the search for a format which might apply in many multicultural settings, and in the search for implementation models for global systems design. (JCV Plan, p. 1)
>
> Conversations have been taking place within a cultural setting using English as the primary language. Yet, if the ISI is to advance towards its ideal of facilitating global societal change through a process of systems design, it must move positively to promote conversations in non-western cultural settings using other than English. The transcultural design group believes that human and cultural diversity are authentic conditions deserving sensitivity and respect. Thus, ethnic groups have a right to honor, maintain, or alter their cultural values through design, and create the way of life they choose to practice. ISI cannot assume that its conversation protocols, which are based on fundamental assumption of equality and openness, will immediately transfer into, other cultures. Inevitably, ISI adaptations will be required and made by other cultural groups. (JCV Plan, p. 2)

7.1. Sensitization Process: Orientation for the 1997 Japan Conversation

In the JCV Plan, Roberta Snow discusses the proposed sensitization process for the orientation of the Japan Conversation participants.

> In designing a conversation that will bring together Japanese and non-Japanese participants issues of cultural difference must be addressed. Because the conversation is a western model of participative design to be carried out in an eastern context, the primary orientation must address two key differences: (1) Japanese participants must become familiar with the conversation as a form of group work, and; (2) non-Japanese, primarily Western, participants must become familiar with the norms and values that guide group behavior in the Japanese context. Without this preparatory time the initial portion of the conversation will be spent in a state of mutual 'culture shock' and the inability to address differences might impede or undermine the overall process. Therefore, in the months prior to the meeting, all participants will be sent two sets of orientation materials. (JCV Plan, p. 9)

7.2. ISI Implicit Rules and Proposed Japan Conversation

Gordon Dyer discusses the ISI Conversation Rules and their roles in the proposed Japan Conversation, which will be based on non-western, Japanese culture.

> The first set of (ISI conversation) "rules" were seen as being equally applicable to Japan as in the West. We felt that this set of rules probably applied for conversations in a very wide range of cultures and we offer these for consideration and evaluation as a core set of rules for multicultural conversations. This may be too ambitious a position but we believe it gives us a basis for design of the opening Phase of a conversation in Japan. It was then agreed that we would test how far into Set 2 it was possible to proceed by a sequence of four conversations as shown below. Thus the overall conversation design would be based on phased incorporation of ISI features and the evaluation of the experiment. . . . (JCV Plan, pp. 10-12)

7.3. Expected Outcomes of Japan Conversation for the ISI

The JCV Plan, set forth expectations for the Japan Conversation,

> The (Japan) Conversation plan is expected to provide the first experimental information on a genuine transcultural conversation. It will provide

evidence on the extent to which another culture, in some respects similar but in many respects different to that of the West, can move towards the ideals espoused by ISI in seeking systems design and global change through conversation as a form of dialogic democracy. (pp. 10-12)

8. CONCLUSION

When we consider transcultural communications, the issues of concern include not only differences in nationality, culture, and race, but also transnational issues such as any generation gap, and differences in knowledge level of systems thinking of participants. The TCC hopes to open up the ISI conversations to a more diverse group than systems-knowledgeable academics and educators who are familiar with the ISI sub-culture. Such openness could take adaptation of ISI conversation to non-western settings as the next step of our conversation journey. Such an idealized design is aimed at proposing the ISI conversations to be held in various western and non-western countries with participants from a wide range of cultural, social, racial and other backgrounds.

REFERENCES

Allen, D.T., 1990. C.I.S.V. village guide. In C.I.S.V. Parent's Manual. C.I.S.V. Japan.

Dyer, G.C., 1996. Enthalpy as metaphor for the chemistry of conversations. *Systems Research*, **13** (2): 145-58.

Dyer, G., Horiuchi, Y., Pointe, G., Lee, I., Minati, G., Osakabe, A., and Snow, R., 1996. Formulating a transnational design inquiring system using Japanese cultural values in a Japanese setting: Report of the Transcultural Design Group, 1995 Asilomar Conversation. *Review of Administration and Informatics*, **9** (1): 1-15.

Samovar, L.A., Porter, R.E., and Jain,, N.C., 1983. *Understanding Intercultural Communication*. (Tsukasa Nishida et al. trans.). Seibunsha, Tokyo.

Shimada, H., et al., 1991. *Ibunka-to Communication.* (Transcultural Communications), Nihon-Hyoronsha, Tokyo.

Ting-Toomey, S., 1999. *Communicating Across Cultures*. Guilford, New York.

The Transcultural Council of the International Systems Institute, 2000. Minutes of the Sixth Annual General Meeting Held at Asilomar 8, November, 1999. *ISI Newsletter*, January.

Chapter 10

THE CRITICAL ROLE OF DIALOGUE IN EMANCIPATORY SYSTEMS DESIGN

LARRY A. MAGLIOCCA* and KAREN E. SANDERS**
The Ohio State University, The Compact LLP***

1. INTRODUCTION

This Chapter examines the concept of *Dialogue* within the context of our emergent understanding of its meaning and use within organizations and groups. Dialogue plays a critical role in the emancipatory systems design of the CogniScope™ System. From its Greek origins, dialogue has meant a "flow of meaning", and in applied use further differentiated with this chapter as "deliberative dialogue", "generative dialogue", and "focused and open dialogue". Current researchers share a framework of inquiry that invites the investigation of assumptions, safety of expression, and active listening. They describe dialogue as free- flowing natural language that is self-organizing and inclusive. However, when dialogue has been used in emancipatory systems design activities, limitations and challenges have been discovered in its application. Unshakeable human burdens must be lifted during dialogue in the form of limits of human cognition, group pathologies, and unequal power relations. Successful strategies for overcoming these burdens are provided based on documented use with the CogniScope™ System approach which, as been extensively tested in more than 200 applications. The authors conclude that, in the process of emancipatory systems design, dialogue must play a substantively deeper role than other concepts of dialogue to lift the human burdens of engagement.

Dialogue has become a fashionable word. It is used in our everyday conversation to mean a free exchange of ideas. When choosing the word dialogue, you would expect the speaker to mean at least an equal opportunity to converse. This implies an obligation to listen while the other is speaking. There would also be an expectation that the purpose would be to shed "light"(reason) on the point of conversation rather than "heat"(emotion). This tradition has survived as the truly human aspects of democratic government from the ancient Greek's use of dialogue in the Agora.

Current practitioners and researchers have taken up this ancient art of conversation. A great deal of consensus has arisen about the framework and conduct of dialogue in groups and organizations. There have been some trouble spots in its use. Researchers refer to these trouble spots by pointing out how intensive or frustrating dialogue may become. In the CogniScope™ System methodology explored in this Chapter, these trouble spots have been identified as the unshakeable human burdens of interaction surrounding complex systems design situations. Three decades of research have revealed a number of important strategies to relief these burdens. This will be the primary purpose of this Chapter.

2. WHAT IS DIALOGUE?

From its Greek origins, dialogue has meant a "flow of meaning". As defined in current usage, " . . . an exchange of ideas or opinions on a particular issue...with a view to reaching an amicable agreement or settlement . . . " (Random House Dictionary, 1987, p. 547). We may infer from its origins to the present day usage that dialogue is a deeper form of conversation than discussion or debate. Where two persons or more meet to exchange ideas, their movement forward to achieve a jointly held consensus requires that individual perspectives must transcend their differences to achieve the confluence of a new meaning. Conversation such as discussion, and particularly debate, emphasize the individual's efforts to win others over to one's point of view. The origin of the meaning of debate means to "beat down". Whereas discussion may primarily focus on asserting one's point of view and debate emphasizes winning over one's "opponent", dialogue is best described as an inquiry process whereby an emergent shared meaning of participants enables the possibilities of a collaborative decision to be made.

When invoking dialogue as a conversational approach, participants should expect the ensuing process will differ from ordinary conversation in several significant ways. There is an anticipated diversity of perspectives in dialogical situations so potential conflicts exist; therefore participants should feel *safety* in speaking freely and directly to the points of conflict. *Active listening* should be the rule for each of the dialogical partners. In the spirit of an *inquiry process*, evaluative comments or the display of negative attitudes to others' conversational contributions would be discouraged. Understanding and/or agreement are the desired outcomes. Most authors when using the word "dialogue" to describe a particular type of communicative interaction share these characteristics.

An inquiry framework to dialogical conversation is an explicit assumption of most approaches to dialogue. From this common framework and other similar expectations, it is important to see the distinctions of other perspectives and the CogniScope™ System approach. In Table 1, three other current approaches to dialogical communications are presented in juxtaposition to the CogniScope™ System approach.

Current usage, as noted earlier, also indicates dialogue has the contextual expectation of reaching an amicable agreement, consensual decision, or settlement. Thus we see the qualifications of the basic process of this form of conversation in Table 1 as "deliberative dialogue" (Matthews, 1999),"generative dialogue" (Isaacs, 1999), or "focused and open" dialogue (Christakis, 1996) in the CogniScope™ System. Matthews (1999) finds the distinction in the public forum context necessary between "dialogue" as a search for mutual understanding around conflictual issues and "deliberation" as a process of framing the issues as options and working through individual limitations to reach consensual decision-making. The convenor of deliberative dialogue prepares extensive informational alternatives of the issue to inform the participants of the possible choices during deliberation. Isaac (1999) proposes generative dialogue as an art of thinking and reflecting together in a philosophical mode. The convenor of generative dialogue describes a four-stage process for participants to expect that evolves from politeness to breakdown to inquiry to generative dialogue.

Table 1. What is dialogue?

Mathews (1999): [D]ialogue . . . means conversation aimed at mutual understanding . . . While deliberation is for deciding…[it is] very dependent on mutual understanding. (p. 223)
Ellinor & Gerard (1998): Dialogue has become a distinct way of communicating in groups with the specific intention of becoming conscious of the shared meanings that we co-create. (p. 28)
Isaacs (2000): Dialogue . . . is about shared inquiry as a way of thinking and reflecting together. (p. 9) . . . [and] to uncover a base of shared meaning that can greatly help coordinate and align our actions with our values. (p. 19)
Magliocca & Sanders (this Chapter): Focused and open dialogue is the pluralistic-systemic inquiry of individuals to develop new social contracts through the collective construction of a consensual linguistic domain when the unshakeable burdens to human interaction have been minimized. (Please see comprehensive definition below.)

Within the CogniScope™ System dialogical conversation is the key interactive process of stakeholders in emancipatory systems design. Christakis and Dye (2001) provide a comprehensive definition of focused and open dialogue below:

> . . . Participants are convened that represent the full diversity of viewpoints with equal representation for each viewpoint (rather than a population based representation). They . . . individually articulate their observations concerning a constrained domain of common interest and . . .collectively assert relationships amongst everyone's individual observations. This

> structured languaging . . . maintains the boundaries between stakeholders by forbidding judgmental commentary and avoiding collaborative authorship. In so doing it protects individual authority, authenticity, and authorship of observations. It also preserves the autonomy of an individual within a pluralistic set of purposes. While it affirms the legitimacy of individual worldviews there emerges at the collective level a consensual linguistic domain [Maturana & Varela, 1988], which becomes a situational-context-specific language in which a social contract can be articulated. . . [T]his new socially constructed language enables them to evolve their articulation of their observation within the context of other participant's observations. Through abductive reasoning . . . a consensus on the most salient observations on which to form the contract is achieved. This shifts the context of inquiry . . . to the identification of highly leveraging directives to be enacted by the stakeholders, and that becomes the social contract . . . The methodology . . . minimizes individual cognitive burden, and maximizes productivity of the human interaction in the group. (Personal Communication, 1/20/2001)

In the CogniScope™ System, focused and open dialogue is fundamentally authentic languaging by the individual as part of disciplined inquiry. The convenor (a facilitation team) seeks through a discovery process of context building to develop a triggering question that focuses the dialogue. The convenor (a facilitation team) ensures that the all observations of the individuals in the dialogue are contributed in an authentic voice.

All of the dialogical methodologies begin with the creative production of diverse perspectives. The stakeholders are encouraged to actively listen to others' perspectives, to suspend judgment in the spirit of inquiry, and to examine one's own assumptions. The demands on the individual stakeholder are minimal; most find the creative experience enjoyable. How these diverse perspectives are transformed into knowledge creation is far more demanding of the stakeholders. They must be able to retain and relate multiple ideas as they occur, understand the meaning of the various ideas, participate constructively in an interactive group situation that requires some communicative competence, and contend with the invisible effects of power relations. Suggestions to provide a "container" to create personal safety for interaction (Isaac, 1999) or for the moderator to model active listening for participants (Mathews, 1999), for instance, are not sufficiently prescriptive for convenors of dialogue to ensure consistent outcomes of consensual decision-making in complex design situations. These suggested methods of dialogue closely resemble the well-known "muddling through" strategies of many organizations (Lindblom, 1956; Braybrooke & Lindblom, 1963). Without a framework of disciplined inquiry that permits organizational learning in an explicit way, particularly within the design contexts of work of the CogniScope™ System, the theories-in-use by individuals engaged in dialogue resembles the undisciplined, muddling through strategies of human languaging that attempt to avoid threat or embarrassment, as aptly demonstrated by Arygris (1993).

Rather than creating knowledge, stakeholders may elect because of fatigue or intimidation to go along with those members of the group who have managed to dominate the verbal interaction. Lacking a disciplined inquiry framework, the authenticity of the stakeholders' voices is in doubt.

3. THE UNSHAKEABLE HUMAN BURDENS OF DIALOGUE AND THEIR SOLUTIONS

In the CogniScope™ System approach (as derived from Interactive Management (Warfield & Cardenas, 1994), dialogue developed as disciplined inquiry because of the rigorous demands of systems design activities of considerable situational complexity at Batelle Memorial Institute (Columbus Laboratories) in the early 1970s (Warfield, 1976). For instance, a project called DEMATEL (decision making trial and evaluation laboratory) was initiated in 1971. Forty-eight world problems were identified that threatened the long-term future of humanity and required international cooperation to resolve them. Leaders from many nations shared their appraisal of the importance of the problems and how the problems impacted on each other. Applying many methodologies to analyze their responses, no consensus emerged from these world leaders of a leveraged resolution to these issues, i.e., no problem appeared to be more fundamental than any other problem making it impossible to initiate work on a significant subset. However, as Warfield (1976) concluded, the most important learning from the study was the unexpected diversity of understanding of complex systems by world-class experts; a phenomenon later quantified and analyzed during design sessions that became identified as "spreadthink" (Warfield, 1995).

Three decades of research have revealed unshakeable human burdens that undermine successful dialogue. The first two types of burdens include well-documented research into human limitations of cognition and group pathologies (individual, group, and collective). The third burden relates to the current focus of critical theorists: the impact of power relations (Ulrich, 1983). This third burden was explicated in the work of Magliocca and Christakis (1998, p. 2001) in the contributions of the CogniScope™ Systems approach to leadership in groups. In describing each of these burdens, we will also describe how they are explicitly confronted in the CogniScope™ System. The third burden, the impact of power relations, is the least understood issue. The convenors of dialogue may overlook its influence since most organizations assume the hierarchical power structures of positional authority of the industrial era. Organizational managers in the upper levels of authority are assumed to know more about what should be done. They expect and are given deferential treatment in design situations by most dialogical processes. In a similar fashion those individuals with superior communicative competence dominate group processes by intimidating others through the ease and quantity of their verbal facility. In the systems theory and design literature, the question becomes who is designing the system? When systems are designed by experts or those with authority, what is the role in systems design for those who must live within the system?

An effective mechanism for mutual exploration and consensus around leveraged approaches to complex systems became the impetus for a new science of complexity (Warfield, In Review). Many of the concepts in this Chapter are informed by Warfield's manuscript, *Understanding Complexity: Thought and Behavior*. This new science had to understand and resolve the obvious human limitations in situational design efforts, particularly as they escalated in scale. Dialogue was the metalanguage of engagement for situation design efforts of stakeholders, but it was discovered to be insufficient to meet the demands of complexity. As we review these burdens, we will identify the CogniScope™ System strategies for lifting these burdens.

3.1 The Limits of Human Cognition

Complex design situations are non-linear by nature; the spoken language of dialogue is linear and incompatible with the demands of complex design tasks. By itself, dialogue as unaided natural language–"metalanguage"–will not be comprehensible to the stakeholders once more than three or four ideas and their relationships are articulated. Miller (1956) initially established the limitations of human information processing capabilities in his famous formula of "7 $\pm$ 2". In design work, Warfield (1988) interpreted this principle to mean 3 or more ideas and their relations. As Simon's (1974, p. 1982) research indicated, the tendency is for people to "satisfice" once the quantity and non-linearity of the ideas is sufficient in number. That is, if stakeholders simply depend upon the natural language of their dialogue, the tendency will be to limit the inquiry and select whatever seems satisfactory. In complex situational design work, this might prematurely shutdown the generating and relating of ideas well before an appropriate design is conceived (e.g., falling short of meeting the need for requisite variety in the conceptualized design as articulated by Ashby, 1956).

The solution to this particular problem within the CogniScope™ System approach is to provide two additional languages. The first language, a graphic-prose hybrid language (displayed in visual form), was conceived to support the ongoing linear nature of dialogue. The graphic-prose language displays an unambiguous relational schema so that many ideas may be displayed and understood within the limitations of human processing. It provides the requisite variety for stakeholders to visually comprehend the relations of their generated ideas.

In addition, Interpretive Structural Modeling (ISM), yet another language in software form, was invented to assist participants to relate their diversity of ideas through formal reasoning with real-time products that enabled dialogical exploration of their interrelated ideas. ISM begins with the individual ideas of each stakeholder. Stakeholders clarify their ideas through dialogue in several ways. Through the ISM software, they are able to relate pairs of ideas as a group activity. This task is well within the limits of human processing capabilities. ISM permits multiple comparisons to be structured so that an influence map of relations is produced. This influence map enables the stakeholders to identify the leveraged subsets of ideas that influence the entire system.

To develop their individual ideas, stakeholders are asked to write their ideas silently in response to a triggering question. When each stakeholder articulates their ideas (which are printed and posted on the wall in large print), the authors of each idea are asked to explain its meaning. After the author's meaning has been given, clarification questions or comments are encouraged from others. Research has shown that stakeholders will begin their dialogue believing that others see the design situation as they do. As they are asked in a CogniScope™ session to explain their contribution so that it can be recorded, they will simply say there's no need for an explanation; everyone knows what they mean. Other participants invariably indicate they are not sure what the author means.

The importance of respecting the limits of human cognition should now be apparent during dialogue. Other researchers implement dialogue as a naturally flowing, self-organizing process of conversation within an inquiry framework of active listening and safety. Dialogue may be romanticized as a "magical solution" to human design issues once a safe space is provided and evaluative comments by others are withheld. These dialogical methodologies appear to work when they are applied as generative strategies; that is, stakeholders may feel progress is being made when a diversity of ideas are produced (as documented in the "spreadthink" phenomenon). However, because of well-documented human cognitive limitations, they may revert to focusing on the most recent ideas (forgetting previous ones) or focus on ideas with the least salient characteristics, (e.g., they are faddish or repeated often). They will certainly have considerable difficulty in relating more than several ideas and their relations into a coherent system of ideas. The strategies of synthesis in dialogue are critical to the stakeholders' design thinking. Both generative and synthetic strategies in dialogue are important in the CogniScope™ System. Synthetic strategies of dialogue in particular are critical to nurture and achieve group learning and knowledge creation.

3.2 Group Pathologies

Our human fallibilities leave us with inadequate capabilities for dealing with non-linear, complex design situations. Natural conversation language that is unaided during dialogue is simply not rigorous enough to meet a threshold of adequacy for complicated issues. This fallibility is compounded in group situations exponentially. Each individual carries into the group their own unshakeable burdens of human cognition. However, individuals will also confront emotional and cultural pathologies that may compound individual problems of capabilities.

Group dialogue work may become seriously frustrating when several types of social-emotional problems occur. Bales (1951) identified how individuals can disrupt the work of groups through expressing negative social-emotional behaviors. For a host of reasons (beyond the purpose of this article) individuals find group situations as a way to vent their anger and frustrations, perceive the situation as a threat to their self-interests, use the situation to get attention, dominate the group or follow some inappropriate strategy to meet a social or emotional need. Most groups find it difficult to confront these types of issues without assistance.

Tuckman (1965) characterized a typical pattern of group activities consisting of four stages:

- Forming (group members begin to develop initial stages of group identity);
- Storming (the inherent conflicts of differing views and approaches to the task surface);
- Norming (consensual arrangements permit the group to proceed); and
- Performing (group members may now contribute to the group task).

Group dynamics is a developmentally fragile process by itself. Many groups do not develop beyond the storming stages. This fact has contributed to the increased use of facilitators for group meetings. An "outside" facilitator may be successful in reaching the norming stages of group process since that may be the implicit justification for bringing in a facilitator. However, putting aside the individual fallibilities for the moment, a group may pose significant hurdles to reaching the performing stage even when the goal of changing the situation is embraced by all.

Since most group dialogue work is done either within or between organizations, the context of change is fundamentally important. What are the operational beliefs about change expressed or implied by the top manager(s) of participating organizations? Do these managers see change as systemic, episodic, or expedient? If systemic, the context of change may be more comprehensive and inclusive of stakeholders. The time scale for design will be adjusted in acknowledgement that systemic change takes time. If episodic, the clear message to the group is "getting on" to a solution. A group will feel rushed with considerable pressure to hold questions or objections of weaker ideas. If expedient, group dialogue may center on "What does management want?". A group may believe that management knows what it wants, but is going through the motions of securing others' inputs. We will revisit similar issues when we deal in the next section with unequal power relations.

Arygris' (1993) longitudinal study of an organization struggling with change is instructive. In spite of the fact that the CEO and the Directors of an organization wanted to put aside counter-productive strategies for dealing with each other to become a true learning organization, they needed considerable assistance. The learning organization framework would allow the participants to discuss and learn from their many different perspectives, including the ones that are potentially threatening or embarrassing. Even when the individuals in the group were shown how they contributed to the counter-productive strategies, they seemed unable to change their own behaviors. Argyris constructed an action map (see Figure 1) of these organizational defensive patterns in four "orders" of the consequences of group dynamics that build upon one another (pp. 98-99):

First order	*Second Order*	*Third Order*
advocate positions; convince others	exhibit low confidence in group effectiveness	hold bilateral conversations . . .
do not encourage inquiry into own position	have mediocre resolution or no resolution of wicked problems	build coalitions among directors
smoothover	build false consensus . . .	engage in lobbying, horse-trading . . .
request candidness; become upset when it comes	distance from acting responsibly on errors	Exhibit . . . bad-mouthing of directors to those outside the directors' group
become polarized		. . . do not ask for help; if you offer help, expect it will not be accepted.

Figure 1. The four orders of defensive patterns

At the *First Order* of consequences, the Directors in this organization experience considerable difficulty conversing about issues of common concern. By the *Fourth Order* (not listed above) of consequences, the group is tied up in an organizational deadlock. The *Fourth Order* is "Experience doubt and cynicism about problem-solving and decision-making effectiveness". As Arygris concludes, many of these strategies are undiscussed by the group and, worst they are undiscussable. Organizational cultures may completely dominate an unproductive framework of interaction.

Two related works explore the impact of organizational culture and group decision-making: "Groupthink" and "Clanthink". "Groupthink" (Janis, 1982) are the sets of interactions that suppress individual opinions on important issues so that deference is given to group recommendations that few, if any, members of the group may feel are appropriate. Typically a group may be under external pressure to make a recommendation. There are several identified "policing" strategies within the group that suppress dissension; silence by individual members is considered consent. In addition, the group culture may support the idea that a particular organization is invulnerable to making the wrong decision.

"Clanthink" (Warfield & Teigen, 1993) shares similar attributes of "Groupthink", but here the members of the group all believe something to be true when it is not. The culprit in this type of group interactions is the undiscussability of assumptions that underlie decisions. The group forms a closed system that is impervious to outside information that is contrary to their beliefs. Clan beliefs don't allow for examination of what is held to be true. Thus no corrective feedback loops exist to raise concern about the accuracy of particular beliefs. Warfield and Teigen (1993) conclude that deference to image is given in lieu of substance as a standard operating procedure.

The primary detrimental effect of group pathologies is that they stifle the potential for learning in a group. Without the possibilities of learning, knowledge creation through synthesis of this emergent learning is very limited. In the CogniScope™ System a number of strategies are combined before and during dialogue to break the closed cycle that creates stalemates for groups. We will begin with the individual stakeholder and then describe strategies on the group and organizational level that prevents or minimizes group pathologies as much as possible.

As previously indicated, it is fundamental to the dialogue methodologies that have been cited to provide safety for its participants. This is usually done by the convenor of dialogue expressing the desirability of open expression of thoughts. The safety of this procedure is usually demonstrated by asking participants to refrain from making evaluative comments of others contributions in the initial stages. However, once the dialogue is in full swing most convenors have few protections from several types of pathologies, which may be quite subtle in their effect.

The CogniScope™ System insists that all stakeholders at the table have a voice. If top managers or other persons of authority or expertise sit at the table, their views are given no more weight than others. Many times the facilitation team will discourage charismatic top managers from being active participants; they may observe or play the role of supportive participant to the dialogue, which is carefully monitored by the facilitation team. Prior to design sessions the facilitation team works very closely with top managers in several ways. Top managers must be certain in their minds and hearts that the process will only work if people are able to contribute their best in a candid fashion. The facilitation team works closely with them exploring how they may not believe this proposition; the facilitation team tries to convince them that this belief will show through. They won't be able to help themselves.

Active, hands-on managers can play a powerful, less intrusive role. Senge (1990) describes the powerful role of leader as designer. He uses the metaphor to describe this role by asking the question, "Imagine that your organization is an ocean liner, and that you are 'the leader'. What is your role?" (p. 341). Most will answer the captain of the ship, or the navigator, or the helmsman. Yet another role eclipses them all in importance, but is rarely mentioned. Senge points out the invisible role of the ship's designer may be most important. It's an important lesson as a leader to know when the most powerful role is establishing a framework so that others may creatively design. This metaphor conveys very quickly what the nature of this role should be.

The context stage within the CogniScope™ System approach is an extremely important process for preventing individual or group pathologies during design work. Many times the context work by the facilitation team requires more time than the actual design sessions. As noted above, top managers must understand their role as designers. Key constituents and stakeholders may be interviewed in depth to explore the scope and boundaries of the systems design issue. One key product of the context work is a triggering question(s) that reflect the goals or outcomes of the design target. Relevant documents may need to be summarized and reviewed with stakeholders. A white paper may be completed for stakeholders to read and study before the actual design sessions. Dagord, Magliocca, Tyree recently enacted another context method, and Runyan (2000) to use the Future Search technology (Weisbord & Janoff, 2000) as a group based way for several major state agencies in Iowa to visualize a platform for change. Context work should prepare all of the stakeholders to engage in the dialogue on an equal footing. In the Iowa case study, the Future Search method complemented the specific need of the CogniScope™ System to prepare the participants for equitable participation.

Nominal Group Technique (NGT) (Delbecq, Van de Ven, & Gustafson, 1975) is used as the means to solicit individual contributions to the triggering question. This method is slower than more free flowing methods such as "brainstorming". The slower pace is deliberate. It establishes a process of thoughtful and reflective dialogue. Individual stakeholders will vary considerably in the level of communicative competence. This is particularly important for individuals who are less verbal and reticent to express their viewpoints. Many group pathologies creep into group dialogue when some individuals begin to monopolize the group's attention by producing long lists of ideas and expressing them in a loud or forceful way. In NGT, each person is allowed to contribute one idea at a time in round-robin fashion. From the beginning of all design sessions, we incorporate a method of contribution, which respects and supports individual differences in communication style.

Setting the stage for individual distinctions or observations is the primary way in the definitional stage of CogniScope™ System work that pathologies are countermanded. It seems simple but the effect is profound. All individual contributions are accepted, recorded and posted as the author indicates. No one is permitted to publicly interpret what the author means other than the author. By writing the contribution and displaying it in real time the facilitation team is demonstrating the importance of individual distinctions. In almost all sessions it has been noted that other members of the stakeholder community who will attempt to explain to the author what they meant; the facilitation team halts this immediately. Another intrusion occurs when another stakeholder decides that they want to co-author a contribution. We find these efforts, however well meaning, co-opt or distort individual distinctions. The facilitation team must actively guard them against.

3.3 Unequal Power Relations

Dialogue is not substantively possible in a group situation where unequal power relations permeate the consciousness of the group. Power for the individual will be equated with the capacity to act. If the individual's perception is one of powerlessness, their involvement will be superficial and their commitment to any actions inconsequential. In the CogniScope™ System, equitable power relations are built through *context setting*, a *structured process that provides a transparent map* from the individual ideas to the final design of an action plan, and *role definitions*.

We have previously discussed *context setting* in several ways. One important context is the organizational structure and its culture and the attitude of top managers to shared power. Groupthink and Clanthink, as described earlier, are significant examples of disenfranchising organizational members. Both phenomena are rooted in top-down hierarchical structures of power relations. This structure of power relations will be invoked in a design situation as a part of the organizational culture. It begins with top management of the organization. When there are oppressive or coercive strategies of power relations within an organization, invoking dialogue as a process will not change that condition. However, as we will soon see, unequal or oppressive power relations in a group do not necessarily have to be organizational in origin.

How is it possible for a group to construct a relational pattern of influence that supercedes the influence of authority, personality, "expertness" or even conflicting beliefs in confronting change? One major organizational change that is impacting our concepts of power is the emergence of post-industrial organizations and societies (Bell, 1976). The rate of change for organizations is unprecedented; they must be flexible and adaptive to change or their survival is in jeopardy. Knowledge workers populate these post-industrial organizations. They are highly skilled and work collaboratively in teams. So the organizational texture of power relations is changing. Decision-making is essential lower in the hierarchy to support rapid responses to change.

Michel Foucault's work, *Power/Knowledge* (1980) presents a liberating vision of power relations as a constructive, positive force. Power occurs in a relational mode through the events of discourse; discourses which construct the social reality of the participants. Power is an emergent characteristic of a complex strategic relationship. To the extent that knowledge is shared and created, the relational mechanism is empowering to all. Power is not about "agency"; it is relational mechanism for constructing a social reality of participation and commitment through discourse. Through this group construction of reality, individuals may be transformed. As Foucault states it:

> . . . in thinking of the mechanism of power, I am thinking rather of its capillary form of existence, the point where power reaches into the very grain of individuals, touches their bodies and inserts itself into their actions and attitudes, their discourses, learning processes and everyday lives. (p. 39)

In the CogniScope™ System approach to dialogue, equitable power relations will have a transforming effect on the stakeholder. This can't be done by posting "rules of equitable power relations" on the wall or lecturing a group on how to be empowered. It must become the social fabric of the dialogue process. The stakeholders must *experience* equitable power relations. This reality begins by following the eight steps in Table 2. This structured process allows the facilitation team to describe the scope, boundaries, and products of the dialogical process as a *context setting* process. Thus, expectations of both managers and stakeholders are very clear as to how dialogue will be conducted, what contributions can be made, respect for individual voice and diverse perspectives, and acceptance to a consensual action plan as a product.The context setting strategy of the importance of the power of distinction-making in an organization is explained further by Tsivacou (1997):

> The actors that distinguish and have the chance to dictate the selection of the dominant explanatory path, immediately put themselves into the position of powerful, reducing the others involved into the position of powerless. Independent of their social status and role, those who control the information distinctions in a given situation acquire power and restrict the autonomy of the others. (p. 25)

Table 2. Structured Process That Provides a Transparent Map

1	Enabling the observers to draw distinctions in response to a "triggering question" that has been carefully framed in preparation for the conversation;
2	Providing the opportunity to all observers to explain their distinctions. i.e., "unpacking the meaning" of specific observations, by selecting the explanatory path that is more meaningful to the praxis of their life;
3	Constructing patterns of observations by employing efficient methods of inquiry for discovering through focused and open dialogue relationships among the diversity of observations generated;
4	Interpreting patterns of observations and discovering categories of similarities and dissimilarities among observations through a systematic and systemic mode of inquiry;
5	Interpreting patterns of observations and discovering interdependencies and the strength of interactions among them;
6	Selecting specific paths of observations as alternative pathways to action in a multidimensional design space;
7	Evaluating alternative "action pathways" and selecting the preferred one for implementation;
8	Designing an action plan by exploring the temporal relationship among the observations selected for inclusion in the preferred action pathway.

A *structured process that provides a transparent map* reflects the safeguard against a phenomenon of many organizations: the control of information about how decisions are made. As Tsivacou (1997) indicated above, inequitable power relations occur when members of the stakeholder group dictate the selection of the dominant explanatory path during dialogue. Information as a source of power is a primary strategy of hierarchical power relations. Many times this type of power relation strategy is invisible or opaque to the stakeholders. They are asked to dialogue and reach consensus on a set of recommendations about an important design issue. However, a major power-relations strategy is the lack of transparency about how stakeholder information is aggregated, consensual selection among alternative design pathways, and the final commitment to action is to occur. The CogniScope™ System employs the structured process in Table 2 that makes the entire process transparent. At each step there is comprehensive documentation of what ideas were generated, the meanings of those ideas, the records of voting, alternative design pathways explored and the final consensual action plan. This is known during the early stages of the process by everyone. The temptation of top managers to subvert such a process by simply "shelving" or "reinterpreting" the decisions are minimized.

Finally, *role definitions* in the CogniScope™ System between "process" and "content" seem to be unique to other approaches to dialogue. Negotiations of the entire design process with the sponsor and the stakeholder group begin with a clear distinction of what the process role is. The facilitation team plays the process role. They are the process experts; that is, they have identified a structured framework for engagement and approach dialogue as disciplined inquiry. As process experts, they will not provide any opinions on the validity of any content suggestions whatsoever. Their role excludes them as participating in the dialogue as a stakeholder member. They offer no suggestions related to content. This excludes the facilitation entirely from assuming a dominant role in the dialogue around content.

The stakeholders are identified as the "content experts". They engage in the generation of ideas, the distinctions among ideas, and the selection of design pathways to action. They are free of the obligations inherent in initiating and maintaining an equitable group process. They can expect to speak freely to the triggering question without fear of censure or criticism since the facilitation team will block this from happening. They will also be free to consider the opinions of "technical experts" and consider the importance of their statements on the design situation. However, in the end, the stakeholders have complete control over the content of the design. They also must accept that they will not be able to influence how the process will be conducted.

Process control is another subtle strategy of inequitable power relations that is averted in the CogniScope™ System. A stakeholder or a small group of stakeholders can attempt to control the design process by inserting their own process. This would of course set up a control situation that the other stakeholders would find almost impossible to surmount. We have had numerous experiences in design situations that are described in other writings that identified the nature of these power relation strategies (Magliocca & Christakis, 1998, p. 2001).

4. CONCLUSIONS

In the process of emancipatory systems design, dialogue must play a substantively deeper role than other concepts of dialogue to lift the human burdens of engagement in complex systems design situations. In the CogniScope™ System approach, dialogue is the means to achieving three important outcomes: (1) group learning that transcends self-referentality to develop a new systems-centered perspective of the design situation, (2) knowledge creation by the stakeholders that results in synthesized products of group learning, and (3) the emergence of a social contract by the stakeholders that embodies commitment to action toward achieving a collective purpose.

As noted in the introduction to this Chapter, dialogue has become a fashionable word. Recently, a high school student (son of the first author) described a community event of one of the candidates for U.S. President at his local high school in Ohio. The meeting was billed as a "town meeting". The candidate would engage the audience in "dialogue" around his specific policies regarding education. One female student courageously asked an incisive question on a specific education policy based on the candidate's past record. The candidate quickly disputed the student's input and declared the student to be wrong. The high school student recounting this experienced was disappointed by the candidate's response. Obviously, public dialogue in these situations has come to mean listening to a public officials point of view with the possibility of asking a question to see if the policy is understood correctly. Dialogue in such situations draws citizens into such forums because they have the expectations of honest engagement. The concept of "spin" has entered our everyday language to reflect the substitution of an attractive concept for a another concept which is less appealing. "Town meeting" and "dialogue" may have become attractive "spins" for staged events that mimic true dialogue. How often does this happen?

Several prominent researchers have identified their approach to dialogue. We share fundamental similarities with their work in regard to what dialogue is about—distinct from debate and suppositional discussions. However, because of its unique history of dealing with complex design situations, we have come to recognize that there are unshakeable human burdens that must be addressed if stakeholders are to succeed. We have identified three major burdens and general strategies that have developed to assist with these burdens.

REFERENCES

Arygris, C., 1993. *Knowledge for Action: A Guide to Overcoming Barriers to Organizational Change*. Jossey-Bass, San Francisco.

Ashby, W. R., 1956. *An Introduction to Cybernetics*. John Wiley & Sons, New York.

Bales, R.F., 1951. *Interaction Process Analysis*. Addison-Wesley, Cambridge, MA.

Banathy, B.H., 1996. *Designing Social Systems in a Changing World.* Plenum, New York.

Bell, D., 1973. *The Coming of Post-Industrial Society*. Basic Books, New York.

Braybrooke, D. and Lindblom, C. E., 1963. *A Strategy for Decision*. Free Press, New York.

Christakis, A.N., 1996. A people science: The CogniScope™ System approach. *Systems*, **1**(1): 16-19. A copy can be obtained from the CWA Website (www.cwaltd.com/index1.htm).

Christakis, A.N., and Dye, K.M.C., 1999. Collaboration through communicative action: Resolving the systems dilemma through the CogniScope™ System. *Journal of Transdisciplinary Systems Science*, **4**(1).

Christakis, A.N., and Dye, K.M.C., 2001. *Personal Communication*.

Dagord, C., Magliocca, L.A., Tyree, R.B., and Runyon, C.W., 2000. *Visualizing a Shared Platform for Change in the Children's Mental Health System in Iowa*. Iowa Depts. Of Education, Human Services, and Public Health, Des Moine, IA.

Delbecq, A.L, Van de Ven, A.H., and Gustafson, D.H., 1975. *Group Techniques for Program Planning: A Guide to Nominal Group and DELPHI Processes*. Scott Foresman, Glenview, IL.

Ellinor, L., and Gerard, G., 1998. *Dialogue: Rediscover the Transforming Power of Conversation*. John Wiley & Sons, New York.

Foucault, M., 1980. *PowerlKnowledge: Selected Interviews and Other Writings,* 1972-1977 (translated by C. Gordon). Pantheon Books, New York.

Isaacs, W., 1999. *Dialogue and the Art of Thinking Together*. Doubleday, New York.

Janis, I.L., 1982. *Stress, Attitudes, and Decisions*. Praeger, New York.

Lindblom, C., 1956. The science of muddling through. *Public Administration Review*, **19**: 79-88.

Magliocca, L.A., and Christakis, A.N., 1998. A transformational process of leadership through the voice and spirit of the people. *Journal of Management Systems,* **10**(4).

Magliocca, L.A., and Christakis, A.N., In Press. Creating transforming leadership for organizational change: The CogniScope™System approach. *Systems Research and Behavioral Science*.

Mathews, D., 1999. *Politics for People: Finding a Responsible Public Voice*. University of Illinois Press, Urbana, IL.

Maturana, H. R., and Varela, F.J., 1987. *The Tree of Knowledge: The Biological Roots of Human Understanding* (Revised Edition). Shambhala, Boston, MA.

Miller, G.A., 1956. The magical number seven, plus or minus two: Some limits on our capacity for processing information. *Psychological Review,* **63**(2): 81-97.

Random House Dictionary of the English Language, 2nd Edition, Unabridged., 1987, *Author*. Random House, New York.

Senge, P. M., 1990. *The Fifth Discipline: The Art and Practice of the Learning Organization*. Doubleday, New York.

Simon, H.A., 1974. How big is a chunk? *Science*, **183**, 482-488.

Simon, H. A., 1982. *Models of Bounded Rationality*. MIT Press, Cambridge, MA.

Tsivacou, L., 1997. The rationality of distinctions and the emergence of power: A critical systems perspective of power in organizations. *Systems Research and Behavioral Science,* **14** (1): 21-34.

Tuckman, B.W., 1965. Developmental sequences in small groups. *Psychological Bulletin,* **63**(6): 384-399.

Ulrich, W., 1983. *Critical Heuristics of Social Planning: A New Approach to Practical Philosophy*. Haupt, Bern.

Warfield, J.N., 1976. *Societal Systems: Planning, Policy, and Complexity.* Wiley, New York.

Warfield, J.N., 1988. The magic number three--plus or minus zero. *Cybernetics and Systems,* **19**: 339-358.
Warfield, J.N., 1994. A *Science of Generic Design: Managing Complexity through Systems Design* (2nd edition). Iowa State University Press, Ames, IA.
Warfield, J. N., 1995. Spreadthink: Explaining ineffective groups. *Systems Research,* **12**(1): 5-14.
Warfield, J. N., In Review. *Understanding Complexity: Thought and Language.*
Warfield, J. N., & Teigen, C., 1993. *Groupthink, Clanthink, Spreadthink, and Linkthink: Decision-making on Complex Issues in Organizations.* IASIS, Fairfax, VA.
Warfield, J. N., and Cardenas R., 1994. A *Handbook of Interactive Management.* Iowa State University. Press, Ames, IA.
Weisbord, M., and Janoff, S., 2000. *Future Search: Action Guide to Finding Common Ground in Organizations and Communities.* Berrett-Koehler, San Francisco.

SECTION III

MODALITIES OF DESIGN CONVERSATION

Chapter 11

DIALOGUE AND DESIGNING OUR FUTURE

Conversation as Culture Creating and Consciousness Evolving

PATRICK M. JENLINK*# and BELA H. BANATHY#**

Stephen F. Austin State University, International Systems Institute#, Saybrook Graduate School and Research Center***

1. INTRODUCTION

In this chapter, first, we begin with an exploration of the meaning of dialogue. Then, we examine the relationship of conversation to culture creating and consciousness evolving within society. Dialogue is introduced as a form of conversation that enables our species to connect within and across cultures, forming and sustaining communities through intersubjectivity and cultural creativity. The importance of dialogue to transcending existing social systems and existing levels of consciousness is explored, and the need for genuine discourse that is also relational is presented. The chapter will examine why dialogue is important as well as how dialogue may be used to create a collective evolutionary consciousness essential to designing our own future.

2. THE MEANING OF DIALOGUE

While we are by our nature a communicative species, not all communication is dialogue. What distinguishes dialogue from other forms of discourse conversation lies, in part, in understanding dialogue as a culturally and historically specific form of social discourse accomplished through the use of language and verbal transactions. It suggests community, mutuality, and authenticity–an egalitarian relationship. So understood, dialogue provides a meeting ground, *communitas*, and manifests itself in a variety of spontaneous and ritual modes of discourse in which nature and structure meet (Turner, 1969, p. 140). In this section we examine the meaning and nature of dialogue.

2.1 Etymological Meaning of Dialogue

Etymologically, dialogue means a speech across, between, though two or more people. Dialogue comes from the Greek *dialogos*. *Dia* is a preposition that means "through," "between," "across," "by," and "of." *Dia* does not mean two, as in two separate entities; rather, *dia* suggests a "passing through" as in diagnosis "thoroughly" or "completely." Logos comes from *legein*, "to speak" Crapanzano (1990, p. 276). Logos means "the word," or more specifically, the "meaning of the word," created by "passing through," as in the use of language as a symbolic tool and conversation as a medium. As Onians (1951) points out, *logos* may also mean thought as well as speech–thought that is conceived individually or collectively, and/or expressed materially. Consequently, dialogue is a sharing through language as a cultural symbolic tool and conversation as a medium for sharing.

The picture or image that this derivation suggests is a "stream of meaning" flowing among and through us and between us. Etymologically, dialogue connotes a flow of meaning through two or more individuals as a collective, and out of which may emerge new understandings (Bohm, 1996, p. 6).

2.2 Dialogue as Collective Communication

Dialogue may be transformative or generative in nature, as well as strategic. That is, it may be seen as transformative in relation to the creative actions of individuals through collective communication, the sharing of thought and knowledge of individuals as the generative materials to transform existing beliefs as well as create new innovations and cultural artifacts. It may also be seen as strategic or positional in relation to implementing an innovation or introducing new thoughts and knowledge into a cultural setting. In collective communication, as Bohm (1998) explains, the basic idea is to suspend opinions as well as judgment of what others share, trying to understand.

2.3 Dialogue as Relational

Dialogue is not something we do or use; it is a relation that we create and sustain by conjoint agreement and through shared discourse. As a relation, dialogue is characterized by inclusion and a reciprocal sharing, such that the individual's become one in and with each other. Gadamer (1976) is instructive in understanding the nature of dialogical relations:

> when one enters into a dialogue with another person and then is carried further by the dialogue, it is no longer the will of the individual person, holding itself back or exposing itself, that is determinative. Rather, the

> law of the subject matter is at issue in the dialogue and elicits statement and counter-statement. And in the end plays them into each other....We say that we 'conduct' a conversation, but the more fundamental a conversation is, the less its conduct lies within the will of either partner....Rather, it is more correct to say that we fall into conversation, or event that we become involved in it. (Gadamer, H-G, 1976, p. 66)

A dialogical relation will show itself in authentic discourse, but is not composed of this entirely or only. Shared silence as well as shared speech forms the relation that connects individuals through dialogue. Once created, the dialogical relation continues, even when the individuals are separated by space or distance, "as the continual potential presence of the one to the other, as an unexpressed intercourse" (Buber, 1965, p. 97). The fundamental tension underlying a dialogical relation is that participants need to be similar enough to share in genuine communication, but different enough to make it worthwhile (Burbules & Rice, 1991).

2.4 Dialogue as Genuine Discourse

Dialogue, wherein each individual conjoins with the others to share through conversation, suspending personal opinions and judgments to listen deeply, "derives its genuineness only from the consciousness of the element of inclusion" (Buber, 1965, p. 97). In this expression of genuine dialogue, each participant regards the "other" as the person he is, becoming aware of the "other" and that s/he is different from the person. In such relation through dialogue, one accepts the "other" setting aside the need to sway by opinion or judge the "other" so as to form a reciprocal relation that is genuine on both an individual and collective level. Buber explains,

> There is genuine dialogue–no matter whether spoken or silent–where each of the participants really has in mind the other or others in their present and particular being and turns to them with the intention of establishing a living mutual relation between himself and them. (1965, p. 19)

The genuineness of dialogic discourse resides in creating and sustaining a "living mutual relation" that enables all participants to share a common space, a community of creative possibilities. Fostering genuine dialogue requires that participants create what Buber (1988, 1992) referred to as the interhuman–a social sphere in which person meets person. Bohm (1998) suggests that one of the first steps toward dialogue is for people to engage in dialogue together, without trying to solve any problem" (Bohm, 1998, p. 117). If we are to be genuine in our dialogue, we must first come together and create a social sphere, without concern for outside problems. In the next section, we examine dialogue as culture creating, identifying what is problematic in culture and the conditions necessary to cultural creativity.

3. DIALOGUE AS CULTURE CREATING

Dialogue, as a connection between the subjective individual consciousness and the socially institutionalized structure of society, offers the opportunity to understand the influence of existing cultures and the differences that distinguish one culture from another, and a people in one culture from the people of other cultures. Equally important, dialogue offers the possibility of creating new cultures across differences, using difference as the very energy that fires social and cultural creativity.

3.1 Dialogue and Culture–Implicate Order

Implicate order, or the enfoldment of everything into everything, explains Bohm (1998) in his book *On Creativity*, means that everything is internally related. Applying this metaphor to culture, there are patterns of values, beliefs, and assumptions implicate within cultures, patterns enfolded one into another. Equally important, enfolded into a single pattern is the whole, such that, like a hologram, the whole image–in our case cultural image–is enfolded into specific elements or patterns. As such, within culture is language or symbol systems that enable individuals to communicate within and across communities of difference, and which transmit the implicate order of a culture.

Dialogue begins with the belief that there is implicate individual wholeness that can be made explicate (Bohm, 1996). Dialogue consists of meaning that requires a shared "field" of experience and attention. Dialogue, as a medium for making the implicate order explicate, is conceived "as part of the process of the coming into being of meaning" (Gadamer, 1982, p. 147). Unfolding that which had been enfolded in culture and the individual consciousness of members of a culture begins with sharing meaning through discourse.

3.2 Dialogue and Cultural Creativity

When conducted dialogically, the direct relation between person and person within society fosters social creativity and it can "generate frameworks of common discourse between different, often disparate, sectors of society" (Buber, 1992, p. 16).

Understanding the nature of intersubjectivity in the human experience, and its relation to cultural creativity–the conditions necessary to social and cultural creativity–for Buber (1992) centers on authentic intersubjective social relations. He believed that these conditions exist to some extent in all cultures but that their fullest development and fulfillment rarely occurred. One might interpret Buber's interest in dialogue as a concern for how to mediate the problematic nature of culture that, for him, made cultural creativity a rare experience.

The problematics of human creativity in general, and cultural creativity in particular, can be found in Buber's (1992) ideas on the essence of culture. He conceived of culture as constructed around several poles, around several contradictions, from which he signaled out four basic components of duality or polarity in culture:

> a) . . . There are two aspects of culture: creativity and tradition. On the one hand, all cultural life is based on personal creative production. Culture derives its vitality from the plethora of creativity, and when in any culture the flow of innovation ceases, its power is annulled, since that culture lacks any power if it does not have the power of innovation, the power of constant renewal: or self-renewal. But on the other hand, none of these productions succeeds in developing a social character; that is to say, does not become an integral part of that culture, unless it enters into the process of give and take; if is does not become material which can conveniently be passed on and be joined to all productions created throughout the generations to become something paradoxical: a form of generality. There are two basic sides to culture: revolution and conservatism, i.e., initiative and routine existence. Each one alone has great historical value, but only the two together have cultural value.
>
> b) . . . Cultural activity is characterized by a basic duality. First of all it gives to life itself form and permanence, restriction, and elaboration, molds people's behavior, raises the standard of their association and develops social relationships through selection and concentration. Secondly, it creates over and above life, or at least beyond it, a world of matter in the same way as nature is a world of matter, a world of beings independent of each other, like creatures of nature which are bound to each other by invisible bonds: this is the second world, the unique world of mankind . . .
>
> c) there are two basic elements related to the crystallization of culture: the development of form and the development of awareness. Both of them, form and awareness, exist within man's experience as a matter of potential. Form grows, as it were, of its own volition . . . but awareness can also grow, as it were, within us on its own, but it remains within us and does not wish to leave us. . . .

d) every culture that is in a state of full development tends to produce a number of cultural types and areas which are totally independent, each of which has its own domain and immutable laws, that is, it has a tendency to pluralism of spiritual spheres. . . . (The Face of Man, pp. 383-386; in Hebrew). (Eisenstadt, 1992, pp. 9-10)

While Buber believed that the tensions giving rise to the problematics of cultural creativity were necessary, he also believed that the central characteristic of situations conducive to creativity is the existence of dialogue, a communicative openness. For Buber, dialogue was both an intersubjective relation between an individual and others, and between an individual and God. It is in these intersubjective relations that the development or "crystallization of a common discourse could occur, and it…is essential for holding a society together, for meeting conditions conducive to cultural creativity and for counteracting the possible stagnative or destructive forces that are endemic in any society" (Eisenstadt, 1992, p. 11).

5. DIALOGUE AS CONSCIOUSNESS EVOLVING

David Bohm (1998) explains that the basic idea of dialogue is to be able to communicate while suspending personal opinions, not trying to convince the "other" but simply trying to understand. This is an important step to understanding how consciousness evolves through discourse. The realization on the part of each person that s/he has a perspective–the evolution of perspectival consciousness. The evolution of conscious awareness of perspective, through dialogue, begins with all individuals' capacity to "perceive all the meanings of everybody together. . . . That will create a new frame of mind in which there is a common consciousness. . . . a kind of implicate order, where each one enfolds the whole consciousness" (Bohm, 1998, p. 118).

5.1 Consciousness as Implicate Order

David Bohm, in his book *Wholeness and the Implicate Order* (1995), is helpful in understanding consciousness as implicate order, an enfoldment of thought, perspective, worldview. An individual's consciousness is an enfoldment of many thoughts and perspectives over time, creating implicate patterns or relationships. Not dissimilarly, a collective or societal consciousness such as that represented in a particular culture or people may also be understood as being implicate order. When one individual's thought or perspective is enfolded on that of another person's, then patterns of thought or perspective become enfolded into the cultural fabric. Implicate in these patterns are the values, beliefs, and assumptions of individuals within the culture, and likewise these patterns are implicate in each individual.

Language, as an artifact of a culture and its people, is an enfoldment of symbols and meanings that create an implicate order. Meaning enfolded in the words and structure of language creates implicate patterns of meaning through the use of language to generate thought and action. As individuals engage in communicative relationships, the meaning implicate in language is unfolded in the social or cultural groups through the discursive interactions. As meaning unfolds through communication, the implicate nature of meaning is made explicate, creating opportunity for the participants to generate common meaning through sharing. Such sharing moves from the individual consciousness level to a collective consciousness level. Thus, as Bohm and Peat (1987) note, "there is an internal relationship of human beings to each other, and to society as a whole" (p. 185). What is seen in society–the explicate order–on one level is seen as enfoldment inseparably within the consciousness of each individual member in the society. Therein, implicate order "is the *content* of the culture, which extends into the consciousness of each person" (p. 185). Consciousness as implicate order, is made explicate by engaging in the unfolding of what individuals, culture, and society has enfolded. Making the implicate explicate requires a evolving of consciousness–a dialogic consciousness.

5.2 Dialogic Consciousness

Dialogic consciousness refers to a "way of being in the world, that is characterized by what Schachtel (1959) calls "allocentric" knowing…a way of knowing that is concerned with both "*the totality of the act of interest*" and with the "participation of the total person" (of the knower) (p. 225). It requires an attitude of profound openness and receptivity–a trust relation. It involves, Schachtel (p. 181) explains, a temporary eclipse of all the individual's egocentric thoughts and strivings, of all preoccupations with self, and self-esteem. For Bohm (1996, 1998), this is the suspension of personal opinion and desire to judge the opinion of the "other." For Buber (1992), this is the creation of an authentic intersubjective relation made possible by genuine dialogue.

Dialogic consciousness is where one is turned toward other (human or nonhuman) "*without* being in need of it" or wanting to appropriate it to achieve something. The latter would point to preoccupations with self that antedate the experience of "I" as separate from the world and block full perception of other (p. 177). Such participatory consciousness is possible through genuine dialogue wherein the participants yield to the will of the "mutual living relation," and create a shared collective consciousness.

The importance of dialogic consciousness to culture creativity rests in genuine dialogue that gives way to cultural creativity. Importantly, evolving to a level of dialogic consciousness means recognizing that differences within and across cultures are implicate in the individual consciousness of each participant. It also means recognizing that differences translate into thought, which "may establish distinctions," but the distance "*between* those distinctions–between people" is mediated by dialogic consciousness (Bohm, 1996, p. 89). As Burbules and Rice (1991), explain, "if dialogue across difference is to succeed, sensitivity is required to the various kinds of diversity one may encounter" (p. 407).

5.3 Dialogic Consciousness and Differences

At the present there are great differences that define society, and many of these appear non-negotiable. Differences in perspective–Buber's dualisms–contribute to the problematic nature of culture and therein limit the potential for cultural creativity. Likewise, differences concern the evolving of consciousness understood as necessary to cultural creativity. It is important to understand that when these differences are enfolded in society–then dialogue may run "up against deep linguistic, cultural, or paradigmatic uncommensurabilities" (Burbles & Rice, 1991, p. 408).

Differences are not necessarily always problematic; rather it is in the differences that Buber (1992) believed that the creative tensions necessary to cultural creativity resided, in part. Respecting difference as a defining aspect of *genuineness* also enables the participants to recognize

> the constructed world-view and subjectivity of the persons who enter a dialogical relation; thus difference (or its absence) cannot always be inferred or assumed from the outside. (Burbules & Rice, 1991, p. 407)

Where difference in thought, perspective, beliefs, values, and assumptions are enfolded at a number levels in society, what "may not be apparent to, or salient for others, may be paramount in the minds of the individuals at hand" (Burbules & Rice, 1991, p. 407). Dialogue–evolving to dialogic consciousness–offers a path to establishing intersubjectivity and a "mutual living relation." Genuine dialogue is recognized and enables participants to create "a degree of understanding across (unresolved) differences" (p. 409). Recognizing this carries those engaged in the communicative act

> beyond the conception of dialogue as a single, convergent method aimed toward Truth. Dialogue can also serve the purpose of creating partial understandings, if not agreement, across differences. Complex understanding

> and total incomprehensibility are not the only two alternatives—indeed, both of these are quite rare. (Burbules & Rice, 1991, p. 409)

Importantly, dialogue does not eliminate differences; rather, through dialogue participants create a consciousness of differences that can sustain differences within a larger social compact of toleration and respect. Genuine dialogue enables the evolution from individual consciousness, to a level of conscious awareness of differences, to a level of dialogic consciousness.

6. CLOSING THOUGHTS

Homo Sapien Sapien, as a social species, is communicative by nature. We are also sentient beings. Placing within our social and evolutionary grasp the capacity to be equally creative and destructive, whether through discourse or social action. When we consider, in relation to being a communicative species as well as sentient beings, that we are by nature extremely diverse, we are filled with great potentials, potentials waiting to be unfolded through genuine dialogue. Whether we choose, as sentient beings, to be creative or destructive is a consequence of our choices in matters of communicative action within and across our societies, nationally and globally.

The fact that we display great diversity–that there are differences in how we see and experience and interpret the world as well as in who we are as cultural, racial, political, intellectual, moral, religious beings–figures largely into the potential for either creative or destructive actions, of either the betterment or the detriment of humankind. Always present is the question of whether or not we take responsibility for our actions, individually and collectively, with and cross the geographical and spatial boundaries that define our cities, our nations, and our world; whether we take responsibility for humanity, all of it? Undertaking this responsibility will shape our actions, and, when enacted through communicative action that is a collective dialogue, in turn will shape our collective future. The realization of our potentialities as a species rests in no small measure, in our capacity for cultural creativity and in our capacity to achieve new levels of consciousness.

REFERENCES

Banathy, B.H., and Jenlink, P., 2002. *Dialogue as a Means of Collective Communication*. Kluwer Academic/Plenum Publishers, New York.
Bohm, D., and Peat, D., 1987. *Science, Order, Creativity*. Bantam Books, New York.
Bohm, D., 1995. *Wholeness and the Implicate Order*. New York: Routledge.
Bohm, D., 1996. *On Dialogue*. Routledge, New York.
Buber, M., 1965. *Between Man and Man*. Collier Books, New York.

Buber, M., 1988. *The Knowledge of Man: Selected essays*. Humanities Press International, Atlantic Highlands, NJ.
Buber, M., 1992. *On Intersubjectivity and Cultural Creativity*. The University of Chicago Press, Chicago, IL.
Burbules, N.C., and Rice, S., 1991. Dialogue across differences: Continuing the conversation. *Harvard Education Review*, **61**, 393-416.
Crapanazno, V., 1990. On Dialogue. In T. Maranhão (Ed.), *The Interpretation of Dialogue* (pp. 269-291). The University of Chicago Press, Chicago, IL.
Combs, A., 1996. *The Radiance of Being: Complexity, Chaos, and The Evolution of Consciousness*. Paragon House, St. Paul, MN.
Eisenstadt, S.N., 1992. Introduction. In M. Buber, *On intersubjectivity and cultural creativity*. The University of Chicago Press, Chicago, IL, pp. 1-23.
Gadamer, H-G., 1976. *Philosophical Hermeneutics*. Berkeley, CA: University of California Press.
Gadamer, H-G., 1982. *Truth and Method*. Crossroads, New York.
Onians, R.B., 1951. *The Origins of European Thought About the Body, the Mind, the Soul, the World, Time Fate*. Cambridge University Press, Cambridge, UK.
Schachtel, E.G., 1959. *Metamorphosis: On the Development of Affect, Perception, Attention and Memory*. Basic Books, New York.
Turner, V., 1969. *The Ritual Process*. Cornell University Press, Ithaca, NY.

Chapter 12

THE MAKING OF A NEW CULTURE

Learning Conversations and Design Conversations in Social Evolution

ALEXANDER LASZLO and KATHIA CASTRO LASZLO
Syntony Quest and the Monterrey Institute of Technology (ITESM)

1. A POSSIBLE DIRECTION

Even a cursory glance at the impact humankind is having on the life support systems of planet Earth makes patent the unsustainability of contemporary cultures of consumption. Creating a new culture through learning and design conversations is not a quest of foolish arrogance – it is the survival imperative for sustainable co-existence of humankind with planet Earth. Societies all around the world are currently experiencing a period of rapid and extensive transformation. The signs of change are pervasive, and the rate of change is itself changing and accelerating, speeding contemporary societies toward a critical threshold of stability and engulfing the individual in a confusing blur of behavioral choice. Global flows of information, energy, trade, and technology are swept up in massive economic reforms and political reorientations with the result of creating a disorienting and disrupting vortex of social and cultural change on both local and global levels.

Nevertheless, we need not be victims of change, destined for one future or another according to either a predetermined plan or random chaos. Both individually and collectively, *we can learn how to have change happen <u>through</u> us, not <u>to</u> us!* But we must find out how to look, listen, and learn – to *really* see and hear and understand the underlying patterns of change so that we can distinguish between those dynamics that are destabilizing and those that forward sustainable futures. The sharp discontinuity between where we – as not the most unobtrusive species on Earth – are going and where we *should be* going is underscored by the need for new ways of thinking, new ways of learning and new ways of conversing. Lester W. Milbrath (1989) notes in his book, *Envisioning a Sustainable Society: Learning Our Way Out*, that

> As a society, have to learn better how to learn – I call it *social learning;* it is the dynamism for change that could lead us to a new kind of society that will not destroy itself from its own excesses . . . for we must share a vision for a new society before we can realize it. Designing a better society and maintaining a good life require deep thought and sustained effort by all of us. Reasoning together is the only way we can bring it about. (pp. 6 & 1)

Reasoning together, conversing together, searching together. The challenge is nothing short of the collective consideration of a radical transformation of the social systems, which embody our attitudes and dispositions. "Our goal," as Milbrath saw it over twelve years ago, "will be to design a new society that provides a decent quality of life while coexisting in a long-run sustainable relationship with the natural environment that nourishes it" (*ibid.,* xi). Not only is this goal entirely relevant to current design conversations around the world, it has increased in urgency as the years have passed. Indeed, it must no longer be considered a side conversation, relegated to conferences and classrooms. This is *The Conversation* for being and becoming with our world.

When we engage in conversation with each other, if we do so authentically and inclusively, we end up also conversing internally – with ourselves, as well as externally – with the more-than-human world of which we are a part. Through multi-faceted reflection on where we stand, where those who surround us stand, and where we would like to be, we are brought inexorably to a consideration of our ethics. We may find that our dearest (and not so dear) acquaintances, and we tend to be more of the take-make-waste worldview than of the syntony-quest worldview. Although, as we will see, this may be neither pleasant nor reassuring, such awareness marks the first step toward transcendence. Here, at the threshold of conscious evolution and the capacity to creatively contribute to evolutionary consonance, we need to step back, take a look at what is happening in the "big picture," and find ourselves somewhere there. How are we contributing to that big picture? Are we over there with those who are heedlessly stamping down this earth, or over here, with the mindful walkers and insightful listeners?

Carolyn Merchant (in Hinman, 1996, p. 516), author of *Environmental Ethics and Political Conflict,* distinguishes among three approaches to environmental ethics:

> An *egocentric ethic* is grounded in the self and based on the assumption that what is good for the individual is good for society. A *homocentric ethic* is grounded in society and based on the assumption that policies should reflect the greatest good for the greatest number of people and that, as stewards of the natural world, humans should conserve and protect nature for human benefit. An *ecocentric ethic* is grounded in the cosmos, or whole environment, and is based on the assignment of intrinsic

> value to nonhuman nature. This threefold taxonomy may be useful in identifying underlying ethical assumptions in cases where ethical dilemmas and conflicts of interest develop among entrepreneurs, government agencies, and environmentalists.

There is also a fourth stage, a truly transcendent and evolutionary level of ethical consideration that should serve as the basis for self-directed sustainable change efforts. It is what can be called evolutionary ethics. Without a doubt, ecocentric ethics assigns intrinsic value to "the whole environment, including inanimate elements, rocks, and minerals along with animate plants and animals" (Merchant in Hinman, 1996, p. 524). But it is still synchronic, considering "the big picture" only at any one point in time. An evolutionary perspective needs to infuse this ethic to make it sustainable in the long run. Otherwise it is just optimizing *what is,* not working in stewardship of *what should be.*

Years ago, C. H. Waddington anticipated the challenge for conversations based in an evocentric ethic. He pointed out that

> we have found ourselves faced by a series of problems – atomic warfare, the population explosion, the food problem, energy, natural resources, pollution and so on – each complex enough in itself, but then it turns out that each of these is only one aspect of, as it were a Total Problem, in which all aspects of the world's workings are inter-related. (in Merry, 1995, p. 78)

This is what others, such as the Club of Rome, have termed the *global problematique,* and as Waddington suggests, it must be considered as a continually unfolding condition. An ecocentric ethic simply will not bear up to the challenge of dealing with it (much less a homocentric ethic, while an egocentric ethic can only make it worse). The time is nigh for societal design conversations based on an evocentric ethic.

1.1 The Vision of a Sustainable and Evolutionary Future

Culture cannot be purposefully planned. It arises out of the confluence of values, beliefs, ideas, and forms of expression characteristic of interpersonally aligned individual cognitive maps. That is to say, it arises from a community. Through the development and application of Evolutionary Systems Design (in all its dimensions of philosophy, theory, and practice), action oriented systems thinkers of today can engage in condition creating conversations that set the stage for the emergence of tomorrow's culture of design. This conversation objective leads down a path of inquiry framed in a normative evolutionary perspective. The approach to bringing forth a design culture draws on the principles that underlie the patterns of change described by all complex dynamic systems with a throughput of information and energy. As such, a sustainable design culture is conceived to emerge when conditions for individual and group empowerment are consciously created in a framework that incorporates evolutionary understanding. Evolutionary Systems Design (ESD) is key to the gaining of such understanding since it is firmly based on the acquisition of evolutionary competence: the state of self-actualization (of individuals and groups) that is marked by the mastery of the knowledge, abilities, attitudes, and values required for co-evolutionary actions, and therefore, for the pursuit of sustainable modes of being.

An evocentric conversation dynamic leads through four stages of evolutionary learning: 1) creating AWARENESS of the need for a design culture through promoting Evolutionary Consciousness; 2) constructing an UNDERSTANDING of evolutionary design through developing Evolutionary Literacy; 3) generating a sense of RESPONSIBILITY that is matched by the change competence of RESPONS*ABILITY* as we learn to affect purposeful, positive, co-evolutionary change in the communities with which we work/play/learn, and thereby gain Evolutionary Competence; and 4) learning how to become CATALYSTS of desirable evolutionary change in social systems through engaging in Evolutionary Praxis with communities open to transformative learning. (See Figure 1 for representation of the stages.)

1. Evolutionary learning is a core aspect of ESD conversation. Recent development of the above portrayed operational learning framework for the stages through which individuals and groups pass as they become evolutionary change agents has provided a scaffolding for Evolutionary Systems Designers to gauge their progress. (A. Laszlo 2000)

With its focus on the development of evolutionary competence, the Evolutionary Learning Community (ELC) serves as the vehicle of choice for such searching together conversations. By engaging in participatory processes of learning how to learn in harmony with the dynamics of its physical and socio-cultural milieu, the ELC does not adapt its environment to its needs, nor does it simply adapt to its environment. Rather, it adapts *with* its environment in a dynamic of mutually sustaining evolutionary co-creation. Through such a process, individuals and groups can self-empower for the creation of responsible ecosystemic transformation in whatever community they choose to join. When one or more Communities of ELCs (or ELC Ecosystems) begins to appear, this in effect, marks the emergence an authentic Design Culture. Through this process of future-creating conversation a vision of a sustainable and evolutionary future can be developed.

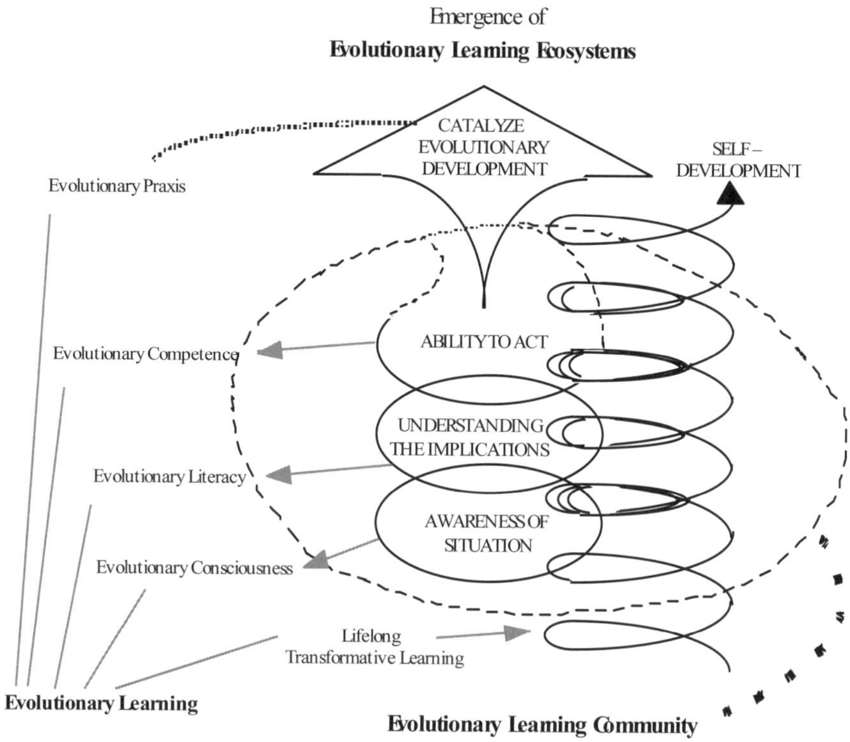

Figure 1. The four learning stages of evocentric conversation

2. HOW DO WE GET FROM THERE TO HERE?

The evolution of evolutionary learning community is a purposeful process that starts with the creation of a Healthy and Authentic Community from any Human Activity System (Checkland, 1993) that is a willing participant in the learning and design process. The formation of a Healthy and Authentic Community is facilitated through a process of generative dialogue and is dedicated to creating the appropriate context for collaborative learning and design. When the community is ready to make the commitment to become a Learning Community, the members engage in a process of learning-how-to-learn that includes the development of evolutionary consciousness and evolutionary literacy, as described above. By subsequently developing evolutionary competence, the learning community can design itself into a Designing Community capable of continuous autopoietic re-creation as a community. This process of evolutionary learning creates the conditions for evolutionary praxis. Through the integration of ideals and actions, the individuals and the

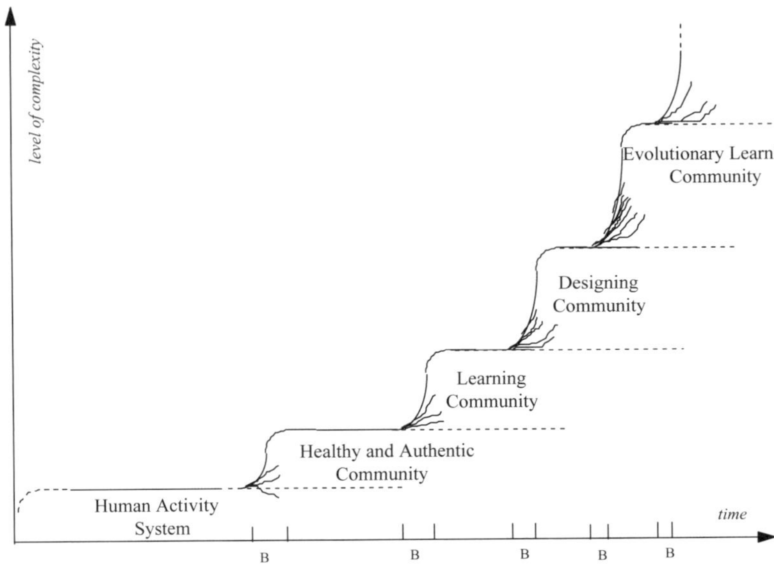

Note: B refers to Bifurcation

Figure 2. The evolution of Evolutionary Learning Community

community become fully empowered as stewards of their ongoing evolution. The evolutionary path from Healthy and Authentic Community to Evolutionary Learning Community can be represented as follows:

Because the conversations focus on issues of learning and applying in real-world contexts, ELCs serve as the catalyzers other systems of syntony. These range from communities of syntony, to social systems of syntony, and eventually to entire ecosystems of syntony. Figure 3 portrays how ELC serves as the vehicle for the generation of these additional stages. The first type of ELC is the community of syntony. Such communities are dedicated to learning how to learn and consciously seek to do so in ways that are evolutionary. At the next level, social systems of syntony are formed from various sorts of individual ELCs, each collaborating to create a subculture of syntony. Those social systems of syntony that have developed their full evolutionary potential have attained a degree of evolutionary competence that permits them to engage with their more-than-human world in the creation of greater ecosystems of syntony. These form evolutionary learning ecosystems (ELEs).

Groups of people engaged in purposeful ESD conversations form an evolutionary learning community, and such communities can foster the emergence of other systems of syntony. At the level of the ELE, people no longer 'come first' – the whole ecosystem comes first. All aspects of the ELE, from psycho-personal and socio-cultural to bio-physical and process-structural – all are 'actors' with a voice in the creation of evolutionary consonance or syntony (*cf.*, A. Laszlo 2000).

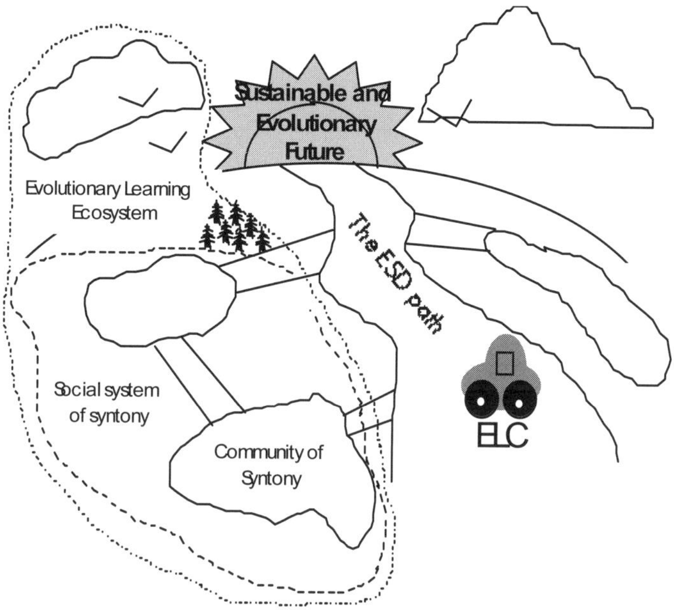

Figure 3. Nested Systems of Syntony

All listen to and create with one another. These ELEs are, of course, communities as well; ones in which people act as stewards of their own futures in syntony with their dynamic surroundings. Figure 4 depicts how ELC facilitates the emergence of systems of syntony within and beyond itself.

By manifesting an evolutionary consciousness, ELEs are embodiments of syntony. They draw on an expanded conception of self that leads to empathetic identification with others – including non-human others. Such an inclusive self-concept fosters understanding and love for other people, species, and future generations. It helps us learn to 'think like a mountain,' in Aldo Leopold's wonderful phrase. Evolutionary consciousness of this sort creates less need for ethical guidelines on protecting and preserving nature – it would be just natural to do so. This kind of thinking leads to actions that are empathetic and inclusive, giving voice to such sentiment as John Seed once showed: I'm not working to protect the rainforest, I'm part of the rainforest, in human form, protecting itself. To be human and to express such more-than-human identity and volition in future creating design conversations suffuses the processes with an ethic of appropriateness that marks a truly evolutionary state of consciousness.

ESD provides a clear presentation of the importance of learning for the purposeful design of the future. As a conversation approach for realizing the vision of a sustainable and evolutionary *learning* society, ESD offers a means

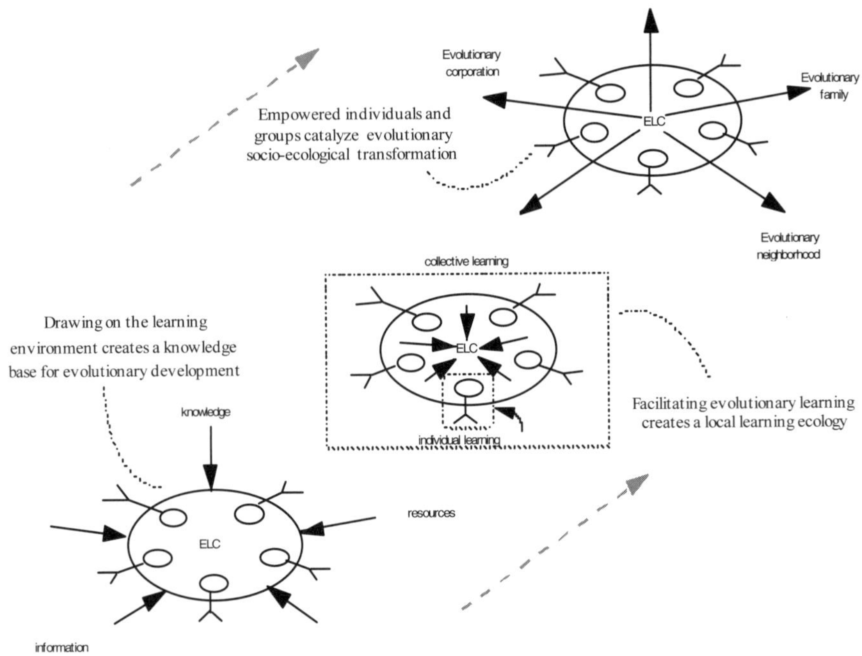

Figure 4. The Process of ELC Empowerment

of recreating the ancient Greek ideal of a *paidea* – a society where the promotion of lifelong learning and the achievement of the human potential in the broadest sense is a central priority (Milbrath 1989, 94). The *paidea* of the future are social systems of syntony where the vehicle of conversation for facilitating evolutionary learning and transformation is the ELC.

2.1 The Role of Evolutionary Systems Design

While Social Systems Design (SSD) can be been characterized as a form of soft systems thinking primarily serving Habermasian practical interests (Jackson 1991), Evolutionary Systems Design (ESD) is conceived as an attempt to evolve SSD into a form of critical systems thinking by also serving an emancipatory interest. This means that ESD draws from wellsprings of soft systems thinking, critical systems thinking, and emancipatory systems thinking in addition to General Evolution Theory and lifelong evolutionary learning orientations. The result is a humanistically oriented systems approach comprised of a meta-methodology that facilitates the critical application of various systems perspectives to real-world situations (Laszlo & Krippner 1998, 59).

ESD defies classification as constituting either an epistemology or ontology. Rather, it is reminiscent of the Greek notion of *gnosiology* concerned with the holistic and integrative exploration of phenomena and events. There are aspects of ESD that are ontological and aspects that are epistemological, and aspects that are at once both and should not be circumscribed to either.

Much as with TSI (Total Systems Intervention, *cf.*, Flood & Jackson 1991), those engaged in ESD conversations must select or design appropriate approaches for addressing their particular purposes. Rather than consider the application of ESD on philosophical grounds, designing communities face practical challenges for socio-ecological survival and must learn to move "toward what will work to provide answers where no reliable guides exist" (Salner 1996, 8). This does not mean that ESD is methodologically eclectic or that it disregards the need for a coherent body of theory to inform its practice. By empowering evolutionary agents neither as activists nor as theorists, but as a synthesis of the two, it offers a way – an integral path – for human becoming in partnership with Earth. Figure 5 illustrates the key conceptual influences upon ESD.

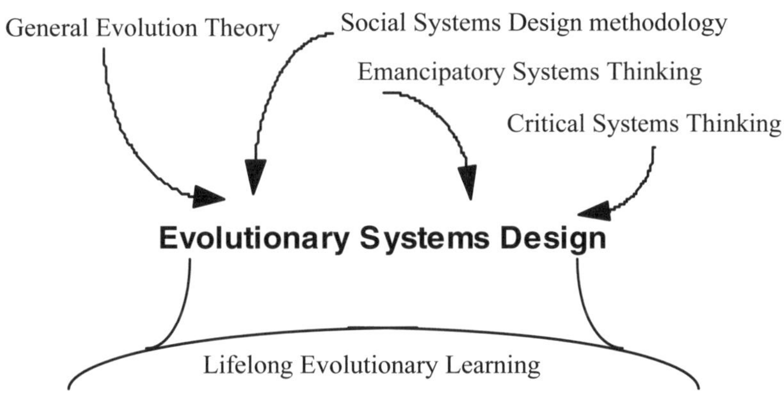

Figure 5. Conceptual Influences on ESD

Various thinkers have identified complementary learning objectives relevant to the creation of sustainable and evolutionary futures: Donald Michael's (1973) psychological barriers to be overcome for future oriented societal change; C.A. Bower's (1993) educational changes for embracing sustainability; Alfonso Montuori's (1989) ideas on the creative elicitation of the future; Dean Elias's (1997) learning for expansion of consciousness; Lester Milbrath's (1989) learning tasks for a sustainable society; and Robert Ornstein's (1991) and Ornstein and Ehlrich's (1989) ideas on conscious evolution. By drawing on the orientations put forth by these thinkers, and infusing them with Bela H. Banathy's (1996) agenda for evolutionary learning, it is possible to engage in future creating design conversations that create operational outcomes in line with the four stage learning framework described in section 1.1, above (*cf.,* A. Laszlo, 2000: "The Epistemological Foundations of Evolutionary Systems Design").

3. CONVERSATION: THE HOLDING PROCESS

Two complementary modes of dialogue comprise design conversation: generative dialogue and strategic dialogue (Banathy 1996, 218). One provides a process through which individuals become friends and partners and a community generates common meaning in learning conversations. The other focuses on particular tasks, in the creation of solutions for a specific social circumstance in design conversations.

Generative dialogue can be considered as the core transformative process for a group to become an authentic community. Banathy suggests that the involvement in generative dialogue "will lead to the creation of collective consciousness, collective inquiry that focuses on the thoughts, values, and worldviews of the group and creates a flow of shared meaning, shared perceptions, a shared worldview, and a social milieu of friendship and fellowship" (1996, 219). Once the community has bonded, and if there also exists the intention and the commitment, they can then enter into strategic dialogue through a focus on communal activities future oriented design. When dialogue consists of collective learning and coordinated action, paradigm exploration, cultural healing, and collective creation inspired by a shared vision of the future (Isaacs in Banathy 1996, 217), the community is on its way to become one of *learning and design.*

As with the entire ESD conversation process, the willing adoption of a learning orientation is an essential aspect. To seek the transcendence of our differences and to promote co-creation is a path toward higher complexity and wholeness. It is a syntony quest, that is, a creative aligning and tuning with the evolutionary processes of which we are a part, and as such, a contribution to the evolution of consciousness and the creation of new possibilities for the future. This is lifelong learning. In face of the planetary challenges of our time, this is the most extraordinary common conversation in which we can engage.

3.1 From Traditional Community Through Surrogate Community, Learning Community, Healthy and Authentic Community and Evolutionary Learning Community to Community of Syntony, and Other Models of the Evolution of Designing Communities Through Conversation

In its most fundamental conception, community can be considered "a group of two or more individuals with a shared identity and a common purpose committed to the joint creation of meaning" (Laszlo & Laszlo, 1997, p. 6). Authentic communities are able to enhance their own development while at the same time enhancing that of each individual in the community, thereby promoting both freedom of personal choice and a sense of responsibility for the whole. In such communities, the operating principle is that of unity in diversity.

There are different types of communities and attention to their distinct characteristics helps clarify the particular orientation of ELC. Four types of communities are relevant for this consideration: Traditional Community, Surrogate Community, (simple) Learning Community, and Evolutionary Learning Community (McCormick, et al., 1998).

3.1.1 Traditional Community (TC)

A closed, stable system where the individual's identity is determined by a collective identity rooted in transmitted myths, values, norms, rituals, and beliefs. That is, an individual born within this kind of community is socialized into the local culture. Many indigenous communities are good examples of traditional communities. They are natural rather than designed communities. In many cases, change within this type of community is slow and gradual unless it is caused by a violent imposition of values from an external dominating group. Traditional communities have been the primary social manifestation of human evolution since the formation of tribal hunter and gatherer groups. Within them, humans developed relationships of mutual support in exchange for a sense of belonging, security, and well being. But this kind of fealty did not bind others outside their community. In our current interconnected world, such orientations tend to be limiting. As Ruth Richards astutely points out, "our survival has become strongly dependent upon our commonalties as human beings, not on the differences between each other as individuals and as members of narrow reference groups" (Richards, 1993, p. 168).

3.1.2 Surrogate Community (SC)

A closed, unstable system artificially created to attract and satisfy disenfranchised individuals yearning for community through imposed norms and values. Modern industrial societies have fragmented the traditional experiences of community for which human beings yearn. Surrogate communities are identified as an artificially designed means to satisfy the need for shared identity and a sense of belonging among individuals who would otherwise not have access to authentic forms of community. Surrogate communities are not authentic in the sense that individuals who join them must accept pre-established values, beliefs, and rules — as defined by others — under which the community operates. For instance, there exists an organization ostensibly dedicated to encourage community in the US and abroad. To do so, it sells community workshops — two-day encounters among individuals interested in having a cathartic experience leading toward feelings of empathy and connectedness among each other. After the workshop is over, participants must continue to pay a substantial fee in order to "experience community" again, or else they will "loose community" and life will continue for them much as it did before the workshop.

3.1.3 Learning Community (LC)

An open dynamic system is one in which individuals collectively learn to adapt *to* their environment. The "learning organization" is a case in point. In both organizational and educational contexts the notion of learning community has become a focus of attention in recent years (Senge, 1990; Caine & Caine, 1997). Individuals in learning communities have an explicit common purpose: to learn together. However, in some cases there is no difference between a learning community and a community of individual learners. Even when a learning community demonstrates creative and fluid processes of collaboration and synergy by which to adapt to its environment, it tends to do so in a reactive mode. Such simple learning communities are often excellent means for learning about "doing things right," that is, for increased efficiency and efficacy in a rapidly changing world. And as such, they are ideal spaces for exploring new ways of working, learning, and enjoying life in an integrated way. But they rarely incorporate an ethical futures perspective, such as required for "doing the right things." Ethical futures perspective marks the quest for sustainability and for the sort of evolutionary possibilities explicitly espoused by ELC. In this sense, simple learning communities can be stepping stones toward evolutionary learning community.

3.1.4 Evolutionary Learning Community (ELC)

An emergent (self-designing) learning system demonstrating dynamic stability by adapting *with* its environment and generating developmental pathways that are sustainable in the context of broader evolutionary flows. ELC is a human activity system that strives toward sustainable pathways for evolutionary development in synergistic interaction with its milieu. It does so through individual and collective processes of empowerment and learning how to learn and through an ongoing commitment to evolutionary learning (Laszlo & Laszlo, et al., 1995). "ELCs do not adapt their environment to their needs, nor do they simply adapt to their environment. Rather, they adapt *with* their environment in a dynamic of mutually sustaining evolutionary co-creation." (Laszlo & Krippner, 1998). ELC is not to be considered separate from the environment, but a part of it. Just as the concept "system" is more a pattern than a thing, ELC is best conceived as an ideal image of community that can serve as a beacon for the design of new social systems appropriate for a new evolutionary era. Just as the concept of "system" is more a pattern than a thing, ELC is more an ideal image of community than a particular social arrangement. Loye and Eisler (1987, p. 57) indicate the need for new visions of the future; a need for "a clearer sense of system goal states or prohuman images of the future." ELC is an image that can serve as a beacon for conversations dedicated to forwarding a new culture of design appropriate to the challenges of life in a new evolutionary era.

4. LEARNING CONVERSATIONS: LEARNING TO LEARN TOGETHER

Peter Senge's (1990) vision of the "learning organization" was first presented as an integral part of a systems approach to organizational change in his bestseller, *The Fifth Discipline.* His idea of a learning organization as one, which continually expands its capacity to create its future through increasing people's capacity to learn at all levels of the organization, is most valuable, indeed. Nevertheless, ESD seeks goes beyond those ideas through it's focus on conversation dynamics.

Senge argues that organizations, as institutionalized human social systems, should adopt particular "disciplines" by which to ensure sustainable development. These disciplines are purported to provide the means by which the organization can learn how best to adapt to it's ever changing environment in ways that ensure it's continued operational development and financial growth. However, in a post-modern world of uncertainty and constantly increasing opportunity, a concern for organizational sustainability may not lead down the evolutionary pathways required by the interplay of individual, societal, and ecological demands for life on a small planet. What is needed is an approach to organizational design that emphasizes the joint learning curve of the individual, organization, society, and ecology through mutually adaptive and sustainable orientations. To focus at any one level is to risk a false synchrony with the other levels, or else, to blindly adapt to the changes occurring at that level. Neither outcome provides the competencies needed to harmonize organizational learning with the dynamics of global change in any more than a passive way. A *learning how to learn* approach may provide the means to move beyond mere patchwork adaptive processes: it marks the first step toward making organizational design a conscious future creating process. ESD conversations address the need for, and the ways to, evolutionary management through organizational design based on *learning how to learn* orientations.

Historically speaking, humankind has more or less consciously pursued the strategy of adapting the environment to its needs in order to gain mastery over nature. We have adapted our environment to us, molding and modifying our surroundings however we please in order to be more comfortable. We also have seen fit to do whatever we like with the animals and plants that share our planet. By the dawn of the third millennium, this strategy of adapting the environment to us in accordance with our every whim has brought us to the threshold of sustainability with the life support systems of planet Earth. However, there are alternatives.

Understandably, if a long period of human history has been marked by a certain type of behavior (in this case, adapting the environment to us), then when people finally decide they want change they often swing the pendulum of behavioral response to the other extreme (in this case, adapting ourselves to the environment). This is what many of today's more radical "green" movements advocate. Deep ecology, the basic philosophy in which they are grounded, is sound enough, but unfortunately, it is often taken to extremes by those who are reacting to the legacy of adapting the environment to us, saying we must now fully adapt ourselves to the environment. There must be an alternative to these two approaches. The first one makes us the villains of evolution by casting us in the role of planetary home wrecker. The second one makes us the martyrs of evolution by suggesting that we should safeguard Earth and all it holds by removing ourselves from the scene. If one strategy is adaptation of the environment to us, and the other is adaptation of ourselves to the environment, what's the alternative? Through a *learning how to learn* strategy of co-adaptation it is possible to engage in processes of adaptation *with* the environment.

When we seek to adapt ourselves *with* the way in which something else is evolving, we embark on a syntony quest. As we transit from one historical period to another, across the dividing line of a millennium, we are beginning to explore such ways of fitting our individual melodies together to create sustaining and enduring harmonies. This is more than just a nice metaphor: it is the essence of syntony. As with jazz musicians jamming in an improv. session, we have to learn certain skills, to develop and practice certain competencies, and to manifest a willingness to think and act interactively and responsibly. The notion of "will" – of active intention and passionate purpose – is crucial here. In fact, it is what makes the difference between merely *seeking harmony* and *engaging in a syntony quest.* And so are certain skills and competencies of evolutionary stewardship. The conversations of the new millennium need to be those that forward the attitudes, dispositions, values, and beliefs, as well as the skills and competencies required for such stewardship. These are the culture building learning conversations of evolutionary systems design.

In organizational and educational settings, the notion of learning community has become a focus of attention in recent years. Corporations and educational institutions are recognizing the importance and value of the experience of authentic community within their operational contexts. Individuals in learning communities have an explicit common purpose: *learning together* in ongoing, flexible and self-organized ways for the attainment of common purposes. Nevertheless, learning together means different things for different people, just as learning *per se* holds different meanings in different contexts. In some cases there is no difference between a learning community and a group of students with a professor in a classroom setting. In other, more dynamic and innovative settings, a learning community is a group of individuals who come together to learn in a flexible and self-directed way.

Learning communities can potentially have a very creative and fluid dynamic of collaboration and synergy, to better adapt to their environment, but they generally do so in a reactive mode. They create their own context and tend to partially or completely ignore the broader context in which they operate. This gives rise to a potential for of creating learning communities that do not question the basic assumptions of their purposes given their particular social and environmental contexts. Learning communities can be excellent means for doing things right, that is, for increased efficiency and efficacy in a rapidly changing world. For learning communities to foment also doing the right things implies an explicit ethical stance. This is where evoethics comes in as a distinguishing factor that sets evolutionary learning communities apart from learning communities, as such. In this sense, learning communities are like stepping stones toward evolutionary learning communities. Learning-oriented authentic communities are ideal spaces for exploring new ways of working, learning, and enjoying life in an integrated way. But they do not necessarily imply the ethical future-oriented perspective required for "doing the right things," which is at the heart of societal design conversations related to ethical evolutionary sustainability. This is one of the particular contributions offered by ESD.

5. DESIGN CONVERSATIONS: THE CHALLENGE OF SHAPING OUR SOCIETIES

Syntony is the means by which a sense of appropriateness is gained. The capacity to distinguish what is appropriate from what may not necessarily be so, is a matter of discretion. Life in the natural world teems with examples of "goodness of fit" – between living system and milieu; life process and evolutionary dynamic; organismic form and organic function. These systems of syntony are manifest expressions of the natural syntony sense, which all beings possess. However, this sense has atrophied to the point of vestigial capacity in human beings due largely to our ever-increasing dependency on technologically mediated relations with our environment and our consequent distancing from nature and natural process. Study of grouping behavior among social animals (schooling among fish; flocking among birds; herding among ungulates) provides insight into systems of syntony in which the syntony sense is strong (A. Laszlo, 2001). Certain human societies also demonstrate strong syntonious capacity, although for the most part, they are nested in some form of aboriginal culture. A consideration of the different ways in which syntony is manifest in both the ecosystemic and societal settings permits a greater understanding of syntony as an organizing force in societal evolution. Such understanding may provide a basis for design conversations dedicated to discerning evolutionarily appropriate courses of action, both at the individual and at the collective/community levels.

Design conversations are strategic. They are motivated by a desire to actively participating in the shaping of our common future. Learning conversations are empowering because through them the community develops the consciousness, literacy and competencies required for designing conversations. Design conversations are self-empowering, because through them the community not only continues to develop their individual and collective capacities, but also are able to bring about changes that will approximate their visions of the future.

6. A NEW CULTURE

Evolutionary learning community can be seen as the vehicle for fostering the conditions that propitiate the emergence of a culture of design. However, when used in this way, the notion of ELC serves as a fuzzy guiding image of that which could and should be. As yet, it can provide a general though incomplete outline of an alternative way of being in the world. Therefore, any human activity system interested in becoming an ELC is faced with the challenge of translating this generic image into a concrete vision for themselves. In other words, through future creating design conversation, each community has to create an image of *themselves* as a concrete, authentic, and unique ELC.

ESD conversations of this sort combine evolutionary systems theory, social systems design methodology, and lifelong-learning orientations to individual and group empowerment. They are dedicated to the design of collaborative learning ecologies *with* others. Essentially, ESD conversations seek to create conditions that empower individuals and groups to develop the skills necessary for the co-creation of sustainable, evolutionary futures. Those who engage in such conversation dynamics and who employ the vehicle of ELC to do so see themselves as catalysts of purposeful and creative aligning/tuning with the evolutionary processes of which they are a part; finding and generating meaning and evolutionary opportunity, both individually and collectively. This process of searching together through ESD conversations is what we have been referring to as the syntony quest.

Erich Jantsch considers syntony as "inquiry at the evolutionary level par excellence" (1975, p. 103). He describes this inquiry as the process of cultural organization which "may be helped in an evolutionary sense by furthering cultural differentiation, a pluralism of as many ideas, life styles, and world views as possible. The invention and introduction of new forms of cultural organization ought to become increasingly a matter of conscious design" (Jantsch, 1975, p. 260).

According to Webster's Unabridged Dictionary (1976), syntony can be defined as "in radio, resonance," while 'to syntonize' is "to tune or harmonize with each other." In this sense, syntony is the process of communication and resonance at the psychic and spiritual levels which, in a subconscious way, has united parts of humankind over certain periods of time. In terms of evolutionary systems thinking, syntony connotes evolutionary consonance; the occurrence and persistence of an evolutionarily tuned dynamic regime. More loosely, it can be considered the embodiment and manifestation of conscious evolution; a purposeful creative aligning and tuning with the evolutionary flows of one's milieu.

As the communities of future creating design conversations of the twenty-first century, it is our challenge to learn to *consciously* tune in to the general evolutionary forces that shape us as we shape them. The syntony quest is involves searching together for pathways of cultural differentiation and the purposeful transcendence of social systems through convergent evolutionary dynamics. As Jantsch puts it, "we shall have to learn now to design systems of syntony" (Jantsch, 1975, p. 270).

The creation of the conditions that foster the emergence of a new culture can be brought about through disciplined future creating design conversation. However, it takes more than just good intentions, for as famed systems thinker and economist Kenneth Boulding once said, "intentions are fairly easy to perceive, but often do not come about. Design is hard to perceive. But it is design and not intentions that creates the future" (in Banathy 1996, 71).

ESD conversation involves learning to think together about our values and to use the resulting understanding to co-create pathways for sustainable socio-ecological emergence. Design competence enables people to participate in the creation of a shared image of a sustainable and evolutionary future and to bring that image into being future creating conversation. The challenge is to engage with each other in generative and strategic dialogue on practical ways of living in harmony with nature; to search together for ways to live lightly, meaningfully, and simply in and with Earth, and to realize an extended sense of identity that embraces the world in enlightened self-interest. By learning to create systems of syntony it is possible to elicit such harmonious patterns of change – in ourselves, in others, and in our broader environment – purposefully, consciously, and yes – even intentionally.

REFERENCES

Banathy, B.H., 1996. *Designing Social Systems in a Changing World.* Plenum, New York.

Bowers, C.A., 1993. *Education, Cultural Myths, and the Ecological Crisis: Toward deep changes.* SUNY, New York.

Caine, R., and Caine, G., 1997. *Education on the Edge of Possibility.* ASCD, Alexandria. VA.

Checkland, P., 1993. *Systems Thinking, Systems Practice.* Wiley, New York.

Elias, D., 1998. It's time to change our minds: An introduction to transformative learning, *ReVision*, **20**(1).

Hinman, L., 1996. *Contemporary Moral Issues: Diversity and consensus.* New Jersey: Prentice Hall.

Jackson, M.C., 1991. *Systems Methodologies for the Management Sciences*. Plenum, New York.

Jantsch, E., 1975. *Design for Evolution: Self-Organization and Planning in the Life of Human Systems*.

Laszlo, A., 2001. *The Syntony Quest: Evolutionary vision for change in your world.* [manuscript]

Laszlo, A., 2000. The Epistemological Foundations of Evolutionary Systems Design. Proceedings of the 44th Annual Conference of the ISSS, Toronto, Ontario, Canada.

Laszlo, A., and Krippner, S., 1998. Systems Theories: Their origins, foundations, and development. In J.S. Jordan (ed.), *Systems Theories and A Priori Aspects of Perception.* Elsevier, Amsterdam.

Laszlo, K.C., and Laszlo, A., 1997. Partners in life: Syntony at work. Proceedings of the Ninth International Conversation on the Comprehensive Design of Social Systems. ISI, Pacific Grove, CA.

Laszlo, K.C., and Laszlo, A., et al., 1995. Building a Design Culture through Evolutionary Learning Communities. Proceedings of the Seventh International Conversation on the Comprehensive Design of Social Systems. ISI, Pacific Grove, CA

Loye, D., and Eisler, R., 1987. Chaos and transformation: Implications of non-equilibrium theory for social science and society. *Behavioral Science*, **32**: 53-65.

McCormick, S., François, C., Laszlo, A., Laszlo, K., and Nanay, B., 1998. Designing Sustainable Evolutionary Learning Communities. Proceedings of the Ninth Fuschl Conversation. Austrian Society for Cybernetic Studies, Austria.

Merry, U., 1995. *Coping With Uncertainty: Insights from the new sciences of chaos, self-organization, and complexity.* Praeger, Westport.

Michael, D.B., 1973. *On Learning to Plan – and Planning to Learn: The social psychology of changing toward future-responsive societal learning.* Jossey-Bass, San Francisco, CA.

Milbrath, L.W., 1989. *Envisioning a Sustainable Society: Learning our way out.* SUNY, New York.

Montuori, A., 1989. *Evolutionary Competence: Creating the Future.* J.C. Gieben, Amsterdam.

Ornstein, R., 1991. *The Evolution of Consciousness: Of Darwin, Freud, and Cranial Fire: The origins of the way we think.* Prentice Hall, New York.

Ornstein, R., and Ehrlich, P., 1989. *New World, New Mind: Moving toward conscious evolution.* Touchstone, New York.

Richards, R., 1993. Seeing beyond: Issues of creative awareness and social responsibility. *Creativity Research Journal.* **6**(1&2): 165-183.

Salner, M., 1996. A new framework for human science. *Saybrook Perspectives.* Saybrook Institute, San Francisco, CA.

Senge, P., 1990. *The Fifth Discipline: The Art and Practice of the Learning Organization.* Doubleday/Currency, New York.

Webster's New Universal Unabridged Dictionary., 1979. Second edition. Simon and Schuster, New York.

Chapter 13

THE COGNISCOPE™

Lessons Learned in the Arena

ALEXANDER N. CHRISTAKIS and KEVIN DYE
CWA Ltd.

1. INTRODUCTION

A response to the increasing complexity and risk of designing and planning social systems, especially inter-organizational collaborations, is to ground methodologies on well-researched principles and empirically verify the efficacy of their application. Recent examples from our arena of practice include engaging representatives of fifty organizations in planning a twenty year Global Disease Elimination program, international philanthropy-government-venture capital sponsored industrial-academic consortia in Northern Ireland, external stakeholder-driven modernization of regulatory policy, Federal and State interagency coordination for resource stewardship, and statewide engagement of stakeholders in special education.

The *CogniScope*™ methodology is a codified and tested means of collaboratively defining a complex situation and developing a social contract amongst the situation's representative stakeholders regarding directives to address it (Christakis et al., 1988; Warfield and Cardenas, 1994, Christakis and Dye, 2000). The process imposes a structured discipline of "focused and open dialogue." Stakeholders generate and clarify observations of the situation, collaboratively discern collective challenges, and construct patterns of interaction amongst participant's observations. Principal outcomes are individual learning, integration of the diversity of viewpoints, the discernment of salient priorities for design, and the emergence and codification of a situation-specific consensual linguistic domain that enables communicative action.

Professor P. N. Murthy of Tata, Asia's largest consulting firm based in India, identifies nine criteria of methodologies for problem solving in complex systems (Murthy, 2000), four of which are:

- *Complexity implies multidimensionality. Therefore one must have an approach for identifying as many dimensions as possible, if not all of them. Since it is not possible for a single individual to identify all the parameters, the method should provide for group techniques and constant learning.*
- *Since mutual influences and relationships between constituent elements and parts of the complex system are the basic hallmarks of complex systems, these two must be traced by a cybernetic relational diagram.*
- *The method should allow (a) complementary use of other system methodologies and (b) continuous amendment of the solution to capture as many dimensions as possible and complete the incompleteness.*
- *The method should be very seriously concerned with involving emancipatory approaches to get the best inputs from various kinds of participants.*

A key stricture implied in the fourth criterion is the autonomy accorded to stakeholders in fulfilling the other criteria. This chapter seeks to make transparent the evolution of foundations of methodology that have been developed and tested in the arena of practice over the past half century.

1.1 Background

Previously, the task of working together to design complex systems implied primarily the formation of interdisciplinary and cross-functional teams within an organization. Such concepts as Concurrent Engineering, Total Quality Management, Business Process Re-engineering, Architectural Programming, Design Patterns, and Knowledge Management, were promoted as a means for those teams to "work smarter" in designing tasks and achieving results. However, as has been documented in many papers and studies over the past decade, many of the techniques espoused in these methodologies, such as Quality Function Deployment, are not designed for economies of scale, the techniques do not transfer well from one domain to another, (Chap. 14, Warfield and Cardinas, 1994), the representations employed such as Ishikawa diagrams are often inadequate and at worst misleading in the description of complex problems. The fundamental drawback of most of these methods appears to be that they are mostly driven by "management fads" rather than by scholarly work, systemic thinking, and the careful collection of empirical evidence to dispute the null-hypothesis of their conjectures (Warfield, 1994; Banathy, 1996).

A marshaling event was the Club of Rome (CoR) proposal on the "Predicament of Mankind" in 1970, in which the concept of the Problematique was defined for the first time (Christakis, 1988). Some of the ontological constructs incorporated in the development of the methodology described in this essay were also conceptualized in the CoR proposal for the first time.

In the early 70s, as a follow-up to the CoR proposal, a group of researchers formed the Academy of Contemporary Problems with the sponsorship of Battelle Memorial Institute. They began formal observation of the cognitive and behavioral limitations of interdisciplinary teams in designing social systems, such as in large-scale urban planning. A major finding of the researchers observing the deliberations of interdisciplinary teams was low productivity. Each member would come to the interdisciplinary team meeting with proposals drafted with other members of their discipline. For example, an economist presented results of an employment projection model for a city. Meetings involving only economists were very productive and effective because they used a language common to their discipline. Breakdowns emerged during interdisciplinary dialogue in integrating disciplinary contributions for systemic design of the whole city. Frustration would typically escalate, participation would decline, and leaders would turnover. As psychiatrist James Taylor wrote, after observing the deliberations of the interdisciplinary team for over a year (Taylor, 1976):

> . . . there appears to be a pressing, well-recognized need for a kind of social invention: the interdisciplinary team which synthesizes knowledge in order to clarify complex problems. The promise of this social invention is clear, yet in fact no workable model has emerged. The question becomes obvious: why not? What has gone wrong in existing efforts to develop 'meaningful synthesis' of 'pertinent fields of knowledge'?

After a thorough review of extant methods and conducting experiments on complex tasks the researchers realized that a new "scientific paradigm" (Christakis, 1973) and "Domain of Science Criteria" (Warfield, 1994) were needed to guide innovation and testing a process of systems design inquiry. The paradigm was named **DEMOSOPHIA**, a New Greek word which combines the word DEMOS meaning "community" and SOPHIA meaning "wisdom" (Christakis, 1993, 1996). The US National Science Foundation sponsored the launch of "Generic Design Science" (Christakis, Warfield, 1987, Christakis et al., 1988; Warfield, 1994). The codification of the methodology and its practice was named **Interactive Management** (Warfield, Christakis, 1987; Warfield, Cardenas, 1994).

The researchers established centers for Interactive Management at the University of Virginia, 1982, George Mason University, 1984. There, the practice and its science were refined through applications in the arena in a large variety of diverse design domains, subjecting the performance of the method to the peer review literature (Warfield, Cardenas, 1994; Magliocca,

Christakis, 2001). In 1988, Christakis decided to subject the methodology to the discipline of the market establishing CWA, Ltd., and a customized package of methods called the *CogniScope*™.

The reusability and robustness of the approach has been verified in the domains of Environmental and Natural Resource Stewardship; Social, Economic, Educational, Professional, and Cultural Development; Healthcare Systems, Biomedical Research, Development and Regulation; and Technology, Capital, and Knowledge Management. The applications of collaborative inquiry span a three to twenty year planning horizon and include: Policy Development; Long-Range Strategic Planning; Research and Aggregate Program Planning and Decision Making; Enterprise and Information Systems Reengineering; and Process Design. In the past decade it has become a methodology of choice for rapid launch of large-scale interorganizational, interagency initiatives involving scores of constituencies such as Future Scenarios for Regional Development, Formation and Redirection of Industrial Consortia, Stakeholder Driven Modernization of Regulatory Policy, and formation of Cross-Sector Public-Private Partnerships.

2. STATE OF THE ART IN SOCIAL SYSTEMS DESIGN INQUIRY

In the past few years a more focused approach on the foundations of the *CogniScope*™ methodology has begun named Collaborative Design Process Science (Dye, 1997). This approach conforms to the fundamental tenants of "third phase science" [de Zeeuw 1996].

Third phase science grounds its legitimacy in engaging stakeholders as "expert observers" of the situation in which they are embedded. They are the ones that should decide how to take action in their situation, since they are those most affected by the existing situation and its evolution. This grounding stands in contrast to first and second phase sciences, which assert that "academic experts" or authorities are more qualified to design the "systems" on behalf of the community of stakeholders.

"Third Phase Science" as defined by (de Zeeuw 1996) adopts definitions such as: "Observation" refers to any report of what people claim to have seen, or experience, which is intended to be used in the construction of a high quality observation. "Science" refers to all (research) actions that aim to construct high quality observations, which make it possible to improve on action. "First Phase Science" assumes that the construction of high quality observations can be **fully separated** from actions that are to be improved by their use. "Second Phase Science" assumes that constructing observations **fully depends** on those actions (e.g., first order cybernetics). Third Phase Science **fully includes** the actions to be improved within the construction of high quality observations (e.g., second order cybernetics). "Third Phase

Science" seeks the transfers that support the development into a 'collective' such that the efforts of maintaining the transfers are minimal. "Linguistic structure" refers to the constraints on the words, signs, tokens, and symbols that are needed to implement the transfer that is intended to improve on action. "Getting a voice in science" refers to the need for developing and using linguistic structures in the construction of high quality observations that are different from what the first and second phase of science' embrace.

2.1 Construction of Observations

The construction of observations during a ***CogniScope***™ conversation is founded on six laws. These laws are:

- Ashby's Law of Requisite Variety;
- Miller's Law of Requisite Parsimony;
- Boulding's Law of Requisite Saliency;
- Peirce's Law of Requisite Meaning;
- Tsivacou's Law of Requisite Autonomy in Distinction Making;
- Dye's Law of the Requisite Evolution of Observations.

The first three laws have been presented and discussed in the literature extensively (Christakis 1987, Warfield 1994).

The Law of **Requisite Variety** (Ashby, 1958) asserts that a design must possess an amount of variety that is at least equal to the variety of the problem situation. It implies that an individual or a group engaged in designing a solution to a complex problem situation can gain control over a design only by making appropriate specifications in all the dimensions of the design. One way to violate this Law is not to ensure that all the relevant types of observers have been asked to present their observations during the design conversation.

The Law of **Requisite Parsimony** asserts that human beings can only deal simultaneously with between five and nine observations at one time (Miller, 1956). In other words, however complex a design situation may be, the design conversation should not require the designers to deal with more than nine items simultaneously, and usually should involve fewer (Warfield, 1988). Parsimony should be invoked in conversational settings in order to make sure that the conversation does not inherently try to force people to make judgments that exceed their short term cognitive capacities, or overburden their physiology. In particular, the *CogniScope*™ respects the notion of "bounded rationality" as described by Simon and others (Simon, 1974; Miller, 1956; Warfield, 1988).

Requisite Saliency or importance of an observation relative to others can only be brought into play as a useful concept when one is dealing with sets. Being fundamentally important in making comparisons, saliency can be a very valuable concept in simplifying design choices, for example in constructing alternative scenarios from a field of action options. The impact of ignoring relative saliency is usually low productivity (Boulding, 1966) resulting from a sequence of choices that do not make maximum use of the knowledge available to the observers in directing the resources of their efforts.

Requisite Meaning asserts that it is essential that observations and meanings of the stakeholders be excavated through inquiry into the relational structure amongst observations, as depicted in Charles Saunders Peirce's Lectures at Harvard in 1903, (Turrisi 1997):

> . . . All necessary reasoning without exception is diagrammatic. That is, we construct an icon of our hypothetical state of things and proceed to observe it. This observation leads us to suspect that something is true, which we may or may not be able to formulate with precision, and we proceed to inquire whether it is true or not. For this purpose it is necessary to form a plan of investigation and this is the most difficult part of the whole operation. We not only have to select the features of the diagram which it will be pertinent to pay attention to, but it is also of great importance to return again and again to certain features. Otherwise, although our conclusions may be correct they will not be the particular conclusions at which we are aiming . . . The end of argumentation is, of course, the drawing of its conclusion. This conclusion is the intended interpretant of the argument considered as a symbol. When we ask a person what he means by something he has said, we ask him to declare the intended interpretation of what he said.

In particular, the *CogniScope*™ respects the inherent capacity of observers to construct meaning (Apel, 1981) by exploring relationships of affinity, difference, influence, and temporality among their observations (Warfield and Christakis, 1987).

Requisite Autonomy of Distinction-Making – Tsivacou's Law – (Tsivacou, 1997) advances a scheme of inquiry, based on ontological elements of action, that is, communicative distinctions. These distinctions are responsible for the emergence of power as well as rationality and organizational culture. During observation-making participants explain their experience "in the praxis of living." Their explanations are distinctions because they represent selections from alternative explanatory paths. These explanations, when internalized by the other participants, enable them to iteratively converge on a plausible interpretation. According to Tsivacou (1997):

> The actors that ... have the chance to dictate the selection of the dominant explanatory path, immediately put themselves into the position of powerful, reducing the others involved into the position of powerless. Independent of their social status and role, those who control the information distinctions in a given situation acquire power and restrict the autonomy of the others. (p. 25)

For the power of persuasion to be equitably distributed among the observers, the autonomy of individual distinction-making must be ensured, and monopolies on distinction-making prohibited. The actors eventually produce a consensual linguistic domain by means of communicative action.

2.2 Evolution of Observations in Identifying High Leverage Action

A group of CWA researchers recently discovered an empirically substantiated law, which represents a significant improvement when compared to earlier efforts to describe this phenomenon (Kapelouzos, 1989; Warfield and Staley 1996). The researchers selected as the object of analysis the observations made by participants during the conversations that took place in 50 applications of the *CogniScope*™ system. They analyzed all the available data on "importance voting" and compared these preferences to results obtained when the stakeholders were engaged in "influence voting" among pairs of observations. They have called this law **"Requisite Evolution of Observations"** (Dye and Conaway, 1999, Dye and Christakis, 2001). It states:

> Whenever observations made by stakeholders in the context of a complex design situation are interdependent, assigning priorities for action on the basis of aggregating individual "importance voting" leads to erroneous priorities and ineffective actions. The effective priorities for action emerge after an evolutionary search of interdependencies among the observations through a dialogue focusing on "influence voting".

Figure 1 shows the consensus influence among goals versus two previous rounds of multi-voting in goal setting for a regulatory agency redesigning the drug review process (Good Review Practices). An interesting metric that emerged from exploratory data analysis is the **"Pareto Rule."** This metric essentially says that 20% of the intent influences 80% of the situation. For example, in the case of the GRP, shown in Figure 1, the total number of influences among the 21 goals is equal to 468. The number of influences exerted by the four most influential goals, i.e., Goals #13, #27, #5, #26, is equal to 276, corresponding to 60% of the total number of influences. The implication of this is that making progress on these four most influential goals (i.e., 20%) leverages 60% of the intent, while the other 17 goals only influence 40% of the intent.

For example, goal #1 is the highest on the scale of Weighted Voting for both Counts #1 and #2, while in terms of its influence rating it is about 1/3 as influential as goal #13. On the other hand, goal #13 is one of the four most influential goals as shown in Figure1, but its importance voting is about 1/6 of goal #1.

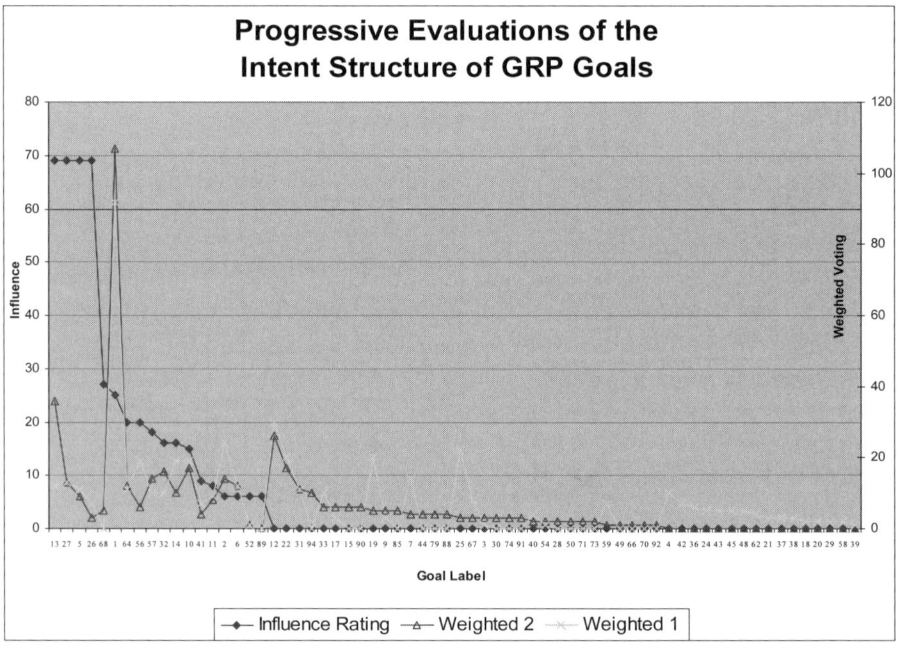

Figure 1. An example of the Evolution of Observations

3. THE APPROACH TO SYSTEMS DESIGN INQUIRY

In this section we present a generic description of *CogniScope*™ inquiry. Each application requires a modular customization of the generic process in order to satisfy the requirement of the particular situation.

The *CogniScope*™ process is based on five components (1) A Design Inquiry and Facilitation Team, (2) Representative stakeholders, (3) A computer-aided Group Design & Decision Support System, (4) A certified set of Consensus Methods and (5) A Collaborative Facility.

An application is launched with the Design Inquiry Team gathering intelligence through a review of extant literature and interviews. This knowledge-base is documented and distributed to the participants. Participants then are asked to address **"What should we do?"** They are asked to clarify their statements through focused and open dialogue. Then the group inductively explores themes, and abductively discerns the most salient challenges in the problematic situation in a "Problematique" (Christakis, 1988, Warfield, 1999). The context of the next stage of dialogue shifts to **"How can we do what we should do?"** The participants generate and clarify options for addressing the pattern of challenges. Participants develop affinity clusters by exploring the relations of similarity among pairs of options, and proceed to select those options that are the most important to them to consider. The most important options are superimposed onto the "pattern of challenges" demonstrating their potential leverage on the situation. Participants design alternative scenarios by focusing on the question **"Which are the preferred options and Why?"** They usually work in small team to identify key options within each of the affinity clusters. They then present their scenario and its rationale in a plenary session. During the plenary session the participants converge on a "consensus action scenario." The Facilitation Team ensures that Requisite Saliency is not violated during the decision process. Finally the participants answer **"When will we do what we can do and who will do it?"** completing the Collaborative Action Plan.

The process of conversations conducted at all the *CogniScope*™ workshops is schematically depicted in Figure 2. The **Definition** Stage begins with a complex, messy situation, depicted graphically in Figure 2(a) by overlapping geometric shapes representing the variety of perspectives among the observers. A review of completed applications indicates that the average number of statements generated is equal to 64, ranging from 30 to 300. Participants respond to a triggering question, such as "...Within the context of the current perception of issues, and based on your experience, what are the major challenges to improving the safe use of pharmaccutical in the USA?"

Supported by powerful "groupware" decision support system, the *CogniScope*™, participants are able to explore relationships among observations and efficiently produce patterns of relationships among observations. Their observations are clarified, selected for further inquiry, assigned to affinity clusters, and structured by means of "influence voting" to show the interrelationships among them. These process steps are shown graphically in Figure 2 (b-I). The influence pattern produced through conversation focusing on the influences among the challenges enables the participants to discover the "deep rooted" challenges, i.e., those that, if addressed, would exert leverage in addressing other challenges.

The second stage focuses on **Design**, i.e., the identification of options, which, if adopted by the community of stakeholders, would meet the deep-rooted challenges and contribute to addressing the Problematique. The options proposed by the participants are clarified, prioritized and assigned to similar categories as shown schematically in Figure 2 (b-g). This kind of *design map* is generated automatically by the *CogniScope*™ It ensures a focus on the options with the highest leverage.

Finally, the Choice stage of the conversation provides the participants with an integrated, systemic model of evaluating alternatives. This is precisely where such a *collaborative decision-making* forum ought to be when faced with the allocation of resources for making progress in the resolution of complex issues confronting most organizations and social systems today. In this way, participants are able to develop truly *collaborative action plans* for guiding *collaborative leadership* in dealing with the most pressing organizational design issues.

In the following section, which describes an actual application, we will attempt to make some of those detailed steps of the inquiry more transparent.

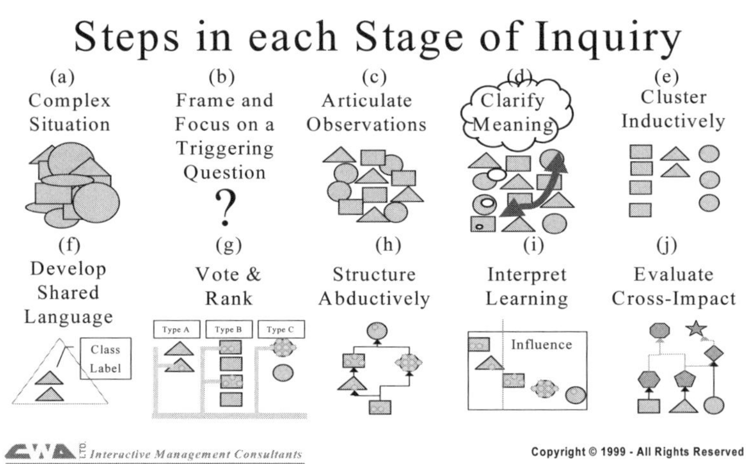

Figure 2. Ontological constructs of methodology

4. EXAMPLE IN PHARMACEUTICAL SAFETY

The case of Safe and Appropriate Use of pharmaceuticals represents a very complex boundary-spanning case (Dye et al., 1999). This application disclosed that at the heart of this very complex problem is a set of value conflicts amongst equally legitimate ethical systems.

The project brought together an unprecedented cross section of stakeholders with forty organizations representing: healthcare clinicians, institutional providers, the pharmaceutical industry, pharmacists, consumer advocates, regulators, policy makers, public health professionals, patients and public interest groups. The goals of the project were:

- Appreciate the diversity of viewpoints regarding problems in the situation.
- Create a shared understanding of the "anticipated challenges" that the community of stakeholders will face in improving the situation.
- Build consensus on an action plan for addressing the "system of challenges."
- Forge a commitment to collaborative leadership.

4.1 The Definition/Anticipation Stage: Understanding the Problem Situation

The first stage of the workshop conversation frames a consensual understanding of the complexity of the situation. Participants responded to the triggering question:

> What challenges do we anticipate the community of stakeholders will face in improving pharmaceutical safety?

Participants generated and clarified 49 potential challenges [see Figure 2 (c) and (d)]. After the challenges had been clarified, the design team developed preliminary affinity groups [see entry (e) of Figure 2]. The participants then voted for those challenges that they considered the most relatively important [see Figure 2 (f) and (g)]. From a subset of 13 most important challenges, the participants created an influence pattern in order to identify those challenges capable of exerting the strongest influence in meeting other challenges. By exploring methodically the influences among the proposed challenges, they discovered those challenges that exerted the maximum leverage on other challenges, and constructed a Problematique or a "system of challenges." The mapping of the propagation of influences was generated by the participants themselves, aided by the *CogniScope*™ software which uses an inference engine and a strategy for selecting subsequent pairs of challenges [see Figure 2 (h) and (I)].

4.2 The Design Stage: Designing Alternative Action Scenarios

The focus of the dialogue on the second day of conversation focused on the triggering question:

> What are options which, if adopted and implemented by the community of stakeholders, will help in meeting the system of challenges?

Participants generated 41 action options. These options were clarified, inductively classified, and voted upon by the participants, following a format similar to that employed during the first day. Participants worked in small teams toward the end of the second day to identify a set of action options for their scenarios presenting their scenarios at a plenary session, which leads to a Consensus Action Scenario amongst the teams.

5. SOME LESSONS LEARNED IN THE ARENA

In this section we describe some of the lessons learned from over 200 applications of the *CogniScope*™ inquiry in the arena. Our description of the "lessons learned" elaborates on the six laws presented earlier in the essay.

At the start of a *CogniScope*™ conversation legitimacy is aligned with the Law of Requisite Autonomy of individual observations in the spirit of demonstrating openness to diversity and inclusion – a major source of variety. As the conversation evolves, the basis of legitimacy shifts to the collective during consensus determination of relationships among observations, and gradually leads to a determination of saliency by making relational judgments among observations in a set. The enactment of autonomy is based on the commitment to a directive and the self-determination of role within the solidarity of collective choice. In this sense legitimacy becomes aligned with parsimony. Relational inquiry enables the community to: (a) construct authentic, anticipatory and autonomous representations of observations (linguistic structures or patterns), and (b) interpret the meaning and the transfer of these patterns in accordance with the tenants of "Third Phase Science."

As an example of the importance of enforcing Requisite Autonomy we have experienced many cases in the arena in which users of Information Technology (IT) systems were asked to participate in the design of IT systems for their organizations. The IT experts within the organization, as well as the IT contractors, were extremely reluctant to compromise their "technical expertise," and to listen to the **authentic voice** of the potential users of these systems (Banathy, 1999). We have had a number of experiences in which managers of IT programs had to be replaced because of their reluctance to enable the stakeholders to voice and explain the system requirements in an autonomous manner.

As mentioned earlier, Requisite Variety corresponds to a phenomenological statement of the inherent multidimensional nature of designing complex systems and the requirement for observers to conceptualize a variety of observations. Requisite Parsimony, on the other hand, corresponds to a physiological or psychological limitation of human beings and limits of communication in groups. These two laws impose contradictory requirements for collaborative inquiry. Requisite Variety asserts the significance of espousing variety, even at the expense of cognitive overload. Requisite Parsimony asserts the significance of respecting human limitations, even at the expense of variety. The *CogniScope*™ inquiry has demonstrated in the arena its capacity to reconcile the contradictory nature of these two laws.

Another interesting facet of the *CogniScope*™ conversation is the capacity to normalize the role of power in drawing and communicating distinctions among the stakeholders. In Habermas' (1984) monumental work *Theory of Communicative Action*, he proposes three forms of action orientation: (1) instrumental action in a nonsocial action situation, (2) strategic action in a social action situation, and (3) communicative action. The first two are oriented to achieving success, either through direct intervention (instrumental action), or through influencing decisions of opponents (strategic action). In either one of these two forms, action is imposed by one party over the other. The *CogniScope*™ conversation is most applicable in those situations where communicative action is required because, by its very nature, it precludes such imposition since both parties are interested to learn and converge on a mutual agreement, where the only arbiter of disputed normative validity claims is the power of persuasion and not the power of manipulation or deception.

This phenomenon manifested itself in a very prominent way within a Federal agency in an application focusing on the Agency's International Position. One stakeholder, with considerable political power because of his contacts with the White House, had been able in the past to dominate the distinction-making function in the international affairs arena. This individual became very antagonistic to the process and challenged its relevancy and applicability. The facilitation team offered to abandon the disciplined inquiry approach imposed by the *CogniScope*™ and let the workshop participants continue their conversation without the presence of the team. The leader of the Agency intervened at this point and asked that the conversation continue in a disciplined and equitable manner.

Participants in *CogniScope*™ sessions describe the process as creating an atmosphere of serenity, equity, authenticity, and empathy. Most feel empowered to make their observations without the fear of criticism or intimidation by other observers. Others consider it a humbling experience depicted in their feelings. As one participants stated: "grieving the death of their world view followed by an expanded sense of the situation I could only describe as a new gestalt." Creating a behavioral climate, in which all stakeholders, independent of rank or authority, are able to legitimize their distinctions and observations, is significant when compared to conversational experiences most stakeholders have in the majority of settings. In essence the *CogniScope*™ system espouses and promotes "the power of persuasion through dialogue, as opposed to the persuasion by power."

Because of this characteristic those observers that have been able to accumulate power and control within an organization are occasionally antagonistic to the deployment of the process. This phenomenon is manifested in applications in which the "leadership" of the organization does not clearly favor democracy and self-organizing. It is becoming increasingly apparent, however, that the trend towards organizational democracy, in the face of escalating complexity, does not appear to be reversible in the post-industrial paradigm (Christakis and Dye, 1999). The *CogniScope*™ inquiry has demonstrated in the arena its capacity to promote the emergence of a type of "transforming leadership" when stakeholders are engaged in focused and open dialogue (Magliocca, and Christakis, 2001).

Christakis (1996) identified the fundamental mechanism of how the *CogniScope*™ conversation is supportive to group learning of moral and spiritual consciousness. The necessity for coping with a complex situation forces the stakeholders to confront their weaknesses as observers, and to come out from their self-referentiality by means of their interactions with other observers. They must transcend their self-referentiality and embed themselves in the role of observers and 'learners from each other' – and as learners, to draw distinctions. Each distinction is a choice made by the learner in a great number of possible opportunities for learning. Through focused and open dialogue, human beings are enabled to break the bonds of their self-referentiality precisely through the development of recursive communicable explanations and representations, in harmony with other human beings.

This phenomenon was observed in a number of instances in the arena. In the majority of cases the participants were able to transcend their self-referentiality and construct collaborative action plans in harmony with the other members of the community. There were, however, some cases that the need for transcendence was very painful to some observers. For example, in one particular case a member of a pharmaceutical company therapeutic team, who was an MD and had to undergo a triple by-pass heart operation a year after the conversation, confided later to other participants "that the scars and the pain from the experience of participating in the *CogniScope*™ conversation were more painful than the heart operation." It turns out that this individual after few years has been finally transformed to a humble and cooperative member of the community of stakeholders, which tells us that transformational leadership does not materialize quickly.

Another lesson is that the implementation of the Law of Requisite Parsimony slows down the pace of the deliberations by the participants. Our experience has been that some stakeholders, especially those that are accustomed to move with speed, which in most cases creates a cognitive overload on other stakeholders, find this characteristic of the process problematic. On the other hand, our experience has also been that high-speed conversation is sometimes used by some as a strategy for dominating the distinction-making process during the conversation, at the expense of autonomy and variety. There is substantial evidence in the literature that high speed in the pace of the conversation increases the risk of underconceptualization, primarily because it violates the Laws of Requisite Variety and/or Requisite Parsimony (Warfield, 1994; Bausch 1999).

Finally, our field experience has demonstrated that in addressing the complexity of large-scale social systems design it is imperative to accommodate the plurality and diversity of disciplinary perspectives and worldviews rather the predominance of the First and Second Phase Science paradigms' objectivism and authority which have proven to be totally inadequate.

ACKNOWLEDGEMENTS

The following people have engaged the methodology in the arena of application and continue to be a source of inspiration as reflective practitioners in its community of science.

Marian Godfrey, Pew Charitable Trusts; Doug Bauer, Pew and Smith-Kline Beecham; Julie Carlston, Alexander Fleming, Margaret Porter, and Janet Woodcock, US Food and Drug Administration; Bela Banathy, Far West Research Laboratory, Saybrook Graduate Institute; Bela Antal Banathy of International Society of the Systems Sciences and the Technology of Social Systems Design program at Saybrook; David Mackett, National Oceanic and Atmospheric Administration; Wilfred Shearer, The Franklin Group Inc., Eleanor Vogt, National Patient Safety Foundation; LaDonna Harris and Laura Harris, Americans for Indian Opportunity; Larry Magliocca, Karen Sanders, and Cesar Da'gord, The Center for Populations with Special Needs of The University of Ohio; Nikos Paritsis, University of Crete; Theodoros Tsekos, Hellenic National Center for Public Administration; Alexander Lane and William Darrow, Schering Plough Research Institute, Robert McDonald of the US National Forest Service, and Marios Michaelidis and Noni Diakou of The Cyprus Academy for Public Administration. Among associates special thanks are due to Diane S. Conaway, Tom Flanagan, Ali Geranmayeh, Hasan

Ozbekhan, Matthew Shapiro, Dimitri Christakis and Chris Feudtner, University of Washington; Ioanna Tsivacou, University of the Aegean, Reynaldo Trevino-Cisneros, Roxana Cardenas, Instituto Technologica Estudios; Lou Kerestesy, Consensus Systems Inc.; Nicholas Christakis, University of Chicago; Ken Bausch, Ongoing Emergence, Surinder Batra, Center for Interactive Management – India; Gary Alexander, University of Idaho.

REFERENCES

Apel, K., 1981. *Charles S. Peirce: From Pragmatism to Pragmaticism,* University of Massachusetts Press, Amherst.

Ashby, R., 1958. Requisite Variety and Its Implications for the Control of Complex Systems, *Cybernetica*, **1**(2): 1-17.

Banathy, B.H., 1996. *Designing Social Systems in a Changing World*, Plenum, New York.

Banathy, B.A., 1999. An information typology for the understanding of social systems, *Systems Research and Behavioral Sciences*, **16**(6): 479-494.

Bausch, K., 2001. *The Emerging Consensus in Social System Theory,* Plenum, New York.

Bausch, K., 2000. The practice and ethics of design. *Systems Research and Behavioral Science,* **17**(1): 23-51.

Boulding, K., 1966. *The impact of social sciences,* Rutgers University Press, New Brunswick.

Christakis, A.N., 1973. A new policy science paradigm. *Futures*, **5**(6): 543-558.

Christakis, A.N., 1987. High Technology Participative Design: The Space-Based Laser, in *General Systems.* John A. Dillon Jr. (ed.), International Society for the Systems Sciences, Vol. XXX, New York, 69-75.

Christakis, A.N., 1988. The Club of Rome revisited in: *General Systems.* W.J. Reckmeyer (ed.), International Society for the Systems Sciences, Vol. XXXI, New York, pp. 35-38.

Christakis, A.N., 1993. The Inevitability of Demosophia, in: *A Challenge for Systems Thinking: The Aegean Seminar*, Ioanna Tsivacou (ed.), University of the Aegean Press, Athens, Greece, pp. 187-197.

Christakis, A.N., 1996. A People Science: The CogniScope system approach. *Systems: Journal of Transdisciplinary Systems Sciences*, **1**(1).

Christakis, A.N., Warfield, J.N., 1987. NSF DTM Ohio

Christakis, A.N., Warfield, J.N., and Keever, D., 1988. Systems Design: Generic Design Theory and Methodology, In Decleris, Michael (ed.), *Systems Governance,* Publisher Ant. N. Sakkoylas, Athens-Komotini, Greece, pp. 143-210.

Christakis, A.N., and Dye, K.M., 1999. Collaboration through communicative action: Resolving the systems dilemma through the CogniScope. *Systems: Journal of Transdisciplinary Systems Sciences*, **4**(1).

de Zeeuw, G., 1996. Second Order Organizational Research, *Working Papers in Systems and Information Sciences,* University of Humberside, Hull, England.

Dye, K.M., 1997. Collaborative Design Process Science, *Working Papers at MIT,* Boston.

Dye, K.M., Feudtner, C., Post, D., and Vogt, E.M., 1999. Developing Collaborative Leadership to Reframe the Safe Use of Pharmaceuticals as a National Health Priority, *Final Report*, *CWA Ltd.* Paoli, PA.

Dye, K.M., and Conaway D.S., 1999. Lessons Learned from Five Years of Application of the *CogniScope*™ Approach to the Food and Drug Administration, *CWA Ltd. Report*, Paoli, PA.

Habermas, J., 1984. *The Theory of Communicative Action*, Vols. I and II. Polity Press
Kapelouzos, I.B., 1989. The Impact of Structural Modeling on the Creation of New Perspectives in Problem-Solving Situations, *Proceedings of the 1989 European Congress on Systems Science*, Lausanne, Switzerland, AFCET, October, pp. 915-932.
Magliocca, L.A., and Christakis, A. N., 2001. Creating a framework for sustainable organizational leadership: The CogniScope System Approach. *Systems Research and Behavioral Science*,
Miller, G.A., 1956. The magical number seven, plus or minus two: Some limitations on our capacity for processing information. *Psychology Review,* **63**: 81-97.
Murthy, P.N., 2000. Complex societal problem solving: A possible set of methodological criteria, *Systems Research and Behavioral Science,* **17**: 73-101.
Simon, H.A., 1974. How big is a chunk. *Science,* **183**: 482-488.
Tsivacou, I., 1997. The rationality of distinctions and the emergence of power: A critical systems perspective of power in organizations, *Systems Research and Behavioral Science*, **14**(1): 21-34.
Taylor, J.B., 1976. Building an Interdisciplinary Team. In Arnstein, S.R., and Christakis, A.N., (ed.) *Perspectives on Technology Assessment*, Science and Technology Publishers, Jerusalem, Israel, pp. 45-63.
Turrisi, P.A., Ed., 1997. *Pragmatism as a Principle and Method of Right Thinking,* State University of New York Press, New York.
Warfield, J.N., 1988. The magical number three, plus or minus zero, *Cybernetics and Systems,* **19**: 339-358.
Warfield, J.N., 1994. A Science of Generic Design: Managing Complexity Through Systems Design, Iowa State University Press, Ames, Iowa.
Warfield, J.N., 1999. The Problematique: Evolution of an Idea. Systems Research, **16**: 221-226.
Warfield, J.N., and Christakis, A. N., 1987. Dimensionality. Systems Research, **4**:127-137.
Warfield, J.N., and Cardenas, A. R., 1994. A Handbook of Interactive Management, Iowa State University Press, Ames, 1994.
Warfield, J.N., and Staley, Scott M., 1996. Structural Thinking: Organizing Complexity through Disciplined Activity, Systems Research, **13**(1): 47-67.

Chapter 14

NARRATIVE STORY AS DISCOURSE IN SYSTEMS DESIGN

PEGGY BARNES GILL
University of Texas at Tyler, Tyler Texas, USA

1. INTRODUCTION

Narrative story, as a component of systems design, provides a method of exploring systemic change and the design of educational systems. Systems design furnishes the construct within which to consider the change process, while narrative story facilitates the design conversations that develop the processes, pathways and patterns that lead to the unfolding idealized school as a socially constructed reality of the participants.

Rather than deal with discrete events individually, the personal experiences that evolve in narrative story allow people to build larger frames of reference and examine underlying assumptions and beliefs that guide their actions. Stories recreate experience in ways that allow the personal, cultural, and historical ground of the educational organization to remain present. It emphasizes the interrelatedness found within members of a school community. This chapter addresses the questions: How can the stories embedded in the history, context and culture of a school inform the systems design process? How can narrative inquiry mediate the process of moving across boundaries to the educational future? How can narrative inquiry identify symbolic boundaries that exist within an educational organization?

Traditional inquiry methods have focused on improving education by examining the components of the system with the intent to improve or restructure these components within a clearly bounded educational system. This approach has failed to alleviate the current crisis in education. An alternate approach, addressing educational change through systems design, allows disciplined inquiry into "the true nature of education as a complex and dynamic system that operates in ever-changing environments and interacts with a variety of other societal systems " (Banathy, 1991, p. 32).

Educational systems are complex and emergent social structures. Narrative story addresses the on-going contextualization of these emergent social structures. MacIntyre (1997) asserts that we can only understand an organization, if we know the stories that are told about that organization. It is our stories that illuminate the significant truths in our lives (Bakan, 1996). These stories or narratives illuminate the meaning inherent in the design process. Designing educational change is a complex undertaking. If we are not to be simply carried along on the tide of change, it is important to come to terms with the meaning of change to those most intimately involved in the change process and acknowledge and examine alternate perspectives of change. Narratives arising from design conversations, built on dialogue and discussion, allow the examination and exploration of alternate perspectives as a foundation upon which to build creative discourse (Jenlink & Carr, 1996). Within these conversations, narrative story seeks to capture the voice of human experience. It does not seek a single truth, but seeks to provide a feasible way to view the situation. It is a narrative of the possible. It suggests possible coherence, possible connectiveness and possible explanations that help describe an ideal image of an educational system.

2. EDUCATIONAL ORGANIZATIONS AS MEANING MAKING SYSTEMS

Piaget (1995) proposed that systems evolve to a higher level of organization through a process of modifications that seeks to preserve both the parts and the whole of the organization. Thus a system is continually reconciling a need for openness to new possibilities and a desire to preserve the current structures. Luhmann (1995) further refines these concepts to view social systems as made up of systems of communication rather than individuals. These systems of communication provide the method of production for the system. Thus the system is fundamentally a meaning- making structure based on a network of relationships.

Within this meaning system, how does the perception of individual participants implicate the change process? While acknowledging the need to view organizations through many frames and from different holistic perspectives, educators and change theorists have not provided a way to examine the varying perspectives of those involved in the change process itself (Nash & Tucker, 1998; Tedesco, 1998). What meaning is constructed by those in the system? Those who facilitate change, those who are engaged in and supportive of change, and those who are resistive to change have different views of the change process. Examining the meaning constructed by individuals with varying relationships to the system provides valuable insight into the design process of change. A strategy is needed that builds on the holistic nature of systemic change to incorporate the perspectives of all participants in the change process into a unifying narrative.

3. NARRATIVE STORY AS DESIGN CONVERSATION

The purpose of narrative inquiry within the systems design process is to construct the meaning of the process in terms of narrative unities in the lives of the lives of those who participate in the system. The emphasis shifts from meaning constructed by the facilitator and/or leaders of the design process to a mutual interpretation of the meaning and purpose for the individual as well as the system as a whole. Narrative inquiry serves to build feedback loops that include personal, experimental knowledge from the participants and to provide connections between and among specific events.

The dialogic process between facilitator and participants lays the foundation for the initial personal stories. These narratives become a shared space of exploration. In this shared space of exploration, the stories are constructed and reconstructed. Through these reconstructions, a larger story develops that represents the evolving meaning of the organization. It is this narrative that provides the unifying story. This larger or unifying story bids others to explore the phenomenon and consider how the particulars of the story fit their current assumptions and thinking. Others within the system are invited to explore the story and the "emotional and motivational meaning connected to it" (Polkinghorne, 1995, p. 11).

Design conversations occur within a particular community of educators and stakeholders who are engaged in a social process of constructing and reconstructing meaning. Within this disciplined inquiry process, the participants' interpreted purpose and meaning bound the dynamics of organizational change. This meaning and purpose may be focused and delineated through examination of the personal stories of those involved in the process.

Rather than deal with discrete events individually, the personal stories that evolve in narrative inquiry allow people to build larger frames of reference and examine underlying assumptions and beliefs that guide their actions. It is a natural way to explore the meaning systems operating within social constructs such as an educational community.

Narrative story acknowledges that organizational meaning and purpose are individually and collectively constructed and reconstructed as the system continuously builds organizational capacity and engages in further design and implementation processes. The design process of evolutionary change establishes the boundaries of the design inquiry and translates the implications of the purpose, meaning, and shared knowledge of the organization within the contextual, cultural and historical setting of the organization. The process is recursive and comprehensive. Narrative story provides the landscape within which to examine how organizational maps of meaning and purpose are created.

3.1 How Can the Stories Embedded in the History, Context and Culture of a School Inform the Change Process?

The evolving history, context and culture of the organization inevitably and continuously implicate organizational change, a social process that incorporates and negotiates social relationships within a social context. Within this context, the catalyst for change calls for the need to make sense of the transition process. Polkinghorne (1995) identifies narrative inquiry as "the linguistic form uniquely suited for displaying human existence as situated action" (p. 5). Stories recreate experience in ways that allow the personal, cultural and historical ground to remain present. Rosen (1984) proposes that storytelling is the very essence of everyday life. The telling of and listening to the stories of others establishes an interpersonal connection between teller and listener. Those listening to the stories identify incidents and events within the story and relate them to similar occurrences in their own lives. Tannen (1988) argues, "Storytelling is a means by which humans organize and understand the world, and feel connected to it and each other" (p. 92). The stories or narratives emerging within a systemic change process contain the qualities, the complexities and the richness found within a school.

The history, context, and culture of a school may be viewed as constitutive, providing the categories and understandings that enable us to engage in practice, or as regulative, providing the norms and conventions that constrain action (DiMaggio, 1994). Narrative story allows both constitutive and regulative representations to remain. The collective body of personal and professional knowledge of an educational organization is constructed within a complex web of relationships that have evolved and changed over time. The stories from these changing relationships provide insight into current periods of transition. Bridges (1980) reminds that, "You find yourself coming back in new ways to old activities when you're in transition" (p. 8). Stories from those who participated in earlier periods of tension "clarify meaning for us in a way that facts or concise definitions cannot" (Sylwester, 1995, p. 104). These stories embedded in the history, context, and culture of the organization begin a conversation based on particulars that allow others to identify with similar issues and examine current situations with a new intensity of awareness.

3.2 How Can Narrative Story Mediate the Process of Moving across Boundaries to the Educational Future?

Narrative inquiry provides stories that live "willingly with plurality, embracing the power of language to make new and different things possible" (Bochner, 1990, p. 5). Langer in *the Power of Mindful Learning* (1997) suggests the myth that there are right and wrong answers stifles creativity and undermines true learning or personal mastery. The emphasis on a singleness of view locks people into rigid thinking patterns that disallow other views, and may even cause people to actually reject viable solutions because those solutions appear to be counterintuitive to the accepted patterns of behavior or thought (Senge, 1990). Nuyen (1997) suggests the need for educational leaders to enter into the realm of possibilities. Only by considering the alternatives is there a means to identify and actualize the most desirable possibility and move across boundaries.

Implicit in any attempt to effect organizational change is the need to respect individual thought (Short, 1994). It is the human interaction with thought, in a multiplicity of ways from formal to informal and verbal to nonverbal that provides the vehicle for interpretation, evaluation and action within the organization. Dialogue opens the learning space within which to engage in sustained inquiry regarding the fundamental assumptions that support the core values and purposes that define the system. This sustained inquiry allows the organization to form new patterns of relationships and expands the repertoire of possible organizational responses. Narrative inquiry empowers people to work effectively in a changing environment. Meaning that is co-constructed and shared through continuous collaborative learning and inquiry shifts the focus from individual effort to interactive professionalism (Fullan, 1991). Within a common purpose and shared meaning, narrative inquiry encourages natural networks to become the strength of the change effort rather than formal policy and structure (Fullan & Miles, 1992; McLaughlin 1990). Narratives or stories configure the participants' actions into a meaningful whole unified by a plot. Possible actions that would lead to boundary crossings can be understood in relation to the collected narratives. The more varied and extensive one's collection of storied explanatory descriptions of previous actions, the more likely that one can draw on a similar remembered episode for an initial understanding of the new situation. The more one can draw on similar remembered episodes, the more likely one will appreciate and search for the elements that make the new different from the recalled instance (Polkinghorne, 1995).

4. THE JOB INTERVIEW AS NARRATIVE STORY: DEVELOPING INVESTED PARTICIPANTS THROUGH DESIGN CONVERSATIONS

Sherman Independent School District (pseudonyms are used for the school district and individual participants), a rural district in Texas, has been engaged in a systemic change process for the past four years. A high turnover rate among teaching staff has prompted a review of employment interviews. Of particular concern was the question of "employee fit" within the school system. Sherman had traditionally used a system of employment interviews in which applicants completed standard application forms and participated in campus level interviews by the building principal and/or site based committees. Each campus was responsible for developing questions and structuring the interview process. Little attention had been given to the particulars of the process.

In general, the traditional interviews were designed to meet the needs of the school system. Applicants were screened for content knowledge, classroom management techniques and philosophical beliefs about education. These answers were then considered in light of current campus needs. Each member of the interview committee took turns asking a predetermined question. Although time was provided for applicant questions at the end of the interview, the interview committee consumed most of the scheduled interview time. Thus a hierarchical structure was presented to applicants from their initial contact with the district. This was in direct conflict with the systemic change efforts going on within the district. A conflict existed between how a teacher was expected to perform once hired in the district, and how an applicant was expected perform during the interview. Although collective and cooperative meaning making was the expectation within the district, the initial relationship established during the interview was not supportive of this expectation.

Using narrative inquiry, members of the faculties began gathering stories that represented lived practice within the district. What conversations in the district provided insight that might help place the right applicant in the right job? How could the interview process become a meaning-making activity to benefit both the district and the applicant? The larger story that emerged pointed to a need for more equality during the interview. The job interview process was redesigned to encourage a conversation between participants rather than an exercise in trying to guess the answers the school district wanted to hear. The interview would be structured around two factors: the desire to share information with applicants and the need to explore specific skills needed to perform the job.

The goal was to structure the interview as a conversation rather than a series of monologues. A conversation implies a more or less equal relationship in which all parties have equal access to information. Given that the district already had information from the applicant that gave previous school employers, educational preparation and personal/professional references, a vehicle was needed to provide the applicants with equal information about the school and school district. In the new process, when applicants are invited to an interview they are given demographic data about the district, results of ongoing inquiry efforts into district satisfaction indicators, the district web address with current information about the initiatives within the district, and the names and phone numbers of current staff members who have volunteered to share their narratives of the district. All parties come to the interview with some common knowledge on which to build a relationship.

In addition to establishing a climate of mutual respect through sharing information prior to the interview, the interview questions were restructured to allow the interviewee to know the intent of the questions. For example, the need to select staff that would stay with the district resulted in direct questions regarding length of anticipated commitment to this job.

> *As you may have noted from our district profile, we have experienced a higher than acceptable rate of turnover in our faculty. We believe it is important to build a community of educators who share similar goals and aspirations for our students. We hope to hire someone who will teach in Sherman for at least four years. While we know that no one can definitely commit to that length of employment, could you share with us what events or circumstances might occur that would result in your leaving sooner than that?*

Sherman School District serves a rural population that has a high rate of poverty and unemployment. Children tend to have a diverse set of learning and emotional needs. As teachers shared their stories about teaching in the district, one of the unifying themes was the need for teachers to be problem-solvers. A question was designed to see how the applicant problem solved.

> *This district has a diverse population that can present many challenges in the classroom. We want to know how you problem solve. We are going to give you a classroom scenario and ask you how you would respond. Regardless of how good your response is, we will identify a problem that exists with that answer and ask you to give an additional response.*

Sherman Independent School District is experiencing a major change in the demographics within the district. Within the past five years, the Hispanic population has grown from less than five percent to over thirty percent. Many of the children are below grade level and speak English as their second language. Another of the unifying themes was the need for teachers to be creative in their teaching approaches. A question was designed to see how applicants viewed their personal teaching style.

This district has many students who have had limited educational opportunities and may not be fluent in English. Although we have support for these students through our Bilingual and Special Education programs, many of these students will be in your classes. We have found that the traditional textbook may be of limited value to these students. We want to know your teaching style. How do you see yourself teaching if you know many, if not most, of your students cannot read grade level material?

These questions do not hide their intent. They are authentic and invite cooperative exploration of the topics. Because the intent of the questions is apparent, it encourages follow-up questioning. It is appropriate to check what is meant by the response and get a better understanding of the applicant. Additionally, applicants gain a better understanding of the district and are able to make an informed choice to not seek employment in Sherman District if they do not perceive a personal "fit."

The redesigned interview process invites the applicant to join in the school's narrative. It provides entry into the school as a partner in the process and encourages the elimination of boundaries that impede movement to the envisioned future of the district. The conversation process of collaborative meaning making is established for all employees.

It is too soon to completely evaluate the effect of the redesigned interview process. Applicants hired within the new process are completing their first year of service.

However, in the ongoing narrative inquiry within the district comments about the interview process are being woven into the stories of the school. New teachers comment on the positive feelings they had about the process. One teacher commented, "It was a really interesting interview. I felt like they really wanted to know about me, and what I can do. It was just like I was visiting with friends."

5. HOW CAN NARRATIVE INQUIRY IDENTIFY SYMBOLIC BOUNDARIES THAT EXIST WITHIN AN EDUCATIONAL ORGANIZATION?

The narrative inquiry process begins with an interest in the systemic change effort within the organization. Underlying constructs of the organization emerge from the storying and restorying of the participant's experiences. Senge (1990) proposes that organizations must expose key assumptions and underlying structures that affect individual and/or organizational actions. These tacit assumptions create symbolic boundaries that limit the systemic change process.

Dialogue as envisioned by Bohm (1990) examines thought both individually and collectively. Participants are active in both the construction of the content of the conversation and in the examination of the conversational processes. Rather than seeking a common denominator consensus, the process seeks to enlarge the shared perspectives to accommodate multiple perspectives. Within the dialogic process symbolic boundaries are identified and addressed. Dialogue, a part of narrative inquiry, enters into the phenomena of the educational organization and addresses all organizational levels.

Johnson (1998) suggests symbolic boundaries may exist around organizational structures, authority structures, physical structures, decision-making structures, and the division of labor. Narrative inquiry seeks the meaning inherent in these structures and allows the significant symbolic boundaries to become focused against a background of less significant features in the change process. The overarching story of the organization is constructed as narrative interpretations of participant practice are presented and considered. A mutual reconstruction of the organizational narrative emerges.

5.1 Stories from a Department in Transition: Identifying Symbolic Boundaries

In 1997 the Department of Education at Shiloh State University (pseudonyms are used for all people and places) began a newly funded doctoral program in educational leadership. With this addition, the department, which had previously offered undergraduate and masters level programs, began a challenging time of transition from a practitioner-focused culture to a scholar/practitioner culture. Over the past three years, the faculty of the Department of Education joined in a collaborative self-study research process.

The collective body of personal and professional knowledge within the department is constructed within a complex web of relationships that have evolved and changed over time. Stories embedded in the history of the organization have implications for the current transition. One group of stories emerged from the work of three women who were the first females appointed to faculty positions within the department. Each of these women's stories begins a conversation about a particular change that continues to impact the community of higher education. Issues of membership, power, and boundaries resurface in each transition. Through the sharing of these stories, the phenomenon of resonance (Bridges, 1980) allows others to identify with similar issues and examine current situations with a new intensity and awareness.

Pam, Connie, and Deborah have been friends for almost 25 years. In the way of long-time friends, they sometimes finish each other's sentences, or share a look that is followed by peals of laughter. They frequently meet during the lunch hour in one of the university classrooms with conversations ranging from family, to friends, to work. With apples, sandwiches and drinks spread out on the table, they weave the past into the present and suggest a possible future. Symbolic boundaries emerge within the stories they share that have implications for the current transition process.

> *Pam: When we joined the department we didn't really have a voice. We don't really know how decisions were made in the department back then. There were no group decisions. At least, if there were, we were never included. The men met every day for coffee and there's no way to know what went on there.*

> *Connie: I went to the Chair and asked to work in the summer. I really needed to work. I had recently completed by doctorate and needed the extra income. The chair looked at me and said, "I'm sorry, but you've got to remember, these men have families they've got to support." I was a single parent with two children to support! There was a mindset I could not change. Men support families. Most of the men carried course overloads, but that was never a choice for us.*

> *Pam: And they taught the off campus courses too. If you taught an off campus course you were paid an additional stipend. That choice was not open to us. Only men could teach overloads, and only the men could teach off campus.*
>
> *Pam and Connie: Why? Because, because they had families to support. (Laughter).*
>
> *Connie: They got first choice on everything. It didn't matter what you were assigned to teach. If a class didn't make, the women would go home and the men would stay.*

Within this fragment, we begin to see a symbolic boundary that may continue to implicate the department today. Not in terms of the specific gender issues addressed, but in terms of power issues. Long-time members of the department did not examine the implications of having new members of the faculty. No efforts were made to incorporate a new set of needs into the existing system. The constructions of meaning continued to be hose of the entrenched faculty members, the resistors. This raises the question for examination by the current faculty, what unexamined boundaries are continuing to separate and marginalize new faculty members? Are new constructions of meaning being encouraged? How do we ensure inclusion of all points of view? Narrative inquiry allows these questions to surface and be examined. The stories embedded in the history of the department provide the vehicle for developing productive communication networks to explore organizational meaning and purpose.

6. CONCLUSION

Within systems design, narrative story provides a means of examining the multiple meanings constructed by those engaged in systems change activities. Narrative story allows alternate explanations and perspectives to address the contextualization of issues surrounding the time, location people, and events that are part of the change process. System change efforts in education occur within constructions and reconstructions of meaning concerning these contextualizations. It is important to examine and understand these emergent social structures. Narrative story allows individual perspectives to remain while encouraging the development of a unifying narrative.

Sharing stories of practice within a system design process fosters reflection and deeper understanding among participants in the process as well as those outside the organization. Individual perspectives are brought forward and retained in the overarching narrative, maintaining the multiple levels of meaning within the organizations shared vision. As stories are shared, points of tension can be identified. The Sherman School district was able to identify the tension created between the current interview process and the expectations of practice at the campus level. A more authentic experience in the interview process seemed critical to developing their idealized school.

Sharing of stories allows the historical, cultural and contextual ground of the school to remain. The Department of Education at Shiloh University, engaged in a collaborative self-study during a time of transition, examined stories from previous times of transition that identified acculturation and inclusion of new faculty members as a potential problem area. With this information, specific activities to smooth the transition could be planned that allowed new faculty members to join the unifying narrative.

As stated in the introduction of this chapter, narrative story is a component of systems design that facilitates the processes, pathways and patterns to an idealized school. It encourages participation from all levels of the school and allows personal knowledge to be incorporated in the design process. It facilitates the reconciliation between new possibilities and current structures as a unifying narrative is constructed that retains both individual perspectives and the overarching themes of meaning.

REFERENCES

Banathy, B. H., 1991. *Systems Design of Education: A Journey To Create The Future*. Educational Technology Publications, Englewood Cliffs, NJ.

Banathy, B. H., 1988. "Systems inquiry in education," *Systems Practice Vol.* **1**(2):193-212.

Bohm, D., 1990. *On dialogue*. Routledge, New York.

Bridges, W., 1980. *Transitions: Making Sense of Life's Changes.* Addison-Wesley, Reading, MA.

Bochner, A., 1990. *Embracing contingencies of lived experience in the study of close relationships*. Keynote Lecture to the International Conference on Personal Relationships. Oxford University, England.

Clandinin, J., and Connelly, M., 1991. Narrative and story in practice and research. In D. A. Schön (Ed.). *The Reflective Turn: Case Studies In and on Educational Practice*. Teachers College Press, New York, pp. 258-281.

DiMaggio, P., 1994. Culture and Economy. In N. J. Smelser and R. Swedberg, (Eds.), *Handbook of Economic Sociology*. Princeton University Press, Princeton, NJ, pp. 27-57.

Fullan, M. G., 1991. *The New Meaning of Educational Change*, Second Edition. Teachers College Press, New York.

Fullan, M. G., and Miles, M. B., 1992. Getting Reform Right: What Works and What Doesn't. *Phi Delta Kappa,* **73**(10): 745-752.

Jenlink, P. M. and Carr, A. A., 1996. Conversation As A Medium For Change In Education. *Educational Technology*, **36**(1), 31-38.

Johnson, B. L., 1998. Reconsidering the Educational Restructuring Process: An Exercise in Retrospective Sense-Making, *Journal of School Leadership,* **l**(7): 540-568.

King, A., 1992. What is Postmodern Rhetoric? in *Postmodern Political Communication.* (A. King, ed.). Praeger, Westport, CT.

Langer, E. J., 1997. *The Power of Mindful Learning*. Addison-Wesley. Reading, MA.

Luhmann, N., 1995. *Social Systems*. (J. Bednarz trans.). Stanford University Press, Stanford, CA.

McLaughlin, M. W., 1990. The Rand change agent study revisited: Macro perspectives and micro realities, *Educational Researcher* **19**(9): 11-16.

Nash, J., and Tucker, M., 1997. "Understanding restructuring in practice: A study of teachers' perceptions of organizational interventions and core technology on student success, *Planning and Changing* **28**(1/2): 57-73.

Nuyen, A. T., 1997. Education for imaginative knowledge. *Journal of Thought*, **32**(1): 37-47.
Piaget, J., 1995. *Sociological studies*. Routledge, New York.
Polkinghorne, D., 1995. Narrative and self-concept. *Journal of Narrative and Life History,* **1**(2&3): 135-153.
Prigogine, I., 1996. *The End of Certainty: Time, Chaos, and the New Laws of Nature*, The Free Press, New York.
Rosen, H., 1986. *Stories and Meanings*. National Association for the Teaching of English, Kettering, England.
Senge, P., 1990. *The Fifth Discipline: The Art and Practice of The Learning Organization*, Doubleday, New York.
Sylwester, R., 1995. *A Celebration of Neurons: An Educator's Guide to the Human Brain*, Association for Supervision and Curriculum Development, Alexander, VA.
Tannen, D., 1988. "Hearing Voices in Conversation, Fiction, and Mixed Genre". In D. Tanner (Ed.), *Linguistics in Context: Connecting Observation and Understanding.* Teachers College Press, New York, pp. 89-113.
Tedesco, J., 1997. Educational change from the perspective of decision makers. *Prospects,* **27**(4): 533- 540.

Chapter 15

CONVERSATION AS AN ACTIVITY SYSTEM

The Mediational Role of Discourse in Systems Design[1]

PATRICK M. JENLINK
Stephen F. Austin State University and International Systems Institute

1. INTRODUCTION

Systems design as Banathy (1996, 2000) explains, is largely communicative in nature, and depends on design and dialogical discourse as semiotic tools for mediation within the cognitive, cultural, and creative activities essential to overcoming deep sociohistorical patterns of education that are woven into the fabric of society. The mediational role of design conversation and the use of other symbol-based systems in ideal systems design is supported by activity theory which presents a systemic view of design activity (Engström, Miettinen, & Punamäki, 1999). The semiotic nature of discourse and language within communities of design practice enables participants to transcend formal cognitive and cultural patterns that often marginalize and disadvantage voices of difference. Essential to the design of new activity systems is the ability of participants to acknowledge the dialectical contradictions that have emerged in their past or present activity system(s), while also acknowledging the importance of creating dialogical relationships toward the goal of designing new systems.

The power of systems design in educational settings lies in the ability of systems design to mediate and transcend the dualisms that challenge social change. Transcending dualisms such as thought and activity, theory and practice, facts and values, systemic and subjective-partisan views, unity and diversity, and internalization and externalization brings the designer(s) into a dialogical relationship with the human activity system being designed.

Design conversation and systems language underlie the socially constructed process of ideal systems design. The framework of activity theory suggests that mediational artifacts such as language and discourse do not exist inside or outside of individual consciousness, rather they reside on the borderline between oneself as designer and the others who are also designers and users. From this perspective one has to take and use cultural artifacts (i.e., design conversation and systems language), to appropriate these artifacts with the intentions of the designer and the design community. The next section will examine the concept of activity theory and human activity systems.

This chapter will examine the mediational importance of conversation in the design of human activity systems that are educative and intentioned as learning systems. The chapter will first examine educational systems design, viewing design as post-epistemological in nature. Design conversation is then deconstructed to create an understanding of how this type of discourse draws from other genre or types of discourse such as dialogue, discussion, dialectic, ethical and critical discourse. The chapter will then elaborate on design conversation as activity through an activity theory framework, demonstrating the dynamic relationship between participants, purpose, mediational artifacts, community, design work, and socio-cultural rules governing design. The author will conclude with reflections on the mediational importance of design conversation in transcending dualisms that challenge social change.

2. EDUCATIONAL SYSTEMS DESIGN

Educational systems design has been elaborated in significant detail elsewhere in the design literature (Banathy, 1991, 1995; Jenlink, 1995; Jenlink et al., 1998) and therefore the text in this section provides an overview of the more salient points to be considered. Relatedly, educational systems design is considered a form of social systems design. In this understanding of educational systems design, design is a communicative process among individuals that enables collective action(s). These actions lead to the creation of change in and/or of the social system; the transfer of the conception of a new or alternative system into action (Churchman, 1971). Drawing from Churchman's ideas of systems design, Banathy (1996) further elaborates that

> the designer's main tool is subjectivity, which includes social practice, community, interest and commitment, ideas and ideals, the ethics of the system and the moral idea, affectivity, faith, and self-reflection. For Churchman, the issue is not whether we can design systems that are wholes and unique, but whether we can design systems that make us more whole, unique, and self-motivated. (p. 164)

As identified in the examination of activity theory, the embodiment of participant's subjectivity is a critical element of the social change or transformation process. Subjectivity as a design tool enables the object of design to be realized, particularly when the object is to create a complex social system like education, from the ideal.

A defining characteristic of educational systems design is that the design process is inquiry oriented and is a form of discourse, practical in nature and based on a social language of systems. This discourse is communicative in action, seeking to bring participants into inquiry-based activity that is focused on creating an alternative or new system. Banathy (1996) defines social systems design, in the context of human activity systems, as "a future-creating disciplined inquiry," an inquiry that "people engage in design in order to devise and implement a new system" (Banathy, 1996, p. 42).

Educational systems design is based on an ideal systems design approach. Incorporating Banathy's (1996) notion of the "ideal" as a focal point, the new educational system is created by "those who serve the system, those who are served by it, others who have a vested interest in it, and all those who are affected by it" (p. 195). The characteristic of users creating or designing the ideal system–user-designers–is a critical component of designing educational systems. From this perspective of systems design, the subjects of the design activity (see Figure 1) are the same individuals who are the stakeholders in the system. Therefore, educational systems design reflects the authentic participation of stakeholders in "the design because they genuinely and deeply care about the future state of their system" (p. 195).

An examination of ideal systems design as delineated by Banathy (1996, 1998) suggests five interrelated and interdependent design spaces that are critical to designing a new system. Each space represents a design space in which participant inquiry and design conversation are situated, and includes the following: exploration and image creation space, design information and knowledge space, design solution space, evaluation and experimentation space, and modeling space. Jenlink (1995, 1999) provides a contrasting yet complementary view of systems design in suggesting that the process includes: contextualization of the system, design of new system and system implementation processes, implementation of system, and critical inquiry and system learning. Together, these form a multidimensional design space in which design activity unfolds. Also critical in this view of systems design is self-renewal and evolution of consciousness as critical processes implicit and explicit in the design; contributing to the participant's ability to transform social structures and transcend existing systems as the ideal is created and realized through the actions of participants. During the design process, consciousness moves through, or perhaps more accurately along, a developmental and evolutionary path, reflecting different types of consciousness in social action including: perspectival, interpretative, critical, ethical and moral, self-reflective, integrative, creative, collective, self-renewing, and evolutionary (Jenlink, 1999). Each type of consciousness relates to different design activities. With a particular focus on forms of conversation, objects of the design process (see Figure 1), design maturity of stakeholders, particular social language, and sociocultural rules that govern or influence the social action of participants engaged in design.

Importantly, ideal systems design of an educational system requires not only a focus on the ideal, but also on the ethical and moral (this can be compared to sociocultural rules in Figure 1). Issues of equity, social justice, and caring provide a constant tension in the design process to create the ideal educational system. The dialectic boundaries set by interacting activity systems, the concern for continued reproduction versus transformation of social structures, and the need to address issues of moral, intellectual and social responsibility in an increasingly problematic society affirm the importance of using an ideal systems design approach to creating new educational systems.

As an ideal-focused approach, educational systems design is a meta-process (Jenlink, 1995) based on a multivoiced dynamic of authentically engaged stakeholders, situated in the design space (Banathy, 1996) and mediated by using design conversation–a complex system of different forms of social discourse. Design conversation enables stakeholders to critically examine, through multiple perspectives, the existing societal structures, the dialectic boundaries of activity systems, the cultural and cognitive patterns of the participants and their communities, and the sociohistorical and cultural artifacts, while simultaneously contextualizing the design of the ideal educational system.

Educational systems design, as explicated thus far, represents a post-epistemological stance of systems design. Stance relates to the type of design praxis that stakeholders engage in as part of the design activities. Post-epistemological refers to the relationship between knowledge (design knowledge) and knower (stakeholder as user-designer). The relationship in systems design transcends more modern views that design knowledge exists extant from or outside the stakeholders sphere of involvement in the activities. A post-epistemological view moves beyond the knowledge structures often bounded in the controlling social structures. In this view, systems design, based on the subjectivity of stakeholders and its embodiment in a system ideal, is guided by an inquiry-based approach to co-constructing, co-creating a new system that overcomes the cognitive and cultural patterns of society. In the next section, design conversation is deconstructed, presenting the dynamical nature by which different forms of social discourse form meta-conversational patterns of systems design discourse.

3. DESIGN CONVERSATION

Design conversation, as explicated in this section, builds on the work of Banathy (1996), Jenlink (1995, 1999), Jenlink and Carr (1996) and Horn (1999), and draws from the extant literature on systems design, conversation and discourse, and activity theory. Conversation, in the context of educational systems design, is viewed largely as a communicative action, providing a medium through which participants in the design process may engage in a multi-dimensional inquiry leading to the creation of a new system. Design conversation is not a singular type or form of social discourse, but rather a dynamic system comprised of different forms of discourse, each with a particular purpose and mediational importance as semiotic tool in the system design activity.

Design conversations occur as socially constructed processes of communicative action, situated within multiple interrelated design activities. Bringing this social action into being requires something more to be exchanged within the discourse than just those intersubjective understandings (or misunderstandings) that belong to the flow of the discourse. "That 'more' consists of what is being talked about, the referential and semantic contents of communication" (Engström, 1995, p. 195). In the design of an educational system, the social languages and co-constructed meanings as well as the ideals generated serve as the referential and semantic contents of communication.

Banathy (1996) has suggested that design conversation is the combining of strategic dialogue and generative dialogue, forming a type of conversation that enables stakeholders to create an ideal system. Jenlink and Carr (1996), in their typology of conversation, defined design conversation as a "disciplined inquiry grounded in systems philosophy, theory, and thinking and practice" (p. 34). Situating design conversation in the context of educational change, this discourse "focuses on a change that transcends both systemic constraints within the [social system] and the constraints of a narrow, traditional view of how change should happen" (p. 35). In the text that follows, design conversation will be deconstructed through the lens of activity theory. The position for design conversation as "trans- or metadisciplinary" (Bohm & Kelly, 1990, p. 449) and meta-conversation will be presented.

3.1 Deconstructing Design Conversation

When examined through the framework lens of activity theory (see Figure 2), the communicative action, practical discourse, and inquiry-based orientation of design conversation reveals a deeply complex array of rule-based social actions. These actions are mediated and governed by discourse and social language that is politically and cultural charged in the contexts of its origins. The complexity of systems design, as reflected in Banathy's (1996) design spaces and Jenlink's (1995, 1999) elaboration on ideal systems design (presented in the previous section) gives support to the argument for recognizing the need for various types or forms of discourse as stakeholders engage in the design process. Systems design, as a series of communicative and social actions, occurs within the larger dynamic of the design activity system. The design activity system is comprised of interrelated and interdependent events, activities, actions, and processes. Each event and activity of the system seeks to transform a particular object into an intentioned outcome–creating the ideal educational system. This transformation or objectivation, using stakeholder subjectivity as a tool, draws into play issues of social justice, equity, difference, voice, consciousness, and ethical and moral responsibility.

Social systems design, as Banathy (1996) notes, "is a process that carries a stream of shared meaning by a free flow of discourse among the stakeholders who seek to create a new system." In order to understand the communicative nature and mediational importance of the design conversation, "various modes of social discourse are explored to search for the mode that is the most appropriate to systems design" (p. 213). Simultaneously, communicative "actions," voices, by bringing up referential potentialities (social languages), carry out the [design] activity of which they are a part" (Engeström, 1995, p. 201). As the user-designers explore and select the form of social discourse most appropriate to design activities, specific forms of discourse are connected to each activity in realizing designer actions. Design conversation, rather than a singular form of discourse, is a meta-conversation necessitated by the need, at varying times, for discursive, dialectical, critical, creative, dialogic, and post-formal forms of discourse.

The "meta" nature of design conversation reflects a dialogic betweeness that connects various disciplinary perspectives as well as the recognition of differences that populate social systems. Critically evaluating the complex nature of educational systems design, and examining the sociohistorical and sociocultural context in which the design process is situated problematizes systems design. The problematic nature of design is made apparent by drawing critical attention to the fact that a singular form of discourse will fail to enable stakeholders to succeed in the design of an ideal system. Rather, given the socially charged nature of discourse and language that already populates the activity systems of schools, the forms of conversation, communication, and language systems needed for educational systems design must be aligned with the activities within the design process. This becomes even more apparent when considering the need to overcome or transcend the patterns of cultural reproduction that seek to maintain social structures that reify educational systems. In the next section, six forms of discourse are examined in relationship to design conversation as a meta-conversation.

3.2 Forms of Discourse

Jenlink and Carr (1996), in their examination of conversation as medium for change in education, identified four types of conversation. These included: discussion, dialectic, dialogue, and design. Three overarching purposes for conversation were aligned with these types of conversation; transacting, transforming, and transcending reflect the purpose of discussion and dialectic, dialogue, and design conversation, respectively. Banathy's (1996) description of design conversation as a combining of strategic and generative dialogue reflects two additional forms of dialogic discourse to be considered in systems design. Banathy also identified ethical conversation as a form of discourse essential in the social systems design process. Horn (1999), building on the work of Jenlink and Carr (1996) and examining conversation through a post-formal lens (Kincheloe & Steinberg, 1999), posits yet another form of discourse, that of post-formal conversation, that has currency in the educational systems design process. Adding to the array of discourse considerations for social change processes, Isaacs (1999) elaborates on his earlier work with dialogue by adding monologue, reflective dialogue, generative dialogue, and discussion as four types or fields of discourse.

The remaining paragraphs in this section will elaborate on the six primary forms of discourse identified by these authors. In the case of dialogue, discussion, and dialectic, distinctions are made in terms of specific types of dialogue, discussion, or dialectic. For purposes of this elaboration, form of discourse will reflect the primary discourse, with type reflecting a subset of the form. A framework, based in part on activity theory and in part on systems design, will be used to guide the elaboration of each form and type of discourse. This framework will include a focus on rule association, consciousness, language, and conversational patterns.

3.2.1 Monologue Discourse

Isaacs (1999) characterizes monologue as a polite type of conversation, where people coming together exemplify the social rules aligned with those normative conditions which guided their development and which are most comfortable. Monologue, as a field of conversation, might be perceived as nondisclosing, that is, people do not share what they really think, but merely follow the politically appropriate rules. This type of discourse is termed rule following, with sociocultural rules characteristic of non-confrontational and largely based in the social structures of the participant's culture. Individual and collective consciousness is interpretive in nature, that is, the participant seeks to interpret the actions of others without engaging in formal action that might reflect judgement or political posturing. This form of discourse enables the evolution from an interpretive to perspectival consciousness. Sidorkin (1999) suggests that a main characteristic of monologue is the separation of individual from ideas, so that what is shared in the discourse is general and does not set boundaries nor threaten cognitive or cultural boundaries. Rather, the individuals may display emotion and aggressive banter as Isaacs (1999) notes, but they do not say what they really think.

Monologue, as a discourse in the larger meta-conversation of systems design, might serve to begin generating the energy necessary to creating a field and forming the container for more dialogic discourse. The social language of monologue is nonreflective, bound in the social norms that individuals bring with them to the design process.

3.2.2 Discussion Discourse

Perhaps the more common discourse found in social activity, discussion is often more pragmatic, giving way to patterns of advocacy, political posturing, and fragmentary boundaries (Jenlink & Carr, 1996). Discussion discourse is more subjectively influenced by opinion and supposition, and often characterized by patterns of rigidity and being closed to sharing personal or professional viewpoints for scrutiny by others. Discussions are often rule revealing (Isaacs, 1999), positing nonnegotiable viewpoints in adversarial and debate like interactions. Through this relation, social rules are surfaced by the participants as each attempts to win the other over this her/his point of view. Sociocultural rules that often come into play include competition, move-oppose, conversation as battle or aggressive confrontation, non-listening, non-suspension of assumptions, and active judgement. As Isaacs (1999) notes, "the challenge of this space is to change the meaning of the trauma that arises, both individually and collectively" (p. 265).

Patterns of conversation often reflect boundary setting, political posturing, defensive routines, and heated exchanges. The language of discussion is often positional, politically charged with advocacy for personal positions, unilateral control, and aligned with social structures that are familiar and provide safe ground from which to argue a particular position. The consciousness that seems to dominate this type of discourse is positional, fragmentary, and advocacy in nature. Discussion discourse provides a transitional discourse between monologue and the more dialogic types of discourse.

The importance of discussion, as a rule revealing discourse, is that participants, individually and collectively, are brought to a level of conscious awareness of the unique perspectives that each person has. This perspectival consciousness is important to the evolution of the design conversation, particularly as the importance of difference is brought into play in the designing of the ideal educational system. The danger with discussion is that if participants remain in the discussion cycle too long, fragmentation and loss of collectivity is often experienced as rigid boundaries set in motion dialectical opposition to sharing and honoring differences.

3.2.3 Dialectical Discourse

Dialectical discourse is a form of disciplined inquiry, presenting a logical argument for or against a particular position that has been presented as truth or is evidenced by logical support. Where discussion discourse is often influenced by emotion, dialectical discourse is typically posited as a rational argument. As Jenlink and Carr (1996) note, dialectic discourse "might be characterized as closed, that is, participants are often rigid in their beliefs and debate for what they perceive as truths. A type of dialectic that is familiar to the inquiry literature is hermeneutic, associated with critical theory and noted as critical hermeneutic (Kincheloe, 1999). The critical hermeneutic "helps produce a situated knowledge that is aware of the conditions of knowledge production and reception and the power relations" (p. 22). The importance of hermeneutic and other types of dialectic resides in disclosing the nature of the relationship that participants in systems design perceive to exist between design knowledge and the stakeholder as participant in the design process. Importantly, the critical hermeneutic enables participants to recognize what is perceived as truth or accepted knowledge and examine how perceptions inform the user-designer of the power structures associated within the culture. Again, this type of discourse is rule revealing (Isaacs, 1999) and is signified by cultural patterns of social language grounded in truth, boundaries, cognitive positioning, rigidity, power structures, and logic.

Discussion discourse is recognized by its dominant patterns of logical argument, distillation of truths, criticality, and at times factionalization that establishes social and dialectical boundaries. Dialectical discourse reflects an interpretative, logical, and critical consciousness. This form of discourse again plays an important transition role as design conversation moves from monologue to dialogue, with the desire of evolving the design discourse to a more reflective and creative influence in the systems design process.

3.2.4 Dialogue Discourse

Dialogue as a form of discourse ostensibly holds significant importance for the systems design process. Whereas discussion is perhaps the more pervasive form of discourse found in social activity and educational settings, dialogue is crucial to bringing the participants to a level collective and transformational consciousness in the systems design process. Dialogue is differentiated into two types by Isaacs (1999) who sees reflective dialogue as rule reflecting, and generative dialogue as rule generating.

Banathy's (1996) identification of strategic and generative dialogue as foundational to design conversation builds on the notion of generative dialogue as rule generative, noting that generative dialogue "is applied to generate a common frame of thinking, shared meaning, and a collective worldview in a group" (p. 215). In contrast, Banathy (1996) states that strategic dialogue "implies communication among designers that focuses on specific tasks of seeking solutions" (p. 215). Each type of dialogic discourse encourages and sustains relational patterns in the larger conversation, patterns essential to creating an integrative and collective consciousness in the participants and across the social activities of systems design.

In dialogue, the social language reflects respect, diminishing of dialectical and positional boundaries, sharing meaning and knowledge construction, collective identify and acceptance of personal world views. Patterns of conversation move to openness toward others, listening deeply, suspension of judgement, disclosure of personal beliefs and assumptions, caring, concern for equity and justice, and a focus on community. Sociocultural rules are co-constructed and collectively respected as communicative and social action informs the design activity of creating an ideal educational system.

Dialogue as form of discourse, in each of its types, is a critically important social discourse that enables the design conversation to serve as the creative and generative medium through which the user-designers create the ideals. As Isaacs (1999) notes, the generative dialogue, while the rarest of dialogic discourse, "is the one where people cross over into the an awareness of the primacy of the whole … this is the space where people generate new rules for interaction, where they are personally included" (p. 279). Dialogue serves as a critical nexus in the forming of systems design discourse as meta-conversation.

3.2.5 Ethical Discourse

Ethical discourse is a governed by social rules of right and wrong. As Banathy (1996) states, ethical discourse is focused on "values, morals, and ethics…among the stakeholders" (p. 181). Stakeholders, as user-designers, must focus not only creating the ideal educational system, they must engage in explicit discussion, aimed at finding common ground and developing consensus. Ethical discourse replaces the aggressive and often conflicting discussion discourse with an "informed and value-based exchange of ideas and perspectives" (Banathy, 1996, p. 281).

Ethical discourse sets boundaries by mutual agreement as to what the ideal system should or should not embody. The social language of this discourse is characterized by personal and collective codes of right and wrong, equity, social justice, and consideration for difference. The conversation patterns of ethical discourse are reflected in Banathy's (1996) statement that "we each bring with us to the ethical discourse a wide variety of values and moral attitudes. Although this creates a more complex discourse, it also empowers the conversation with the capacity to deal with increased complexity" (p. 181).

3.2.6 Post-Formal Discourse

Post-formal discourse "includes an expansion of the awareness of self in relation to others, and a critical awareness of the communication process in relation to how it emancipates or constrains our relations with others" (Horn, 1999, p. 364). Grounded in post-formal cognition as described by Kincheloe and Steinberg (1999), post-formal discourse is a dialogue about power, " a dynamic investigation of our selves, our relations with others, and the political implications of the type of conversation in which we are engaged" (Horn, 1999, p. 364).

This form of discourse shares a similarity with dialectical discourse in that it is grounded in a critical perspective of responsibility for social and cognitive activity, guided by inquiry into social structures of culture based on a critical hermeneutic of power relations. Post-formal discourse surfaces a critical consciousness on the individual and collective level of design activity.

This discourse shares equally defining characteristics of ethical discourse as it seeks to ensure social justice and caring in the ideal design of an educational system. Post-formal discourse is guided by the four elements of a post-formal structure including patterns, process, etymology, and contextualization.

Conversational patterns in post-formal discourse include facilitative, constructive, collective, critical voice, and a focus on sociohistorical and sociocultural relationships that exist between knowledge, knowledge construction and user-designers. In post-formal discourse, the elements of dialectical and discussive discourse surface as participants engage in examining personal perspectives and individual worldviews. This serves the meta-conversation of design by creating an awareness of perspectives, thus leading to a perspectival consciousness essential to generating a quality and energy essential to ensuring that voices of difference are included in the design of an ideal system.

3.2.7 Design Discourse as Meta-conversation

As a meta-conversation, design conversation embodies each form of discourse throughout the systems design process (process is viewed as interrelated sets of design activities), enabling the stakeholders to engage in the objectivation and embodiment of the participant's subjectivity in the transformation of the ideal into reality. Key in this meta-conversation is the transition from one form of discourse to another, and knowing when each form is relevant to the design activities. In the next section, design conversation will be examined as discursive activity through the framework of activity theory.

4. DESIGN CONVERSATION AS DISCURSIVE ACTIVITY

Design conversation as discursive activity serves a unique function, interrelated to the function of design conversation as mediating artifact in the design activity. Whereas discourse serves to mediate communicative and social actions within the design activity at one level of the creation of an ideal educational system, at a different level design conversation serves the function of designing the design system that will be used by stakeholders to create the ideal educational system. This relates directly to the dual functions that stakeholders play in the system design process.

Stakeholders in their design activity must perform a dual function: they must not only design the ideal educational system but design the process of their design, that is, redesign (or to a greater or lesser extent design) the structures, processes, and artifacts critical to their substantive activities of systems design. Because social structures, as well as design processes and artifacts are themselves products of systems design, they are themselves possible objects of transformation or redesign and so may be only relatively enduring. And because design activities, as social activities are interdependent, social structures may be only relatively autonomous. As Bhaskar (1989) states, "it is important to note that because social structures exist only in virtue of the activity they govern, they do not exist independently of the conceptions that the [stakeholders] possess of what they are doing in their activity" (Bhaskar, 1989, p. 78).

Based on the dual function perspective, design conversation serves to mediate the activities of designing the design system, influenced by what the stakeholders' conceptions are of designing an ideal educational system. Design conversation as a meta-conversation, also serves as the medium through which the design system is created. Therefore, design conversation as activity means that conversation is elevated to a level of social action focused on transforming and creating.

An examination of design conversation as activity, using the framework of activity theory, shifts the focus from one of design conversation as mediational to one of design conversation as transformational and transcendental (Banathy, 1996). The concept of systems design as a meta-conversation takes on new meaning as design conversation is explicated, through an activity theory framework, as a conversational or human activity system. Discourse, as differentiated and examined earlier in this paper, is clearly delineated as a mediational or semiotic tool in the design process. Issues of sociocultural and mediational rules, social languages, participant consciousness, and discourse patterns are situated in the meta-conversation. The dynamic relationship of subject, object, mediational artifact, sociocultural rule, community, and division of labor are reconsidered in the context of design conversation as a meta-conversation or human activity system (see Figure 1).

As an activity system, design conversation focuses on the transformation of the object of the activity within the system. More precisely, since the activity system is an interrelated and interdependent set of activities and actions, designing a design system must consider the object of each design activity. The form of discourse and social language used as a semiotic tool for mediating the social actions of stakeholders (user-designer) is determined in large part by the object of the activity, influenced by the sociocultural rules and user-designer maturity in systems design. Isaacs (1999), in examining the conditions for conversation, introduces the concept of conversation fields in forming a container for the conversation. Because each form of discourse has certain sociocultural rules, language, consciousness, and action patterns, the field is unique to the discourse. Field, as used in this context, refers to "spaces in which there is a particular quality of energy and exchange" Isaacs, 1999, p. 257). This quality and exchange relate to the rules that govern the discourse, the type of language, and the individual and collective consciousness reflected through the communicative and social actions of the designers. Each field of conversation contributes to the evolving container that surrounds or supports the design conversation, and, as Isaacs (1999) notes, "are the relatively observable features of fields" (p. 257).

4.1 The Framework Lens of Activity Theory

Activity theory, with its philosophical and historical roots in the classical German philosophy (from Kant to Hegel), in the writings of Karl Marx, and in the theorizing emerging from the cultural-historical school of Russian psychology most often associated with the research of L.S. Vygotsky, A. N. Leont'ev, and A.R. Luria, presents a framework of understanding activity in human systems. Recent work with activity theory in the fields of human cognition, cultural psychology, and communication through the research of Michael Cole, Yrjö Engeström, and Ritva Engeström draws attention to the similarities in social systems and educational systems design and the sociohistorical and sociocultural foundations of activity theory.

Through the framework of activity theory, in the context of educational systems design, participants in a human activity system are guided by object or motive based expectations of creating an ideal educational system. The creative activity is mediated by use of cultural artifacts that might be any combination of rule-based, role-based, symbol-based, cognition-based, discourse-based, process-based, and technology-based tools. A primary example is the use of ideal systems design technology, systems language, and design conversation in the design of an ideal educational system. Also critical to the framework which guides the systemic change process are sociocultural rules that are aligned with the object or motive based expectations. Essential in this framework is membership in a community of stakeholders seeking to design a new ideal for the educational system—a design community. Membership in the community by the facilitator and stakeholders is balanced through a division of labor that seeks to authentically engage all participants in the systemic change process. A core set of beliefs serves as a foundation or center for this framework, which provides social coherence for the design community.

Activity theory, as a framework for understanding the meaning of human activity systems, is based on a relational dynamic between the subject, object, mediational artifacts (or tools), sociocultural rules, division of labor, and community structure of a human activity system (see Figure 1 for an elaboration)[2]. Community refers to those who share the same general object; rules refer to explicit norms and conventions that constrain actions within the activity systems; and division of labor refers to the division of labor of object-oriented actions among members of the community.

4.2 Design Conversation as an Activity System

Within the framework of activity theory, design conversation is viewed as an activity system, wherein the conversation field is evolved within and through the relationship of subject, mediating discourse and language, and object (see Figure 1). The influence of socio-cultural rules (see **(a)** in Figure 1) on the transformation of the object of the activity via the mediating discourse helps to shape the transformation.

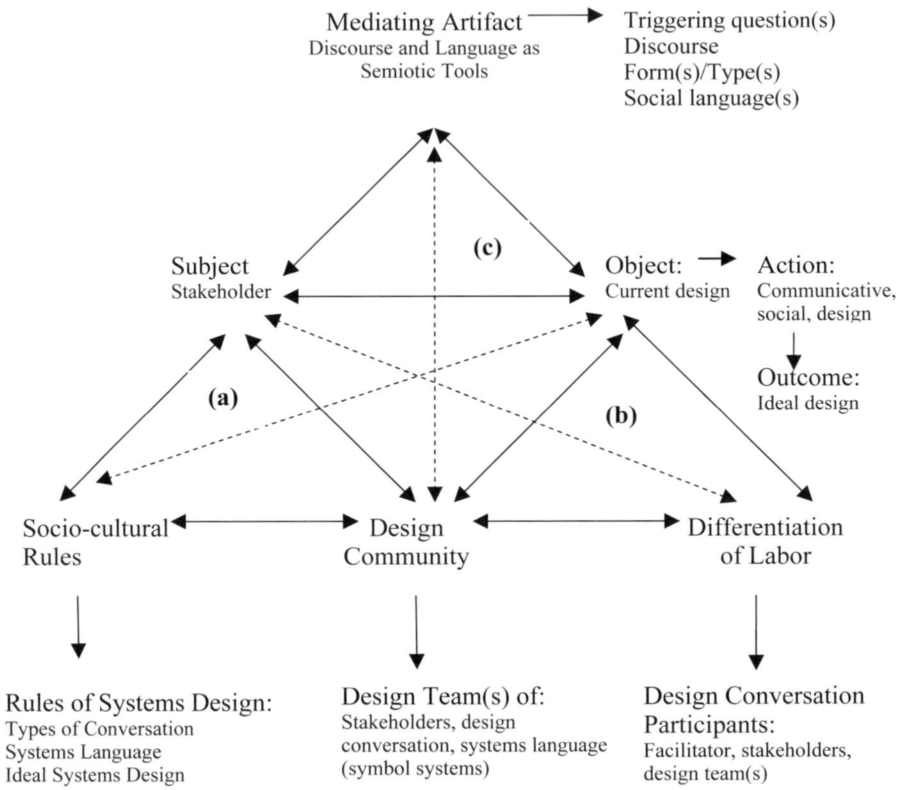

Figure 1. Design Conversation as Activity System

Activity theory claims that an activity, (such as those related to design conversation or educational systems design), is a unit that implies a socially defined goal and the execution of some specific actions that have evolved or have been created to attain that goal (Engeström & Cole, 1991; Leont'ev, 1978; Russell, 1997). As such, activity involves patterns of communication with others related to the setting and the goal, and thus, mastery of a set of symbolic tools (such as systems language), discourse tools (such as communicative action and design conversation), or perhaps process tools (such as educational systems design). In an activity, each of these elements influences the individual's and the collective's actions, practices, and understandings. It is important to reiterate that the various components of the activity system do not exist in isolation from each other. Rather, "they are constantly being constructed, renewed, and transformed as outcome and cause of human life" (Cole, 1995, p. 113).

The top triangle in Figure 1 represents an element of the activity system that defines the subject, object, and mediating artifact(s) relationship. The subject(s) of any activity is the person(s) for whom the activity is created. The object is the motive or intentioned outcome implicit and explicit in the activity. The mediating artifacts or tools are cultural in origin, and serve to mediate the subject's actions and activities as the object is transformed through objectivation. As the object transformed, human subjectivity (social languages, forms of economic and political organization, cultural and ethical norms, ideals for social systems) is embodied in the intentioned outcome or product.

The second triangle in the lower left of Figure 1 represents an interrelated element of the activity system. This element depicts the relationships between the subject(s), the sociocultural rules of the community (related to the object, goal, outcome—see **(a)** in Figure 1), and the designated community made apparent. The third triangle in the lower right of Figure 1, represents the relationship between the object(s) or intentioned outcome(s), the community in which the subject is a member, and the division of labor respective to the particular activity. The division of labor might be thought of as role differentiation by subject(s) within the community (see **(b)** in Figure 1). The community's culture produces, uses and transforms artifacts (see **(c)** in Figure 1) as individuals and the collective engage in activities. Connected, the three triangles form a framework or lens for analyzing and designing human activity systems such as educational systems. In this framework, activity theory is elaborated as a complex set of interrelated and situated relationships that enable participants to accomplish a goal.

Perhaps of more importance is the realization that the type of discourse(s) and language(s) used as semiotic tools, influences the transformation of the object into a design system for designing the ideal education system. Again, the dual function of design conversation as mediational and as design activity, related to the dual function of the stakeholders as user-designers, plays a critical role in understanding design conversation as an activity system.

As a human activity system, design conversation requires that the differentiation of labor enables authentic participation of stakeholders in the discourse mediated activities (see **(b)** in Figure 1). Stakeholder participation as user-designers is a defining feature of the field or container that supports the meta-conversation of systems design. As conversation patterns are constructed through the design activities, the energy and exchange generated through different forms of discourse evolve these patterns, and create a higher level of creative and transformative potential.

The complexity of design conversation as an activity system is recognized by examining the relationship of stakeholders as user-designers to discourse and language as semiotic tools for mediating the actions of transforming the object of creating a design system into reality. The dualism of conversation as mediational and transformational presents a complementary relationship of functions, symbolizing the problematic nature of engaging in dualisms. In the final section of this paper, reflections on the mediational importance of design conversation are presented.

5. REFLECTIONS ON CONVERSATION AS AN ACTIVITY SYSTEM

This chapter has examined the mediational importance of design conversation in the context of educational systems design. Activity theory, as a sociohistorical and sociocultural theory of learning, has been used as a framework to construct understanding about how discourse and social language serve as semiotic tools for mediating the communicative and social actions related to system design activities. Reflecting on the idea of design conversation as a meta-conversation has suggested that original perspectives of design conversation, while suggestive of the function of design conversation, have fallen short of providing a comprehensive explanation of the role that discourse plays in the systems design process. Systems design is a complex social process that is communicative in nature. The communicative nature requires the use of social discourse and social language to effect the social actions of user-designers within the interrelated and interdependent set of design activities. In the larger scope of using educational systems design to create an ideal educational system, the argument has been presented for using a complex system of social discourse to mediate the transformation of the object of each design activity into reality.

As mediational artifact and semiotic tool, discourse is socially and culturally charged with the rules and social languages of the respective contexts of origin. Stakeholder subjectivity is recognized as a primary tool in the generative process of creating an idea system. The critical and developmental role that discourse takes in mediating the creative process is made apparent as human subjectivity challenges the existing beliefs and social structures that represent the old system. Mediating tensions as well as overcoming dialectical boundaries set by interacting activity systems further informs the importance of design conversation in educational systems design. The mediational importance of design conversation is equally supported by the role that educational systems design plays in transforming not only the educational system, but society as well.

NOTES

1 Parts of this chapter were excerpted from a paper presented at the ISSS 2000 Conference, Toronto, Canada.

2 Figure 1 is an adaptation of Engeström (1987, 1995), Engeström, et al., (1999), and Cole & Engeström's (1993) work on activity theory.

REFERENCES

Banathy, B.H., 1991. *Systems Design of Education: A Journey to Create the Future.* Educational Technology Publications, Englewood Cliffs, NJ.

Banathy, B.H., 1996. *Designing Social Systems in a Changing World: A Journey Toward a Creating Society.* New York: Plenum Press.

Bhaskar, R., 1989. *Reclaiming Reality: A Critical Introduction to Contemporary Philosophy.* Verso, London.

Bohm, D., and Kelly, S., 1990. Dialogue on science, society, and the generative order. *Zygon,* **25**(4): 449-467.

Carr, A.A., 1997. User-designer in the creation of human learning systems. *Educational Technology Research and Development,* **45**(3): 5-22.

Churchman, W., 1971. *The design of inquiring systems.* Basic Books, New York.

Cole, M., and Engeström, Y., 1993. A cultural-historical approach to distributed cognition. In G. Salomon (Ed.), *Distributed Cognition: Psychological and Educational Considerations.* Cambridge University Press, Cambridge, UK, pp. 1-46.

Cole, M., 1995. The Supra-individual Envelope of Development: Activity and Practice, Situation and Context. New Directions for Child Development, no. **67**: 105-118.

Engström, R. 1995. Voice as communicative action. *Mind, Culture, and Activity, 2*(3), 192-215.

Engström, Y., 1987. Learning by expanding: An activity–theoretical approach to developmental research. Helsinki: Orienta-Konsultit.

Engeström, Y., and Miettinen, R., 1999. Introduction. In Y. Engeström, R. Miettinen, and R. Punamäki (Eds.), *Perspectives on Activity Theory.* Cambridge University Press, New York, pp. 1-16.

Engeström, Y., Miettinen, R., and Punamäki, R., 1999. *Perspectives on Activity Theory.* Cambridge University Press, New York.

Horn, R.A., 1999. The dissociative nature of educational change. In J.L. Kincheloe, S.R. Steinberg, and P.H. Hinchey (Eds.), *The Post-formal Reader: Cognition and Education.* Falmer Press, New York, pp. 351-377.

Isaacs. W., 1999. *Dialogue: The Art of Thinking Together.* Currency, New York.

Jenlink, P.M., 1995. Educational change systems: A systems design process for systemic change. In P.M. Jenlink (Ed.), *Systemic change: Touchstones for the future school.* IRI/Skylight Training and Publishing, Inc., Palatine, IL, pp. 41-67.

Jenlink, P.M., 1999, June. *Crossing boundaries, Changing Consciousness, Creating Learning Communities: Systems Design as Scholarly Practice in Educational Change.* Paper presented at the ISSS Conference, Asilomar, California.

Jenlink, P.M., Reigeluth, C.M., Carr, A.A., and Nelson, L.M., 1998. Guidelines for facilitating systemic change in school districts. *Systems Research and Behavioral Science,* **15**: 217-233.

Jenlink, P.M., Reigeluth, C.M., Carr, A.A., and Nelson, L.M., 2000. *Facilitating Systemic Change in School Districts: A Guidebook.* The Systemic Change Agency, Bloomington, IN.

Jenlink, P.M., and Carr, A.A., 1996. Conversation as a medium for change in education. *Educational Technology,* **36**(1): 31-38.

Leont'ev, A.N., 1978. *Activity, Consciousness, and Personality.* Prentice-Hall, Englewood Cliffs, NJ.

Kincheloe, J.L., 1999. Trouble ahead, trouble behind: Grounding the post-formal critique of educational psychology. In J.L. Kincheloe, S.R. Steinberg, and P.H. Hinchey (Eds.), *The post-formal reader: Cognition and education.* Falmer Press, New York, pp. 4-54.

Kincheloe, J.L., and Steinberg, S.R., 1999. A tentative description of post-formal thinking: The critical confrontation with cognitive theory. In J.L. Kincheloe, S.R. Steinberg, and P.H. Hinchey (Eds.), *The Post-formal Reader: Cognition and education*. Falmer Press, New York, pp. 55-90.

Russell, D., 1997. Rethinking genre in school and society: An activity theory analysis. *Written Communication*, **14**(4): 504-554.

Sidorkin, A.M., 1999. *Beyond Discourse: Education, the Self, and Dialogue*. State University of New York Press, New York.

SECTION IV

PRACTICAL APPLICATIONS OF DESIGN CONVERSATION

Chapter 16

THE NORTH END AGORA:
Design Conversation at the Neighborhood Level

MATTHEW A. SHAPIRO
Research Associate, CWA Ltd. and President, Mary Parker Follett Foundation

1. INTRODUCTION

Sustained and constructive conversation may prove to have its most powerful societal effect if it can be carried out among neighbors in the context of a common goal or challenge. Neighborhood – and its corresponding forms in the small town and village – has the following characteristics that can make it an ideal setting for conversation that adds value to the community and to the lives of its participants: diversity in a small area; familiarity with those who live around you; numbers that can make a difference; shared concerns that pull people together at a basic level; physical proximity, providing easy opportunity for planned and accidental encounters between people on a fairly regular basis; equality, insofar as there are no formal organizations or structures in which certain people hold authority over others; and a dysfunctionality or non-existence as social systems, which makes them "greenfield" opportunities for innovation and broad-based grassroots participation.

It was in the context of the North End neighborhood of Boise, Idaho that a "design conversation" was conducted over a period of one and a half years, the subject of which was the creation of an institution of communication and community-building called the North End Agora. The North End Agora represents what may have been the first attempted application of idealized social systems design on the scale of a neighborhood. The design conversation, carried out by a small group of neighborhood residents, included both generative and strategic forms of dialogue and was a powerful experience for its participants. It also led to the implementation of one aspect of the design, which was a regularly published "viewspaper" called the North End Community Paper.

The North End Agora experiment survived for only eight months after the initial design work was completed, but the lessons learned were invaluable and the design stands as a powerful artifact of what design conversation can produce. This chapter will describe in detail the experience of the North End Agora, with particular focus on the design conversation and the praxiological tension between design, implementation, and sustainability in the neighborhood.

2. THE NORTH END AGORA

The North End Agora was an institution of communication and community-building for the North End neighborhood of Boise, Idaho, designed by residents or other persons who identify with this neighborhood. Its design began in early 1997 and its implementation began in late 1998. Its lifespan after implementation was approximately eight months. The Agora was designed to have three major facets: a Media facet, a Conversation Pathway, and a Services Umbrella, briefly described as follows:

- The **Media** facet was the first to be implemented. It included the North End Community Paper and a web site. The paper was published monthly to semi-monthly and had a circulation of 5,000. It was supported solely by advertising and was delivered door-to-door throughout the neighborhood. Its length ranged from a 4-page flyer format to a 16-page tabloid. The web site was hosted at no charge by a local Internet service provider and was intended to complement the paper. During its existence it displayed only the markers and core definitions of the Agora.
- The **Conversation Pathway** aspect, which was to be developed according to community response, was intended to encourage and support a broad range of conversation activities (meaning any kind of activity in which people communicate richly and intensively), from block parties and pot lucks to study circles and salons to dialogue groups.
- The **Services Umbrella** aspect was to support the development of other services in support of communication and community-building that neighbors may have wished to create. Under consideration were a conflict-resolution service, a skills exchange, and assistance with facilitation of meetings.

This is the story of the North End Agora. The Agora was the prototype of a neighborhood institution for communication and community-building located in the North End neighborhood of Boise, Idaho. It was created – and was to have been continually re-created – through the process of idealized social systems design, hereafter referred to as "ISD." For us, the emergence of an institution of this kind represents not only the use of a powerful *method* of local creativity and self-empowerment, but the birth of a *design culture* and the seed of an *evolutionary learning community*. This kind of culture and this kind of community has the potential to be replicated with an ever-expanding scope and emergent form, possibly laying the foundation for a *designing democracy* of limitless possibility. But this will be true only if the creators of Agora succeed in cultivating broad ownership and stewardship for Agora that they create.

3. THE DESIGN JOURNEY

3.1 How It Began

In 1995, I was invited to participate in the Societal Design Team, a research group of the Asilomar Conversation on the Comprehensive Design of Social Systems. Through this fellowship, I was exposed to two of the three keys that would eventually lead to the design of the North End Agora. The first was the culture of social systems design, with its associated models and methods. The second was an emphasis on neighborhood. The latter discovery was not an explicit part of the Asilomar experience, but rather a result of reading Mary Parker Follett's *The New State*, which was required reading for the team.

In the summer of 1996 I began to focus more exclusively on neighborhood as the most fertile ground for both personal and societal transformation. Seeking to begin with a project that would help build familiarity among neighbors, I envisioned a neighborhood "atlas" for which residents of blocks would create their own pages. The people, not the physical features, would be the theme of this atlas. But while a number of residents did hear of and take interest in this project through word of mouth, I realized that there was no way to effectively reach all 350 square blocks in the neighborhood in this manner. Later on in the year, I realized that perhaps what we need to begin with was a medium of communication so that efforts like this one could be possible. We needed a neighborhood newspaper – or perhaps more than just a newspaper.

By late 1996 I had become deeply impressed with both the need for the capacity for design and for a revolution in the way we communicated with one another. Influenced by the ideas of Mary Parker Follett, David Bohm, and later Martin Buber and others, I sensed the need for community members to experience something other than broadcast communication and an emphasis upon superficial discussion and debate. We needed an interactive means of communication that had an emphasis on community-building, something that would bring people together face to face wherever possible and serve as an active medium that focused us on the roots of issues, on their interrelatedness, and that itself generated more varied and dynamic modes of interaction. This conception comprised the third key which, implemented in the neighborhood sphere and brought to life through idealized design, could open the door to the future.

By early 1997, I had interested a number of neighborhood residents in the creation of a "newspaper and more" for the neighborhood. None of these people were close friends; I had met them all through informal contacts and in different social contexts. One afternoon in February, eight of us gathered in the living room of one of the participants and introduced ourselves to one another.

We were about to create something that would serve to support communication and community-building for the neighborhood, and we knew that it would include a regularly published paper of some kind, but beyond that we did not know what it would look like. I introduced the concept of social systems design and suggested that if we were seeking something that would be authentic, dynamic, sustainable, and truly owned and authored by the entire community, then we should take the path of idealized design. The group agreed to try it.

The first series of meetings we spent getting to know one another better, clarifying our endeavor, learning the basics of design literacy and figuring out how to tackle some of the tasks that lay ahead. Then our design began in earnest, with the first stage of design: image-creation.

3.2 A Necessary Compromise: Limited Participation Design

One of the key principles of social systems design is that only the stakeholders of a system – those who serve, are served by, and use the system – can design that system. However, in the absence of a means of communication and community-building through which neighborhood-wide design could take place, and in the absence of a design culture or even an inclination toward design, this principle had to be, for the moment, compromised. Participation in the design effort was open to all from the beginning, and we ran an ad in the local weekly newspaper as well as invited people as opportunity arose. Regardless, the group remained relatively small through the implementation of the design. Conscious of this issue from the beginning, the first-generation designers endeavored to create a design that would be as open-ended as possible and would include as a central feature the fostering of a design culture.

3.3 Question-Finding and First-Stage Image Creation

One of the principles of design is that we begin by painting the largest possible picture on the largest possible canvas. Therefore, the first-generation designers of the Agora began with the creation of an idealized image of the North End as authentic community. I drafted the following triggering question, which I gave to each member of the design group inscribed upon a square of paper. Its wording was inspired by that which the Societal Design Team used during their CogniScope work at the 1996 Asilomar Conversation:

> What should be the markers of the ideal image of the North End as authentic community, one which is capable of empowering its members and serving a vital role to Boise, to Idaho, and to society in general?

While the *subject* of the above idealization would seem to be limited to neighborhood, the *scope* is in fact societal, because – as we discovered – the boundaries of local idealization may just as well be the ends of the earth. It seems that just enough of the world's concerns can have an effect in and be traced to a locality of sufficient size and diversity, *i.e.*, the quintessential *neighborhood.* To summarize this point, I hypothesize that *the neighborhood scope is ideal for image-creation because it brings the societal scope down to a human scale.* This said, I do believe that the specific crafting of the triggering question is instrumental in bringing about this effect.

When we were well into our design work, I reflected that there is an art to crafting the triggering question. The question should be designed to allow the greatest range of possible answers within the scope of the subject at hand. I feel that the questions we worked with accomplished that. In retrospect, I must say that crafting a triggering question is an activity that might best be shared by the group. I say this because the question to some extent shapes the answers, and that is a power perhaps best shared by all of the stakeholders.

There was some discussion early on about how to distinguish markers from design ideas. We discussed how a marker is a "what" rather than a "how," although we discovered that some "what's" can also be "how's." We also learned that markers can have different degrees of resolution. In other words, some are more specific than others. This can create a challenge when we are trying to categorize or compare markers, but we got through it.

We went through a round-robin process of articulating markers (core values and core ideas), each author explaining what they meant by what they were saying. It was emphasized that the person who authored a marker had the final say on its wording and its inclusion in the image, so that an atmosphere of safety and inclusion was fostered. By the time we were done we had about one hundred markers on our list. By the time of implementation of the design, the number had grown to 118. The following is a sampling of markers created during this stage:

1. PRESERVATION OF AUTHENTIC COMMUNITY

   Preservation of positive markers that make the North End an authentic community. If it is authentic, let's keep it authentic. This includes what already exists.

2. HIGH LEVEL OF AWARENESS OF INTERCONNECTIONS

   High level of awareness of interconnections (among everything): that we are all connected; everybody has an awareness that we are connected. We are many that create a one, and every individual and collective thought, action and intention has an effect upon that one.

3. GENERAL ACKNOWLEDGMENT OF CHILDREN AND INCLUSION OF CHILDREN IN COMMUNITY SITUATIONS

   Separating children out as little as possible and encouraging their creativity. Everybody can come into contact with children in a positive way. I feel that currently we separate children a lot.

4. YOUTH 14-18 YEARS OLD HAVE POSITIVE SUPPORT DURING THEIR TRANSITION TO ADULTHOOD

   Clear.

5. THE NORTH END IS LIKE A VILLAGE

   The North End has a sense of itself as a whole and that there is a significant degree of internal support. There are systems in the North End which make it relatively autonomous without taking away appreciation for interdependence with other neighborhoods and towns.

3.4 Dimensionality

I suggested to the group that we not begin with a set of categories or dimensions – such as the eight dimensions of Banathy's Evolutionary Guidance System (EGS) – but should instead allow them to arise from the set of markers. In this regard, I was influenced by Alexander Christakis. In my view, we chose to do this in the truest spirit of design. Not only would we create the elements, but we would create the categories as well. This is reflective of the constructivist and fluid nature of all structures and systems. Whatever dimensionality, geometry, structure or form they take is guided by functionality. Those that we created are fluid and could change over time as their functionality is tested.

Following the creation of markers, we wrote each one on an index card and spent one long meeting sorting them according to affinity, *i.e.*, how they addressed a similar aspect of the life of the community. The result was more than 20 clusters, which became the *dimensions* of the first image. The following is the current set of dimensions for this image:

• Validation of Self (featured below) • Family • The Community Family • The Greater Human Community • The Ecological Community • Belonging to Place • Networks	• Information Networks • Heritage • Human Diversity • Generational Relations • Ritual, Ceremony and Celebration • Wholeness Consciousness • Dialogue & Conversation • Basic Needs • Safety • Healthy Lifestyles • Economy & Business Ethics • Education & Learning • Business Ethics • Civic Participation • Democracy

3.5 An Intermediate Step: Exploring Influence Relationships

One of the exercises I was impressed with during our CogniScope experience at Asilomar '96 was the exploration of influence relationships among markers.

I thought that learning which markers have the highest leverage might help guide the rest of our design. I discussed it with the group – who at this point were willing to go along with anything – and we decided to go through this step using a sub-set of our markers. We determined that the number of comparisons would be calculated by n (n-1), and, unlike in a CogniScope session, we didn't have a computer to map our comparisons for us.

We decided that the minimum number of markers to include for sufficient coverage of the diversity of our image would be about 30. To select this set, each of us was allowed seven votes for markers that we felt were most important. 33 markers were chosen in this way. Surprisingly, they were drawn from 83% of the dimensions of our image. We wrote each of these 33 markers on large index cards and for the next month and a half we met once, sometimes twice a week to compare each marker with every other marker, both ways, asking "Would the realization of *this* marker significantly enhance the realization of *this* marker?" 1056 comparisons later, we had a 33 x 33 grid filled with red and blue squares, red representing "yes" and blue representing "no." From this beautiful tapestry that contains complexity beyond comprehension, we were able to glean the relative leverage and "centrality" of our markers.

The follow-up on this exercise was poor, but the true reward was the journey itself. The chart that would illustrate all of the influence relationships that we had explored was never completed, and although the results were interesting, I don't think that anyone felt that they were useful. We realized that this exercise was really an "optional" one for our design experience: it cost us one and a half months and a great deal of energy but it didn't add anything specific to our design. However, I feel that it was worthwhile because it gave us tremendous opportunities for generative dialogue and the exploration of the meaning of our markers. In order to make the pairwise comparisons, we often had to sweep in real-life experiences and imagine different scenarios. The end-result was discovery of relationships we would not otherwise have thought about. There was one marker at the bottom of our chart, which we apparently felt had the highest leverage: "Acknowledgment of spirit – we all are one."

3.6 Time for Consensus: Drafting the Core Definition

It was now time to complete our first-stage image creation. We did not have a great deal of guidance on what a "Core Definition" described in Banathy (1996) would look like. We decided that it would be an integration of markers, something that would transform a list of markers into a coherent description of an ideal while retaining the dimensionality of the image. The markers, dimensions, and core definition would all together then comprise "the image."

At this time I reflected on what a set of markers offers us and what a core definition might do to take us to the next level of design and community. Markers provide the author a space to express what they truly believe. Markers can exist side-by-side without particular relation to one another; they can even be at odds with other markers. But if it is to serve as the foundation for design, then there must be a place for convergence and consensus. This is where moral disagreement (if it exists, which it did not in our group, generally) must be transformed into ethical judgment, *i.e.*, placed into the context of diverse persons who must share, in real life, something they are creating in common.

We went through the markers one dimension at a time, drafting a prose statement for each dimension that would give a voice to every one of the markers within it. This was the first point at which we would have to come to a consensus.

The question of how to go about drafting the paragraphs of the core definition came down to this choice: do we have a series of meetings and all attempt to draft it together, or should one person sit down and make a first draft, with the group revising it afterwards? We chose the latter path. (Somehow this made me wonder how the Declaration of Independence and the Constitution were drafted, and I suspect that the Founders took the same route). I wrote the first draft, and gave each member of the group a copy. They brought their thoughts and notes to the next meeting and we forged major revisions in wording and placement until we were satisfied. We called the finished ten-page work a "Provisional Edition," as it always is. The following is one of the paragraphs from the Core Definition of our first image:

"HERITAGE

In an authentic North End community, we acknowledge and are aware of our history in both a local, recent sense and in a global, long-term sense. We see the past in our present as it becomes the future, and so act with many future generations in mind. We sense this on a personal level, knowing that we live on through our community, so that our lives are not bounded by our own physical birth and death. We keep what is most precious to us and pass it on to the next generation to do with what they will."

When we drafted the core definition, it occurred to us that the dimensions themselves must have some order to them. In other words, there were "meta-dimensions": a geometry through which these dimensions are related to one another. I posited the following meta-dimensions in an effort to provide complete coherence to the image: *wholes* (dimensions which represent things embedded within other things, such as self, family, community, etc.), *connectors* (dimensions which bridge things, such as networks and generational relations), and *actualizers* (dimensions which bring things about, such as democracy and civic participation). Some dimensions, such as dialogue and conversation, were designated as both connectors and actualizers. These designations are noted within the core definition.

3.7 Narrowing the Focus: Second-Stage Image-Creation

With the first image created, we had painted the backdrop. Now it was time to focus on the system itself: an institution of communication and community-building. We would go "back to the beginning" and create a second idealized image, that of the Agora. The triggering question we used, and continue to use, for our second image is as follows:

> What should be the markers of the ideal image of an institution of communication and community-building that can help bring to life our first image (of North End as authentic community)?

With reference to our image of the North End as authentic community, we generated another 100+ markers. The following are examples of markers from the Agora image:

1. THE AGORA HELPS BRING TO LIFE THE COMMUNITY'S IDEALIZED IMAGE THROUGH PRACTICAL SERVICE

   Like any human activity system created through idealized design, the Agora is a bridge between an abstract horizon of ideals and ways that we can approach those ideals in our daily lives. Thus it provides a practical service as well as enables practical service.

2. THE AGORA IS A SOCIAL SYSTEM

   The Agora is not just a medium or a facility; is made up of people and ideas, it has a purpose, a measure of how well it meets that purpose, an identifiable group of stakeholders, a clearly defined social environment and relationship with that environment, and identified designers and decision-makers.

3. THE AGORA OFFERS SOMETHING FOR EVERYONE

   It has enough variety and adaptability so that everyone can find something of interest and value in it. By "everyone" is meant all of its users.

4. THE AGORA ILLUSTRATES THE RELEVANCE TO PEOPLE'S LIVES OF THE IDEAS THAT ARE COMMUNICATED THROUGH IT

   Clear.

5. THE AGORA PUTS INFORMATION INTO CONTEXT AND PERSPECTIVE

   Information is not delivered in an isolated or abstract form. It is made useful by being placed within a bigger picture and in relation to other information.

19 dimensions emerged from this set of markers:

• Forum • Mirror • Systemicity • Service • Conversation • Empowerment • Healing • Story • Ownership & Stewardship • Inclusiveness	• People Becoming • Agora Becoming • Expansive Community • Working with the World • Ethics • Format • Presentation of Information and Issues • Attractiveness • Timing

We then drafted a Core Definition for the Agora image. It had taken ten months to reach this point.

3.8 Designing the Framework: Specifications

The next stage in our design was the creation of a set of specifications. These would be things that someone would need to know in order to build the system. In order to develop our set of specifications, three members of the design group independently went through the core definition of the Agora's image and extracted discrete elements that might be used as specifications. We convened, compared our results side by side, and selected a set of more than 100 specifications. In order to use this as a springboard for the next stage of the design, each specification began with a verb and ended with the word "*by* ... " The following are examples:

The Agora
fosters the articulation of a public philosophy *by* ...
helps us sound out ways to make the neighborhood run more smoothly *by* ...
helps people know the place they live in *by* ...
helps people know the people around them *by* ...
reflects what its community stands for *by* ...

3.9 Creativity and Pragmatism: Design Solutions

Now it was time to open the floodgates of creativity and pour forth the "how's." We worked alone and together to develop more than 300 ideas for how to fulfill the specifications. I reminded the group that we were still idealizing, yet inevitably some pragmatism would have to begin showing itself as we imagined specific ways of actualizing the design. As with the markers, some of the design solutions were of a higher resolution than others. In order to keep track of our design solutions and how they branched off into further design solutions, we created a "specification tree."

Not all of the individual specifications had design solutions attached to them, but there was some overlap and we felt that most of the significant specifications were covered. Once we had developed a reasonably large and broad set of possible design solutions, we began to research their feasibility and build a knowledge base that would later be used during implementation. Not all of the design solutions received equal attention, and in implementation, some remained unfilled. We also didn't spend a great deal of time looking for consensus on each and every proposed design solution, instead focusing on those which were the most compelling or were most pivotal to implementation. For those that did survive but remained unimplemented for any reason (such as lack of resources, or even just lack of initiative), our intention was to attach a notation to each one, explaining the "deviation from the ideal." Design solutions that the community does not feel are desirable would eventually be dropped.

The following is an example of a set of design solutions developed in response to a particular specification, in this case phrased in terms of the monthly newspaper:

fosters the articulation of a public philosophy by...

1. posing questions.
2. utilizing public conversation mapping.
3. linking to and being integrated with conversation groups.
4. asking questions that encourage people of all ages to think critically about the world they live in and themselves.
5. telling the truth and serving as a public voice.
6. conducting polls following discussions and then publishing the results, updating as necessary to reflect learning and change.

Each proposed design solution was written on an index card, and these we arranged according to the function that they seemed to be related to. The resulting clusters became the *function areas* of the Agora. Examples are Agora Design, Resources, Production, Organized Inquiry, and Crossing and Expanding Borders.

3.10 Creating the Organization: The Enabling System

The *enabling system* is the organization and the resources – including people, funding, and other organizations in the environment of our system – that would be needed to carry out the functions. In order to develop ideas for the enabling system, I suggested to our group that we look at each of our function areas and sub-areas and consider the "who, what, where, how, and when" as a guide for idea-generation. Because many of the ideas for the enabling system had already been generated during the design solution stage, it was largely a matter of organizing them appropriately and developing them fully.

3.11 The Complete Description: The Model

Late in the design work we faced the question of how to go about modeling the system so as to have a complete representation or representations to play with and test prior to implementation. The plan was to develop a model or models and to bombard them with questions – solicited from people outside of the group – and scenarios, so as to reveal under-conceptualization.

We decided to converge on a single model rather than create multiple possible models. This may have been due mainly to the limited size and resources of our group, but may also be attributable to the nature of the design approach we took. As for the testing of our model, the momentum toward implementation took hold and we were resigned to a trial by fire: our testing and learning would have to take place in the real world.

We interpreted the term "model" to mean a complete representation of the living system, or at least the system as it is designed to function. It would be somewhere between a Constitution and a blueprint for the Agora, and it would be highly structured for clarity, efficiency and ease of use. It was to describe in full the functions of the Agora and the organization by which those functions would be carried out. However, the written form of the model was only partially developed before the Agora effort ended.

3.12 Implementation

By the summer of 1998 we all knew that implementation was not far off. There was an emotional and psychological momentum building toward this, even if some desirable steps were being compromised.

Implementation naturally began with the Media aspect of the Agora – the newspaper in particular – and with the web site, since these would be the media that would communicate to the neighborhood what the Agora is, how it was/is created, and how it works. It would also help to bring people together to launch the other two major aspects of the Agora: the Conversation Pathway and the Services Complex (or Services Umbrella, as it was later called).

We sketched out the layout of the paper, determined roles that would need filling, designed a "priority cascade" for determining its content each issue, and settled on a legal infrastructure that would fit well with the design-based governance system that we had been designing. We began the legal incorporation process. Several of us stepped forward to fill, if only temporarily, some of the roles that were involved on both the corporate side and the staff side. Rather suddenly we set as our target date for implementation – signaled by the release of the first issue of the paper – the neighborhood's Hyde Park Street Fair at the close of the summer.

In order to meet this target, we had a great deal of work to do in setting up the business end of things, securing advertising revenue, procuring content and organizing layout, production, and distribution. Work on the web site was put on hold, as was the completion of the design documentation. This last step may have been a mistake, as it took some time for completion of the genesis design work to catch up to implementation once the latter was allowed to outpace the former.

3.13 Participation after Launch

After implementation began and resources began to shift into operating the Agora, the original design group lost its purpose. Some of its members were reconstituted in other functions of the Agora, such as the two corporate boards and the Stewardship Board. New people took on active roles. The area of broadest participation was one of the most mundane, yet one of the most vital: the delivery of the North End Community Paper. Within several months, more than seventy-five people were helping in this area. A growing number of people were submitting articles and perspectives for the paper. Some also stepped forward to offer ideas for the Conversation and Services aspects of the Agora. However, after eight months of operation, only a small handful of new people had entered into the design conversation itself. This was one of our greatest disappointments, but we hoped that the numbers would grow as more people came to understand what the Agora and its design were all about.

3.14 After the Agora: Seeking the Next Frontier for Design

The first-generation design group always felt that the design of an Agora (a social system for communication and community-building) by a small group of first-generation designers (necessary because there was no infrastructure – communicative nor cultural – for broad-based design) was an intermediary step toward the emergence of a design culture in the neighborhood. However, I believe that we may have mistaken the nature of that intermediacy.

We thought that as people both used the Agora and participated in its design conversation, the base of the design culture would grow. That growth was expected to be slow – my own hope was for just 100 people within the first year. But even turned out to be too hopeful. Every opportunity was taken to illustrate in the newspaper how the spiral of living design manifested itself in features or policies or conflicts. In every issue we published additional sections of the idealized images. We had posted the first half of the design documentation (including the idealized images and core definition) on the web site. And we provided clear notice of when public design meetings were being held. Yet little comment was received on this aspect of the Agora, and only two new people had come to the monthly public design meetings after three months of operation.

The paucity of participation led to the realization that perhaps the intermediacy of our work was other than what we first thought. The system that we designed seemed to be sustainable, at least; this is "success" for most people, but not for an idealizing system. The Agora was used and it was well-liked, and many people were helping support it. However, the design spiral within it was largely invisible, in spite of our efforts to make it apparent, probably because it seemed to be functioning smoothly to the casual stakeholder-observer. I speculated that there may be some sort of blind spot – perhaps due to low design literacy, or perhaps due to the persistence of the reification of institutions – that hides the relationship between individual and institution, and that this extended even to the Agora.

I realized by the third month of Agora life that it may be unrealistic to expect a functioning system that has been designed by someone *else* to ever be inherited as an idealizing system by its stakeholders, no matter how much the system was designed to be passed on to its stakeholders and forever-designing.

However, as the Agora began to function well, the spiral of living design comprising it could have been used as a designing system by its stakeholders, and in the *next* journey of design – that of some *other* institution – therein may have lain the fulfillment of the role of the Agora for extending design culture. Both the idealized image of North End as authentic community, and the Agora itself, would be essential to this role. And so the Agora would become the infrastructure for broad-participation design that heretofore did not exist. Then, in turn – in the manner of "bootstrapping" – the next design journey would further bring the first designed system (the Agora) into a more authentic spiral of living design.

What would be the subject of this next design journey? It had to be something that was within reach of the neighborhood by itself, because that was the next frontier of the designing community. I saw three near-term possibilities:

- The first possibility considered was the North End Plan. This is an extension of Boise City's Comprehensive Plan, and all neighborhood associations are entitled to re-draft it from time to time. Ours was old and up for re-drafting. Instead of following the conventional planning process, we could possibly use social systems design to create a comprehensive plan/design that is as much social in consideration, content, and impact as it is physical in terms of streets and houses. Unfortunately, the official neighborhood planning process is far too conventional and the idealized design would have had to be scaled back about 95% to be integrated into the City plan.
- A second possibility considered was a charter school, perhaps involving the design of a "community school beyond walls" that would integrate one or more of the existing neighborhood schools. Whether it would be accepted once submitted to the school district for approval was thought to be a matter secondary to the power of a

neighborhood-wide design process. With only one charter school being allowed per year from each area, however, and with the approval of a conservative school board required, the odds of implementation were deemed too long. Furthermore, a notice in the Agora paper about the idea of designing a school generated no interest.

- The neighborhood association, which was ripe for transformation. The idea was that it might evolve into a neighborhood council or assembly of some kind, a focal point for producing and acting upon a common will. With the leadership of the current neighborhood association, and its willingness to be transformed in the manner of a phoenix, this might have been the best opportunity for implementable design within reach of the neighborhood. However, the option was not pursued.

Mid-way through the Agora's life, we still hoped that in time the Agora, or any institution designed through idealized design, could actually be inherited *in toto* by its stakeholders given enough patience and imagination to make it happen. The "bootstrapping" theory seemed to be our best chance, and at least it would accelerate broad-based ownership of the founding system, the Agora. But the energy level of the Agora's founders dwindled and plans to launch new design initiatives were shelved.

By May of 1999, the Agora newspaper had gone from its original format of a 16-page monthly tabloid to a four-page newsletter published every other month. Although a new volunteer stepped forward to take my place as editor, there wasn't enough help in the critical function of obtaining advertising, interest in the design process continued to be low, and the core staff of the Agora became burned out. By the summer of 1999, the Agora quietly stopped operating.

4. EXPERIENCES AND INSIGHTS

4.1 Generative Dialogue

I introduced the concept of generative dialogue to the group very early on, because I knew that it would help to create the "glue" of shared meaning that would hold us together and serve as the foundation for our strategic dialogue. The experience of generative dialogue in our designing group was not relegated to special times or places. It was entwined with our strategic dialogue and entered into spontaneously. It might begin at the end of a design meeting, when we were sitting around exploring some local or global theme or issue, or it might have arisen during the exploration of the meaning of or inspiration for some marker or design solution. Or it might be seen during less formal encounters with members of the design group, at a cafe or stopping by at another's house – sharing a neighborhood makes such encounters easy.

Because of the nature of generative dialogue, it is sometimes hard to recognize when it is happening. I know that we experienced it. However, I feel that we should have spent more time with it, particularly in the latter stages of our time together. We later saw that it was essential that the design conversation be carried on in people's living rooms and around tables in cafes, because such settings are more amenable to generative dialogue than our monthly public design meetings were. Still, even in a formal setting, some generative dialogue was unavoidable when the subject is revision of the design spiral.

4.2 Time Frame

It took approximately five months to finish the first idealized image, that of the North End as authentic community. Creating the second idealized image, that of the Agora itself, took another four months. It took another six months to create the specifications and design solutions, and to begin planning the implementation. Within three months of that point, the Agora was functioning in the public sphere.

We did not keep track how many hours we spent in design meetings. During the image-creation phase, we usually met once a week or once every two weeks, each meeting lasting between two and three hours. After implementation began, the design pace became much slower, although maintaining the day to day operations the Agora was an almost continuous activity.

4.3 The Challenge of Transcendence

The difficulty of transcendence – in design language, the capacity to let go of a current mindset or image and to imagine another one – is illustrated in circumstances surrounding the departure of three of the participants in the first-generation design group. Two of them left after only two meetings. The first was not a resident in the North End but owner of a business in its historical Hyde Park district. At seventy years of age, he was much older than anyone else in the group and said both in private and in the group that he had a difficulty with idealization. He explained that he had lived through the hard reality of the Depression, and that "the world just isn't like that (potential ideal)." He suggested we focus on "how to pay for this thing" from the start, in spite of my best attempt to explain that we should figure out what we want before we figure out how to get it. He also felt that he might be discriminated against for his age, which I reassured him would not happen. But he declined to participate further.

Another early casualty was the former president of the neighborhood association, a scientist by trade who simply did not see the value in idealized design. He was also rigidly fixed on the idea that a good bus system was the only solution for the problems of today's community. He privately expressed to me his view that our effort would fail anyway, and suggested that we simply focus on creating a newspaper if we came together to create a newspaper – "why go through all this trouble?"

A third resident who left the design group was one who participated during much of the image-creation phase. He supported what the group is doing, yet expressed frustration at the gap between the idealization and the current or immanent. He had also adopted the view that conscious or organized efforts to improve the human condition are unnecessary because things will evolve in that direction naturally.

4.4 The Role of Leadership

One aspect of our design experience that needs to be carefully examined is the role of a single individual as facilitator, visionary, and user-designer. Design must be based upon integrated diversity and shared responsibility (stewardship), and to a large extent we achieved this. However, a disproportionate number of the original markers in both idealized images are attributable to my authorship. The chief reason for this, I felt, was the experience I had gained in idealization and marker-creation through my fellowship with the Asilomar Conversation's Societal Design Team in 1996.

It is true that everyone was given equal and continuous opportunity to put forth markers, and that in time the markers become the ownership of the entire community. In fact, we did sometimes forget who the original author of a marker was once it had become a part of our shared image. Furthermore, as a neighbor, I was a stakeholder as well as a facilitator in this design journey. Yet in spite of all of this, the volume of my own input and guidance was a source of concern, at least for me.

One result of my reflection on this issue was a realization that perhaps, in our collective creative and decision-making conversations, we need to focus on "plusquality" as well as on "equality." Individuals are different and bring different kinds of richness to the table. Follett uses the terms "plusquality" and "plusvalence" in her work *Creative Experience* (1924). I would define the former as a kind of status or relationship among different stakeholders or interests which focuses on the interactive, creative, emergent quality of difference rather than on sameness or a balance of interests. It recognizes the emergent quality of human interaction essential to democracy. Plusquality implies integrated diversity while equality implies side-by-side diversity, thus the former is more compatible with the valuing of diversity.

Another aspect of plusquality, I believe, is that there is always a place for situational leadership. It must be said that without leadership *as well as* stewardship, this design journey would not have reached the point of realization.

4.5 Communities of Design Practice

Etienne Wenger introduced the term *community of practice*, which means a kind of community created by the sustained pursuit of a shared enterprise, in whom collective learning is embodied in practices which reflect both the pursuit of its enterprises and the attendant social relations. Communities of practice are discussed mostly in the context of workplaces, but in the designing democracy will be seen at any level of organization.

One of the aspects of a community of practice is that when new persons enter into the community, they initially begin on the periphery, observing and helping rather than participating fully. In a community of basket-weavers in a village, the person on the periphery might be a child who is accompanying an adult while they are going about their basket-weaving activities and attendant social functions. In time they learn through observation and they by practice and social interaction, and become a full member of the community of practice.

In our group we observed the phenomenon of a community of *design* practice when after some months of work a new person joined. She was immediately critical of manner in which we were approaching a particular design step. There was some discomfort in this, as the rest of us had been together for some time and we'd laid the groundwork for this step through prior experience. However, we deferred to her suggestion and tried it. We soon found out that it was not working, and so by consensus we shifted back to the original course. Later on she admitted feeling bad about coming in so aggressively, and from that point on observed and listened more before speaking. In time became a powerful member of the community.

I introduced the concept of the community of practice and *legitimate peripheral participation* (the acceptability of remaining on the sidelines) to another person who joined in later, and during her first meeting she only observed what we were doing, helping carry out some of the support tasks. Shortly thereafter, we began to ask her input on certain questions, and soon after she became a full and active participant.

It is important to remember that the community of practice is not intended to deny fresh perspectives and differing opinions. It is simply an acknowledgment of human nature and the development pattern of a community that works together. An unhealthy or deterministic community of practice will have individuals enter from the periphery, acquire knowledge, become an insider, and then leave or die off to be replaced by another with the same knowledge. In a healthy and heuristic community of practice – what a community of design practice must be – individuals who become insiders must be allowed and encouraged to return to the periphery and gain new perspectives and insights that can be shared with the community. In this way, the culture will remain fresh and dynamic.

4.6 Design Documentation

We gave the name *design documentation* to the artifacts of the design conversation. These include the markers, dimensions, core definitions, specifications and design solutions, and the written model. We called these "artifacts" because the true, living product of the design conversation is the shared meaning among stakeholders and the living system itself.

We had discussed utilizing other forms of media to convey the design, in particular the idealized image. One favored approach was to display it in the form of a mural on the side of a local market, a mural which would evolve as the idealized image evolved. Theater and some kind of oral tradition were raised as additional possibilities.

4.7 An Unbroken Thread

For the design to mean anything, the integrity of a design spiral must be maintained. What this meant for us was that any stakeholder must be able to trace any aspect of the living system back to a marker in the idealized image of North End without skipping over a design stage. What appears "later" in the design spiral (*i.e.*, closer to the working model and the living system) *must* have a predecessor element in the design documentation, or else it may become an aspect of the living design without having a root in the idealized image or in the design conversation as a whole. What appears "earlier" in the design spiral *should* have an element that follows it in order to fulfill it, for the sake of a quality design. However, while it is a shortcoming if it doesn't, it is simply not going to be always possible to take the next step for that element because of limited imagination, limited resources, *etc*. The important thing, I noted, was to make a notation of it and to remember that its "on hold" status is always temporary.

To help in this "tracing of the thread," each element of the design carried a nomenclature. To be locatable within the design spiral, all the element needed was a designation of what type of element it was (markers, core definition dimension, specification, design solution, function area) and which specific element preceded it.

4.8 The Spiral of Living Design

Once we have an unbroken thread established from the idealized images at the core to the system that has actually been implemented – the living system – the design spiral is brought to life. We called this the *spiral of living design*. It works in both directions: the living system will change as the preceding layers are revised, and the preceding layers are revised as experience, imagination, dialogue, *etc*. in the realm of the living system give rise to initiatives for change.

Within the first several months of implementation, proposals for revision to the design had been made and circulated to the neighborhood for review. Among these proposals were several new markers for the Agora image, changes in the wording of two markers, a change in the delivery policy for the paper, and a proposal to drop a feature in the paper. The latter became as changes in Design Solutions when they were adopted.

5. LEARNING FROM OUR MISTAKES

The concept of the "mistake" has limited usefulness in our particular case because I believe that each "mistake" we made was observed and learned from. Thus I preferred to call them "opportunities for learning."

One such "opportunity for learning" was the use of secret voting during the early design process when we might have been better off with an open voting or consensus approach. After having created the initial set of markers of the idealized image of North End as authentic community, we were going to perform a pairwise comparison of markers in order to explore influence relationships and the relative leverage of markers. We couldn't use all 100+ markers, because that would involve almost 10,000 pairwise comparisons, so we sought to reduce the set by voting in secret for what each of us felt were the seven most important. This resulted in the selection of 33 markers, surprisingly covering 83% of the 22 dimensions that we'd created for them.

In retrospect, we could have relied upon our *demosophia* ("wisdom of the people") if we'd placed all of the markers up on a wall, within their respective dimensions, and taking turns selecting markers until we'd reached the desirable number to work with. That way we would have eliminated redundancy and achieved 100% coverage of the dimensions.

Another improvement, in hindsight, might have been to employ the three systems "lenses" (Banathy 1992) to fully explore and develop each design solution. We might also have utilized some kind of formalized choice-making process, such as trade-off evaluation, to prioritize our design solutions, and then employed a process to help us determine the best action sequence for our design solutions. These last two steps in particular would have helped us to develop a stronger implementation plan.

Probably the biggest "opportunity for learning" occurred when we went into implementation of the Agora without having completed the initial design work. A date for the annual neighborhood fair appeared and we saw this as an appropriate target date for distribution of the North End Community Paper, the first implemented aspect of the Agora. At this point, most of our energies were directed toward establishing the paper and producing the first issue. Unfortunately, there were several critical things were not yet finished:

- We had some not finished confirming the complete integrity of the design spiral by making sure that there were specifications and markers for every design solution being implemented.
- We had not gone through the design solutions that could not be implemented and made notations as to why they were not feasible.
- We hadn't finished drafting the Model, which was to represent the culminating description of the Agora as it would be implemented.
- The legal aspect had not been solidified: we were short two board members and had not yet approved bylaws of the two Agora corporations, and we hadn't yet applied for tax-exempt status for the parent corporation.
- We hadn't created the Stewardship Board yet, which, as it turns out, would have been very useful during a conflict faced soon after implementation of the paper. In fact, the Stewardship Board was never fully constituted.

- We hadn't prepared a city-wide publicity and education effort, taking advantage of earlier interest by members of the conventional media.
- We hadn't placed the design documentation (markers, images, specifications, design solutions, model) on the Agora web site, so that a large number of people with access to the Internet could begin to review the design spiral of their Agora.

Another major deficiency was the lack of an active feedback mechanism. We implicitly assumed that feedback would work through residents' simply expressing what they want to see, and that these desires would be incorporated into the living design. This did not happen, at least not through the formal channels we had created. In hindsight, we should have designed a feedback mechanism by which we would directly ask individuals whether they understood the concept of the Agora, what they felt about the markers and other design elements we publicized, whether they used and benefited from the Agora's features, what they would like to see in it, and how they might be encouraged to get involved. Steps were later planned to introduce such a feedback mechanism into the design, but they were not implemented.

Six months into the implementation of the Agora, it had become clear that implementation itself was our weakest link. We underestimated the number of people and the amount of energy that it would take to sustain key roles needed to maintain even minimal functionality for the Agora, let alone expand to the potential reflected in our idealized image. Due to this pitfall, the monthly publication has had to scale back dramatically; however, because the Agora had several other facets, we did have the option to shift energies into another facet and see how things progressed.

6. EXPANDING UPWARDS WOULD MEAN EXPANDING OUTWARDS

We recognized early on that "a rising boat must lift the tide." Just as the empowerment of neighbors depends upon the empowerment of their entire neighborhood, the empowerment of one neighborhood requires the empowerment of all neighborhoods. To this end, if the Agora experience succeeded, there would be an impetus within the Agora to extend the practice of design to surrounding neighborhoods. This might have been facilitated through the city's Neighborhood Alliance, through direct contacts with other neighborhood members, and through the Agora's own planned "Neighborhood Extension Service." We had provided copies of the of the North End Community

Paper to the members of the Neighborhood Alliance on a regular basis, receiving strong congratulations from several representatives.

There appeared to be at least two ways of expanding a community of design practice to other neighborhoods if and when it was created in the North End. One approach would have been to help to facilitate the design from scratch, from the first step of announcing an opportunity to join a "first-generation" design group and crafting a triggering question, proceeding through image-creation, core definitions, specifications, design solutions, etc. This approach was seen to be optimal in the sense that the new neighborhood would experience the unfolding of their design in a complete and original sense.

The second approach would be to transfer to other neighborhoods the image that we created in a manner that allows for authentic ownership and unique design. The other neighborhood might go through the markers one by one, discussing each in turn, clarifying them for themselves, and then choosing to retain or drop them. Then they might add markers of their own. The same pattern might be followed all the way up to the living design, or there might be a point at which they take off completely on their own. At which stage they do this – be it during the drafting of core definitions, the creation of design solutions, or later – would depend upon their inclination and our guidance as peer-designers.

The second approach acknowledges the likelihood that neighborhoods within the same small city will quite likely share similar ideals and will have reviewed the originating neighborhood's design spiral, thus being pre-disposed to "borrow." It also has the advantage of more easily producing a third, shared image that is founded upon elements common to both. This third image might be the foundation of a "super-agora" among neighborhoods.

6.1 Why Neighborhood?

6.1.1 Dysfunctionality/Afunctionality

The role of neighborhood as integrating social system is devastated or non-existent. This makes it more of an *opportunity* for design than an existing dysfunctional system, a system functional within its own context (as are most), and especially a system that is successfully functional within the larger community context.

Because neighborhood is a place which has by and large lost the village function, it is without structure and process, thus making it fertile ground for design and democracy learning. We just don't recognize the neighborhood situation as "crisis" because other forces have mediated neighborhood out of our lives. Television, the automobile, and distant jobs that are too demanding in our lives are among these factors.

6.1.2 Diversity

Enough people live in a neighborhood, in close proximity, to sweep in a critical level of diversity for rich co-creation and democratic learning, even if the neighborhood is relatively homogeneous in terms of ethno-cultural background and income level.

6.1.3 Commonality

Neighborhoods with identity do tend to reflect similar or at least compatible values, political attitudes, income levels, family units, and ethno-cultural composition. While we value diversity as strength, this commonalty provides a base for building based on what we already share.

6.1.4 Proximity

In neighborhoods, we live close enough to each other for convenient physical contact and the sharing of physical concerns related to transportation, water, air quality, refuse, aesthetic conditions, and challenges of nature such as natural disaster. We also share critical local institutions such as schools and market places. Encounters planned and accidental are also more convenient.

6.1.5 Equity

In the neighborhood there is a lack of hierarchy, in part currently because of a lack of social organization. While there may be wide differences in income level or ethno-cultural or home-owner status, there is a relatively level playing field when compared with existing social systems such as schools, health care systems, workplaces, and governments.

REFERENCES

Banathy, Bela H. *A Systems View of Education*. Englewood Cliffs, NJ: Educational Technology Publications, 1992.

Banathy, Bela H. *Designing Social Systems in a Changing World.* London, NY: Plenum, 1996.

Bohm, David. *On Dialogue.* London, NY: Routledge, 1996.

Follett, Mary Parker. *The New State.* London: Longman's, Green and Co., 1918.

Follett, Mary Parker. *Creative Experience.* London: Longman's Green and Co., 1924.

Wenger, Etienne. *Communities of Practice: Learning, Meaning and Identity*. Cambridge University Press, 1997.

Chapter 17

CONVERSATION AS THE COMMUNICATION METHOD OF CHOICE

Designing New Agoras for the 21st Century

DOUGLAS C. WALTON* and HALIM DUNSKY**
Saybrook Graduate School, San Francisco, California, USA
Bainbridge Graduate Institute, Brainbridge Island, Washington, USA

1. INTRODUCTION

Interest in the idea of civil society seems to have resurged lately, along with a corresponding interest in dialogic methods as the vehicle of its manifestation. This idea of civil society pertains to the hope for an ethical ideal of the social order that can offer an arena for harmonizing diverse individual interests with the social good (Seligman, 1992). Especially since humans have acquired the power to destroy the world, the imperative to establish the conditions for civil society seems increasingly evident to many individuals. Moreover, this interest seems to be intensifying, as more and more people become concerned that a haphazard cacophony of particularized interests is being played out in the public sphere—perhaps ultimately resulting in grievous and unrecoverable outcomes.

A number of social theorists are now urging humanity to take up the intentional guidance of its own cultural evolution, rather than allowing it to unfold randomly at the whim of self-interested desires (Banathy, 2000a; Elgin, 1993; Hubbard, 1998). At heart, this notion of *guided evolution* suggests an organization of public discourse that could engage diverse perspectives in the creation of a vision and design for social and institutional systems that might realise a widely acknowledged vision of a better society. It is a vision of a dialogic mode of discourse that could facilitate not only citizen-to-citizen collaboration, but citizen-to-government collaboration as well. Thus, proponents further conjecture that such discourse may result in a deliberative democracy where citizens, politicians, and governmental officials participate more intimately to create mutually satisfactory outcomes.

Banathy (2000a) has likened the establishment of a large-scale, intentional public discourse to the metaphorical restoration for the 21st century of the Agoras of ancient Greece. These *New Agoras* could embody a loose functional structure animated by an evolutionary design conversation. The design conversation, which would be conducted fundamentally within small groups of individuals, would serve to postulate an ideal future and to guide movement toward it. Simultaneously, an ever-widening consensus would be created among disparate groups regarding what that future might comprise and how it might be realized. The popularity of this notion has, in fact, been increasing in locations throughout the world.

2. THE ANCIENT GREEK AGORAS

The ancient Greek city-states are often cited as one of the first known examples of democracy. In these ancient settings, the citizens would often gather together in a public sphere, or *agora*, where discussion was held on matters of import to the society.[1] During that era, circa 500 B.C., citizen participation in the design and governance of society was both encouraged and expected. Held (1987, p. 18) explains that "the process of government itself was based on what Pericles refers to as 'proper discussion', i.e. free and unrestricted discourse, guaranteed by *isegoria*, an equal right to speak in the sovereign assembly."

This early Greek form of public dialogue became difficult to sustain once the population grew, dispersed, and became politically and economically more complex. Eventually, the Greek democratic system broke down and disintegrated, not to re-emerge for over two thousand years. When it finally re-emerged, it was in the form of the modern representative democracy, where communication was conducted via the proxy of elected officials (Roberts, 1997). Although the modern form of the representative democratic system has become the most successful—now dominant—political system in the world, it is also apparent that global society is facing increasing stress, perhaps to the point of breaking (Greider, 1997). This suggests that guidance systems, such as representative democracy, must evolve to meet the challenges of today's global systems conditions.

A revitalized public sphere may be possible (Banathy, 2000a). The construction of New Agoras becomes foreseeable because many of the scale, complexity, and geographic problems that crippled the ancient Greek democracies can now be mitigated by advanced communications technology. The creation of this new arena may offer much greater potential for citizen participation in social system design, particularly when the technology is used in conjunction with a systemic structure that coordinates the aggregation and systemic integration of the conversation. Moreover, as Ray (1996), Elgin

(1993), and Hubbard (1998) contend, the consciousness of global society may now be shifting, such that there are an increasing number of people who are motivated to engage in designing a better society. If this is the case, the task becomes to galvanize and concentrate the energy of those individuals into a force that can manifest the vision.

3. THE PUBLIC SPHERE AND CIVIL SOCIETY

The public sphere can be conceptualized as the arena where individual agents interact to harmonize particular interests with the universal interest of society at large. Typically, this interaction is played out in social settings. Eberly (2000) states, "Above all civil society denotes that sector of society in which nonpolitical institutions operate—families, houses of worship, neighborhoods, civic groups, and just about every form of voluntary association imaginable" (p. 7). To the extent that these interactions can operate in accordance with ethical ideals, the public sphere can be a setting for a stable, civil society (Seligman, 1992). When a civil society can also effectively adapt to existing environmental conditions, there is a dynamic balance that enables cultural evolution.

If we are to create and maintain the conditions for civil society, then we will have to establish means to continue the synthesis of the particular and universal values. According to Seligman (1992), the emergence of civil society in the West has been enabled by the movement to apply basic civil, political and human rights to everyone worldwide. This universalization of rights has resulted in an abstraction of trust, allowing individuals to relinquish reliance on traditional social groups. Nonetheless, Seligman also contends the abstraction of trust is fundamentally undermining its own foundation for two primary reasons. First, values are increasingly being removed from what is legitimate in public discourse—that is, public discourse, influenced by rationalism and scientism, increasingly strives to be more value-free and objective, with the result that all particularized values are treated as equal, without preference among them. Second, the more generalized trust is, the less confidence an individual has that his or her particular interests will be honored. According to Seligman, "Rights and trust thus stand in a complimentary [sic] tension with the other, where the latter is articulated on general or universal principles" (p. 188). On this basis, Seligman argues that civil society is only possible on the basis of preexisting and historically constructed moral values, and the notion of civil society lacks "any blueprint for the establishment of relations between social actors where such notion does not fully exist" (p. 179). The consequence may be decreasing control over, and increasing impetus for, destructively self-interested action—a specter noted at least as far back as

Plato's *The Republic* (Held, 1987). It may also compel a movement back toward reliance on geographically, culturally, or religiously allied groups.

Habermas (1990) has attempted to compensate for the separation of the private and public spheres via *discourse ethics*. In discourse ethics, discernment between legitimate and illegitimate norms is sought by conducting a rational discourse that seeks universal consensus on the basis of discursively valid knowledge claims. That is, claims are considered valid to the extent that the members of the community of discourse can agree on them. In principle, this assumes the availability of an "ideal speech" situation in which unconstrained, open discourse can occur. A number of Habermas' critics, however, claim that the *de facto* impossibility of the requisite ideal speech situation makes discourse ethics unfeasible; furthermore, Seligman (1992) contends that discourse ethics still does not resolve the fundamental problem. "Trust, as we have seen, arises from precisely those shared affective aspects of the social world which cannot, it would seem, be totally subsumed into the workings of a rational formula of linguistic pragmatics. (p. 195).

Seligman's (1992) concerns may be at least partially allayed by the introduction of *practical discourse* (Ulrich, 1983). As opposed to *theoretical* discourse, which seeks to establish the existence of such things as material phenomena and natural laws, practical discourse seeks to justify a course of action based on intersubjective agreement rather than semantic correspondence to phenomena. Ulrich contends,

> Practical discourse . . . must secure rational consensus on (a) what needs or interests ought to be given a legitimate chance for satisfaction, and (b) what negative consequences or side-effects can rightfully be imposed on those whose needs are not served. (p. 145)

Practical discourse, in Ulrich's (1983) view, also proposes that dialectical inquiry is essential to exposing the perspectives of all affected individuals. This process will correspondingly build trust among those stakeholders. The following subsections will show that evolutionary design dialogue can be viewed as a type of practical discourse. It accordingly provides a mode of inquiry that seeks to build consensus on a shared value base and guides conversation so as to expose and integrate various perspectives.

4. EVOLUTIONARY DESIGN AS DIALOGUE

Evolutionary design is a particular type of practical dialogue that focuses on understanding humanity's current evolutionary circumstance and attempts to envision, design, and implement a better society. Consequently, sustaining dialogue in evolutionary design must address several peculiarities. First, the design conversation may be protracted over many years, and, in principle, it continues throughout the life of the system. Second, goals in social systems design are generally complex, shifting, and interconnected. Third, evolutionary design is not likely to have the clarity of a single, well-defined objective. Rather, an ideal future state is envisioned, but that state is inherently indefinite and may change, as members of the community understand themselves and their intentions better and as system conditions continue to evolve.

The complex nature of the design problem actually calls for the judicious employment of distinct modes of dialogue to facilitate a free-flow of shared meaning regarding complex topics over a prolonged period of time. The use of multiple modes of dialogue enables the emerging meanings to be synthesized into a complex model of a new system. In particular, Banathy (2000a) emphasizes that design conversation contains two primary dialogic modes: generative and strategic. In *generative dialogue*, there is no agenda or topic; individuals seek to find common ground on the basis of shared values, perceptions and worldviews. Finding this common ground is an essential starting point for design.

Once the common ground is established, the team can engage in *strategic dialogue*. Strategic dialogue, when used in design, begins with the intentional development of conceptual frameworks and continues into deepening those frameworks and reflecting on them. The original frameworks are often modified by the subsequent work, creating a type of recursion. Thus, the process is as much transformative as constructive. The challenge is described by Ulrich (1983):

> If our goal is practical, we must first of all provide the social inquirer with helpful conceptual tools for grasping normative social practice . . . the corresponding task is then not to explain the sufficient conditions of perfectly rational discourse, but rather to guide the discourse participants toward reflection upon the inevitable lack of perfect rationality. That is to say, we face not so much a task of theoretical explanation as one of practical design—design for conceptual frameworks and social arrangements that together might provide heuristic support to inquirers in their search for discovery and deception. (p. 162)

The process of deepening the design from the initial abstract notions down to specific models of the future state often occurs in a spiral of increasing definition of the design (Banathy, 1996; 2000). As each loop of the spiral occurs, participants may pass through stages of divergent conversation by generating alternative solutions, then through a converging phase of selecting from the alternatives. These alternations of diverging and contracting conversation occur simultaneously while exploring deepening levels of the design, as discussed below.

- *Guiding image.* The designing team must first seek to establish some general image of the desired future state. This is often a collection of statements that describe the intended outcome, but may also include drawings and other representations.
- *Goals.* The goals of the system should be specified next. The initial goal set can often be divided into sub-goals, as well as related to higher goals.
- *Functions.* Functions carry out the goals. Once the goals are established, the functions that will accomplish the goal are identified. This may occur at multiple levels of definition (that is, functions, sub-functions, and even sub-functions of sub-functions.)
- *Components.* The components are the real-world entities that put the functions into effect. Since form should follow function in purposeful systems, components are not designed until after functions.
- *Guidance system.* The guidance system is constituted by the relational arrangement of the components. Guidance, as a function, may also have its own dedicated components.

In progressing through the spirals, there is a certain recursiveness where each new level of definition may at least partially modify the preceding definitions. Nonetheless, for the most part, progress is forward, and the downstream modifications do not normally radically change the preceding definitions. That is, vague and ill-defined concepts become successively more defined and concrete.

5. DESIGNING THE ASILOMAR STEWARD COMMUNITY

The practice of evolutionary design is perhaps best illustrated through the exploration of a real case. An example, which will be detailed in this subsection, is the design of the Asilomar Steward Community. This design effort occurred during the latter half of 2000 and resulted in a rudimentary "steward system," which was able to begin operation and enter an ongoing process of self-definition.

5.1 The Asilomar Conversation Conference

The design was conducted via a systems research team operating as part of the Asilomar Conversation Conference (ACC). The ACC is sponsored by the Asilomar Design Conversation Community, which is affiliated with the International Systems Institute (ISI). ISI is a non-profit institute whose mission is to promote design education. One of the organization's chief activities is to provide a venue for a new type of conference that is based on conducting design conversation, rather than on the presenting of theoretical presentations in a pedagogic format.

The Asilomar Design Conversation Community seeks to perpetuate one instantiation of this new conference design; the community's main activity is to facilitate an annual design conference that is held at the Asilomar Conference Center in Pacific Grove, California. A related conversation project, sponsored by ISI, is held in at Fuschl Lake, Austria, on even years. Additionally, ISI has coordinated research conversations in Crete, England, Finland, Greece, Hungary, and Spain.

5.2 The Conversation Program

Although participants only meet once a year, the Asilomar Conversation Conference is actually a year-round conversation program wherein a group of systems research teams focus on self-selected design problems. The conversation program contains four phases; each phase occurs approximately during the same months of the calendar year. A face-to-face conversation, traditionally held in the fall, is the highlight of the annual program.

The phases of the program are summarized below (Banathy & McCormick, 2000).

1. *Program announcement and team identification.* During this phase, individuals who wish to start a design effort propose a team objective and solicit members. Also, teams from previous years resubmit their announcements. Depending on whether the team is open or closed, an existing team may accept new members.
2. *Preparation.* Once the teams have established a mission and recruited team members, the team members begin communicating with each other.
3. *The Conversation.* In this phase, teams meet for a five-day, intensive conversation.
4. *Follow up.* After the conversation, teams may continue to work together. At the very least, teams usually submit output papers and may contribute some form of proceedings.

6. PROGRAM ANNOUNCEMENT AND TEAM IDENTIFICATION

All design efforts must start with an initial desire for change and an image of the preferred future state (Banathy, 1996). In the case of the Asilomar Steward

Community, the program was first initiated by Dr. Bela H. Banathy and announced in the formal newsletter of the International Systems Institute, *ISI Conversations*, as "Research Team E: The Agora Project – The New Agoras of the 21st Century." This initial proposal was a product of Dr. Banathy's twenty-year history of advocating social systems design, the guided evolution of society, and the development of evolutionary guidance systems. Specifically, Dr. Banathy proposed a design team that would further the work of The Agora Project, or the definition of "New Agoras," where public dialogue could occur (Banathy, 2000c).

The initial program announcement briefly described the metaphor of the ancient Greek agoras and the suggestion to resurrect them for the contemporary age. "We are seeking to build a community of 'Agora Stewards' who will [each] dedicate themselves to sharing responsibility for the collective design and implementation of the project," stated the announcement. The team's initial activity was targeted to "those Fellows wishing to commit themselves to become Agora Stewards, who, on their return back home, will initiate the establishment of an Agora evolutionary design community" (Banathy, 2000c, p. 4).

This announcement drew the interest of nine participants, all either professors or graduate students. There were seven men and two women. Most of the members were experienced participants at the Asilomar Conversation Conference. All had considerable experience in group work and dialogic methods.

7. PREPARATION

During the summer months, the preparation phase began. A preliminary website was constructed and conversation was initiated. The conversation covered several key topics, including introductions, pre-reading, input papers, and some focused discussions on specific topics. During this preparatory phase, Dr. Banathy coordinated the work.

7.1 Use of the Internet

The initial conversation was mostly "virtual," or occurring using computer-mediated communication. Two primary forms were used.

- *Listserve.* A *listserve* is a type of email list where messages that are sent to a designated email address are forwarded to the individual email addresses of all members of the list. Depending on the type of listserve, the listserve may archive messages and accept various commands from the users. In the case of Team E, the listserve merely forwarded messages that were sent to it.

- *Website.* Team E also established a website (www.21stcenturyagora.org) for presentation of public and shared information. At first, this website was quite limited and simply presented (a) the mission of The Agora Project, (b) a tiny knowledge base of relevant book reviews, and (c) some of the preliminary work that was done by the team.

The use of the website by Team E is mentioned here merely to highlight the possibilities of a website as a knowledge base for the work of evolutionary design teams and to suggest that it can be grown from a bare bones structure. Even at the present time, the website is undergoing considerable revision. The important aspect is that it exists and thus forms a basis for not only communicating existing information but for generating knowledge about effective approaches to using the Internet.

7.2 Virtual Introductions

One of the first team tasks was for members to create short introductions of themselves, their work, and their particular orientation to The Agora Project. Even when team members are already acquainted with each other, as was the case for some members of Team E, the practice of generating snapshots of each individual's current moment in life is an effective way to begin building a base of rapport and trust among all members of the team. Adequate establishment of rapport is essential to the success of the overall conversation.

The introductions varied considerably, both in theme and length. For the most part, they ranged from two to eight paragraphs. Some approaches were more biographical in nature, stating academic history and past experiences, while others focused on current interests, research questions, areas of study, books-in-progress, and so on. The introductions were all sent out on the listserve and most were also posted to the group's website.

7.3 Pre-Reading and Input Papers

The next action for the team was to read some foundational material and create input papers. A list of recommended material and book reviews of the material was posted on the website. These included the works of Hubbard (1998), Banathy (2000a), Elgin (1993), Ray (1996), Chaisson (1987), Salk (1983), and Csikszentmihalyi (1993). The input papers were typically either reviews of the books or relevant contributions from a team member's own research. The intention was to stimulate discussion prior to the face-to-face meeting of the team and to endow team members with sufficient knowledge for the Intensive phase.

7.4 The Systemic Structure of the Agoras

A number of discussions occurred by email prior to the conference, and one of the most influential topics was the initial definition for a "systemic structure" of the agoras. This became important during the intensive phase, as it circumscribed the functional space where Team E would concentrate their design effort. The basic structure of the New Agoras is outlined below (Banathy, 2000b).

- *Local Level.* The local agora is the fundamental grouping of the New Agoras. It is a group of approximately 7 to 12 individuals who are interested in creating a common, ideal future together. The local agora meets frequently face-to-face. Ideally, the group would have an *agora steward*, who brings well-developed skills for building agoras and conducting social systems design. This individual helps develop the evolutionary competence and literacy of the agora.
- *Steward Level.* The stewards of each local agora form a "second level" agora. This agora is a type system that enables the interflow of ideas between local agoras. Steward groups would also meet regularly, face-to-face.
- *Linkage Level.* The Linkage Level is the "top" of the systemic structure. This level exists primarily in cyberspace. It is the space for agora groups to interlink with each other, even though they are for the most part geographically separated. Within this top level are the technological systems and tools required to facilitate large-scale ideal image development, consensus, and action planning.

The systemic structure of the agoras describes a basic functional structure for the supporting dialogue in the public sphere. Each of the different levels performs particular functions and interacts in a loosely defined fashion. Thus, whereas the Ancient Greek agoras were a physical public space located in the center of town, global society, having grown too large for any physical space, must build its agora in a virtual public space.

8. INTENSIVE PHASE

The beginning of the Intensive phase was initiated by the arrival of the team members at the Asilomar Conference Grounds. The members then entered five days of face-to-face conversation. This conversation was critical to the program and was facilitated by a highly synergistic combination of factors, including the setting and agenda, the usage of generative dialogue, the dynamics of group-individual interaction, and the use of strategic dialogue in conjunction with the design methodology.

8.1 Setting and Agenda

Location is an important facilitator of community (Brown, Smith, & Isaacs, 1994), and the Asilomar Conference Grounds in Pacific Grove, California suited the requirement wonderfully. Set on 107 seaside acres of forest, dunes, and beach, the conference grounds mix a rustic, cozy setting with tremendous natural beauty that includes abundant deer and other wildlife, as well as stunning ocean views (Asilomar Conference Grounds).

The rooms are structured into lodges, many of which were constructed from redwood and native stone in the early 1900s under the architectural guidance of Julia Morgan. The lodges have many casual rooms, some with fireplaces, for team meetings; there is a main lodge and a main cafeteria that serve as open community places. The site offers plenty of opportunity for either self-reflective walks by the ocean, informal conversations over dinner, or chance meetings in the lodge while getting coffee. Consequently, the conference center tends to encourage a sense of cloister. The conference generally starts on a Friday evening and ends the subsequent Wednesday morning. A typical agenda is shown in Table 1.

Table 1. Intensive Phase Agenda

Friday Evening	3:00 – 6:00 Check in
	6:00 – 7:00 Dinner
	7:00 – 9:00 Opening meeting
Saturday to Tuesday	8:00 – 9:00 Breakfast
	9:00 – 12:00 Conversation
	12:00 – 1:00 Lunch
	1:00 – 5:00 Conversation
	5:00 – 6:00 Sunset walk
	6:00 – 7:00 Dinner
	7:00 – 9:00 Community event
Saturday to Tuesday	8:00 – 9:00 Breakfast
	9:00 – 11:00 Morning meetings
	11:00 – 12:00 Check out

8.2 Generative Dialogue

The members of Research Team E met face-to-face for the first time on a Friday evening at dusk. As the entire community of the ACC gathered together in the main hall to attend the opening talk, Team E members located each other in the crowd and made introductions. For some, it was the reunion of old friends; for others, it was the first sight of someone whom they had previously communicated with only via email. Of Team E, seven members—four professors and three graduate students—were present.

Beginning the next morning after breakfast, Team E met in the second story room of one of the lodges for the conversation. In accordance with well-established practice, the team began generative dialogue, spending the first few hours of the initial morning offering each person an opportunity to present his or her background and orientation to The Agora Project. While each member spoke, other members listened respectfully and, at points, asked clarifying questions. The rule was to maintain attention on each person in turn and ask clarifying questions only. Team members were watchful to not to "hijack" the conversation by introducing tangential topics.

Although a few hours might seem a long time for such a routine and elementary activity, the investment is part of the essential work of generative dialogue. Experience has shown that a significant number of team members afterward report that the experience of listening to other's backgrounds, and being listened to attentively, is a powerful rapport builder. This facilitates the essential building of common ground and shared understanding. The strengthening of interpersonal ties does not happen nearly as powerfully if one member simply comes in and reads off a set of ground rules.

8.3 Roles, Norms, and Disciplined Inquiry Mode

Once introductions were made, Dr. Banathy suggested that the group begin by considering "What questions shall we address?" He also emphasized settling on a disciplined inquiry mode and organizing group roles. In particular, he mentioned ACC custom, wherein the coordinator of the Preparation phase is removed from the role of coordinator during the Intensive phase. This has the advantage not only of distributing the work, but also of disrupting tendencies to relinquish power to a single person.

Initially, the obvious solutions were either to select another person to act as coordinator or alternate the role from person to person on some established basis. One team member suggested that perhaps the group could collectively manage the role without the need for a formal coordinator. There were a few questions about how this could work, after which the group adopted the proposal willingly.

Additionally, the group also assigned a scribe. The scribing method involved recording on-going notes on a laptop computer. Although creating a complete transcript was not feasible, many key ideas and verbatim statements were captured. These notes were eventually mailed out the group, but no attempt was made to clarify the recorded statements. Another team member agreed to transcribe flip charts into electronic format, and both the flipchart content and the scribe's notes were later emailed out to the team.

Following the assignment of roles, the group began discussing ethical and other guidelines for the group work. The initial topics included the need for total group participation and for respecting the contributions of everyone, as well as the following ethical guidelines:

- "Strive to be of service to humanity and take responsibility for the outcome of group decisions."
- "It is a fundamental human right of people to take part directly in making decisions that affect their lives and guide their destiny."
- "If social life is organized so that people can learn to exercise this right, then they can develop a competence that enables them to guide their own evolution and the evolution of the system in which they live and work toward their envisioned future."

After the establishment of guidelines, the group began a lengthy discussion regarding the creation of purposely evocative, or *trigger*, questions for the next phase of the inquiry. A long list of potential trigger questions was generated on flip charts. The flipchart pages included items such as *What do we want to accomplish within the next 4 days? What do we mean by an agora as an evolutionary design community?* and *What do we mean by 'conscious' and 'evolution,' and what do we mean by conscious evolution?* Consequent discussion about these questions involved defining the words "evolution" and "consciousness," as members of the group struggled to establish a common basis for what was being discussed. This became one of several topical strands that would periodically occupy the group's attention for the next two days.

Another strand involved values. This strand was initiated around the middle of that first afternoon, when Dr. Banathy commented, "The design is based on a common ground of values that must be established up front," after which ensued a wide-ranging discussion concerning whether design could be value free, how a steward group could avoid imposing values, and if universal values existed. None of these questions were resolved; rather, various perspectives were introduced.

At points, when the value discussion seemed to wane, an effort was be made to return to the question of how to proceed. This tendency was advanced in questions that inclined toward *What is our objective?* and *What shall we accomplish?* Eventually, this line of inquiry seemed to spur the introduction of the concept of the systemic structure of the agoras. From there, a third strand was initiated, as the conversation began to focus on what a steward system might do and how it might be of service to local agoras. Yet, at points, this inquiry line was again usurped by questions about values. A team member questioned "the fuzzy area of trying not to dictate values, while nevertheless offering a design that embodies our value system." Another member suggested that common values are found and made explicit, rather than imposed. Dr. Banathy suggested, "We should not look for compromise, but instead common ground. . . . Don't ask anyone to give up a value." Gradually, the group began to establish the notion that every design decision is related to the common ground, and one should not try to go beyond the common ground in design. Rather, a designing community can start with a narrow common ground and enlarge it, always ensuring that the design fits snugly to the value foundation.

Throughout the afternoon, the team groped and wrestled with these ideas as the three strands of conversation interwove back and forth. The process is worth a moment's reflection, for this sort of collective wandering often causes anxiety to individuals. In fact, this anxiety was at points expressed by more than one member of the team. Still, the team was willing to stay in this "creative tension" while endeavoring to reach the shared understanding that would be essential to success. It was nonetheless arduous, and by five in the afternoon, the group was exhausted and the meeting was adjourned.

8.4 Clarifying Purposes and Values

The next morning, having had time to reflect, walk by the beach, eat, and sleep, the group began with a check-in from all members. Each person reflected briefly on the previous day's dialogue and provided his or her current thinking about how to proceed. Emerging from this check-in, there was a renewed sense of clarity and energy, and the discussion tightened to focus more intently on the potential purpose of a steward system and how it would operate in relation to the local agoras. This topic persisted for several hours, until several trigger questions were proposed, including the following:

- "I am now engaged in an agora, what is the experience and what will happen?"
- "We are an evolutionary design community, what do we look like and what is our work?"
- "What will the Steward System accomplish in a year?"

Still, the three strands were not completely resolved, for the discussion of these trigger questions led back to topics related to values, including issues of ensuring that agoras have suitable human values and concerns about the possibility of reconciling values across agoras. Dr. Banathy suggested that "there is set of universal values that can be used and universally required, which are invariant across any agora." Similarly, the group affirmed the underlying assumption of an aspiration toward human betterment, with the caveat that "our values will be an example of, not a prescription for, values that others would hold, although we consider some of these to be generalizable to all groups and might require them for our participation."

Eventually, toward the end of the afternoon, there was an almost magical transition when the purpose and notion of what to do seemed to gel. If one was to trace the spoken conversation that got the team there, it would seem erratic and disconnected, but nonetheless a shared meaning emerged after a period of struggling with the issue. On the basis of that meaning, a general agreement on the purpose of the team was established: *The steward system would gather, filter, and present information that would be useful to creating local agoras.*

Having clearly articulated this definition, it was further decided that the team should be begin generating a set of common values. From that point on, the work developed a more focused nature of having a loose agenda and seeking to advance it. Strategic dialogue had begun.

8.5 Dynamics of Group-Individual Work

The group began a series of iterations that generated the foundation of the design. Each iteration manifest in what might be called a "group-individual dynamic." Generally, this dynamic started with gaining consensus on a trigger question. Once the trigger question was identified, each member, equipped with a self-adhesive flip chart page (or two) and a marker, spent about 30 minutes developing his or her own response to the question. Each individual recorded their personal responses on notepaper or laptop computer, then transcribed them to flip chart page. Once everyone was finished, the flip chart pages were affixed to the wall all around the room. Each member then presented his or her responses to the rest of the group. During the presentation, questions were often raised, and occasionally additions or modifications were made to the charts. Both the overlap and the complementarity among the contributions of the various group members were astonishing. No immediate attempt was made to reconcile them into a single, unified list. The sheets remained on the walls and subsequent responses were added after each trigger question was answered. Later, the flip charts were taken down and compiled, after the conference, by one of the group members, using a review and approval process that involved all participants.

This method of alternating between group and individual work seemed highly effective throughout the process. The technique avoided some of the problems that have been identified with general group brainstorming (Diehl & Stroebe, 1987) and thus allowed each member to develop his or her own ideas without interruption or discomfort about interjecting comments into group discussion. The contributions of each were honored and kept visible as the combined output of the whole, allowing later reference, reflection, and refinement of meanings.

8.6 Values, Functions, Components, and Tasks

The trigger questions that were explored in each cycle of the group-individual method were generated roughly along the lines of the Banathy's (1996) five spirals of the design architecture. Values were the first element to be identified. A sample of values, in their eventual consolidated form, are shown in Table 2. It was noted that the list contained both process and product values, where process values are those that pertain to the process of the operating a steward system and product values are those that pertain to the outputs of the steward system.

Table 2. Sample of Values

Value	Process	Product
Beauty	Self-reflection	Nonviolence
Balance	Self-realization	Reconciliation
Synergy/Symmetry	The common good	Humility
Freedom	Tolerance	Obligation
Responsibility	Morality	Giving
Uniqueness	Community	Honoring
Wholeness	Solidarity	Deep listening
Connectedness	Cooperation	Willingness to take risks
Internal Consistency	Respect	Dialogic democracy
Cultural diversity	Love	Emancipation
Biodiversity	Harmony	Self-determination
Authenticity	Flow	Participation

The next day, Monday, functions and components of the target steward system were developed using the same group-individual dynamic. Sample results are shown in Table 3.

In a further cycle, the group developed a list of tasks that could be performed in the first year of steward system development. Individuals volunteered to accept responsibility for various tasks on the list, although some tasks had to be downplayed due to lack of resources. A sample of the tasks is presented in Table 4.

By the time an action plan was established, it was Tuesday afternoon. As the ACC tradition is for all research teams to present findings to each other during a community meeting on Tuesday evening, Team E spent the remainder of their conversation time creating the presentation. Team E settled on dividing the presentation in to seven sections, one to be presented by each member. The presentation was given later that evening, which was the last group event of the Intensive phase.

Table 3. Sample Ideal System Statements

Steward System	• Works in service to the values of the Agora Stewards, in the individual and collective lives of stewards, and as an Agora Stewardship Group. • Continuously refines and revisions the steward system ideal model in relationship with a changing society and in relationship with Agoras across and within society, and moves toward that ideal. • Continuously derives achievable system goals from ideal system design. • Engages in constant self-critical reflection as individuals and as an Agora Stewardship Group, in order to learn, grow, and evolve. • Maintains itself: – Stays aware of who is showing up and monitors how well that fits the intended audience and system goals. – Conducts an ongoing conversation on the values, ideals, goals, and performance of the steward system. – Enables new stewards to become integrated into the steward system. – Provides mentoring and consulting to stewards. – Recruits, develops, and retains volunteer and/or paid personnel as necessary. – Provides for needed financial and in-kind resources. • Stewards share learning with each other and reflect on what has been learned, what has been created, and how well it is working.
Knowledge Base	• Develops a base of knowledge concerning methodologies and resources in support of the activities of stewards, Agoras, and Agora-like groups. • Identifies, gathers, filters, organizes, synthesizes relevant knowledge/data. • Conducts scholarly inquiry and original research; experiments, develops, tests, and refines knowledge. • Learns from experience and from other groups and schools: interacts with Agoras to learn what they do, how they work, what services/support might add value to their work, how (best) to provide the support, what scales the Agora concept is applicable to, and other related questions. Data types include: – Conversation records. – Case studies of Agoras and Agora-like groups. – Simulations. – Bibliography. – Web references. – Original research and theory reports.

Table 4. Sample Task Statements

1. Develop presentation for systems conference.
2. Develop revised website design and implement.
3. Coordinate the creation and submission of reflection papers to *ISI Conversations*.
4. Develop a proposal for a special journal issue on The Agora Project.
5. Coordinate the compilation and reconciliation of a group value statement and ideal system statement.
6. Develop resources.
7. Work on local agoras.
8. Create starter materials for local agoras by mining existing evolutionary inquiry literature.
9. Continue to develop theories, methodologies, and philosophy.
10. Create a statement for *ISI Conversations*.
11. Develop a richer statement of purpose.

9. FOLLOW UP

After departing from Asilomar, there was little team activity for about a month or so, probably in part because everyone was recovering from the conference—intensive dialogue can be exhausting. However, team members gradually began to work on their chosen tasks. Over the subsequent months, progress was made on a number of projects, including those listed below.

- Occasional status reports were submitted to the listserve by various team members.
- Reflection statements were sent out. Material was collected for a version of the *ISI Conversations*, reporting on the experience to other members of the community.
- The website was redesigned to better serve the functions of the steward community, as identified in the preliminary designs. Although the website remained a living prototype, it moved a step closer to becoming a serviceable tool for supporting The Agora Project. A trial "community space," which provided a password protected environment for sharing of files and hosting online discussions, was also established.
- Members began working together on shared book projects and shared articles for journals.
- Members began giving presentations on dialogue and The Agora Project at various conferences.

Although this work remains in infancy and is constrained by the fact that the members are volunteering their time and have other work, the progress does show ongoing development of the project. For example, what was once one steward team with a vague notion of how to proceed has expanded by Summer 2001 to include two additional teams, one focusing on developing the content initiatives and another on developing the design for the Linkage System.

10. TOWARD DELIBERATIVE DEMOCRACY

Design conversation has been demonstrated as the core modality of social systems design. In particular, evolutionary design has been posited as a type of rational public discourse that can facilitate the synthesis of the unique interests of individuals with an ideal vision for the collective good. Thus, widespread employment of this type of dialogue may contribute not only to the guided evolution of society but also to fostering the conditions for civil society in service of a more robust democracy.

The implication of a more robust democracy, in this context, is not the design of institutions or governmental structures, but of what is often called *deliberative democracy*. As noted by Weeks (2000), deliberative democracy is widely considered a possible remedy to some of the current ills of democracy. "The goal of deliberative democracy is to revitalize civic culture, improve the nature of public discourse, and generate the political will necessary to take effective action on pressing problems," contends Weeks (p. 360). In fact in, on the basis of a review of four large-scale trials using discourse methods to involve the citizenry in local government decision making, Weeks was able to conclude that "Using off-the-shelf social research methods, it is possible to convene a large-scale public deliberative process that enables local governments to take effective action on pressing community problems." This does not mean that perfect methods are available. Rather, it simply indicates that the possibility for effective influence exists, if appropriate methods are used.

In many ways, the work of Team E was highly illustrative of the possibilities for a powerful new dialogue in the public sphere. Although the work was rudimentary, it illustrated the potential for the self-organization of a functional component of the New Agoras, beginning with the usage of a disciplined mode of inquiry based on evolutionary inquiry. Certainly, many specific activities might have been done better or differently, but what is important is that, using skilled dialogic techniques, the team was able to take an abstract idea for human betterment, refine it, and deepen the vision into a design and an action plan. Of course, the plan is not final and neither is the model of an ideal state. Rather, a system was put in place that could continuously hone in on a better future.

Evolutionary design conversation is an important form of practical discourse that may be able to overcome some of the previously noted critiques of discourse ethics. As a heuristic methodology, design conversation seeks to continuously inquire about itself. Hence, it embodies a self-critical nature that can persistently root out sources of self-deception. This enables its usage in less-than-ideal speech situations. Moreover, in response to Seligman's (1992) argument that civil society is fundamentally undermined by the equalization of individual values and the abstraction of trust, the evolutionary design conversation may offer the missing blueprint for continuous trust building. Because evolutionary design groups (local agoras) conduct conversation in-person and identify shared values, they create their own, new bases of solidarity. The intensive dialogue experience then is in itself rapport building and conducive to value reconciliation and cohesion. The trust and good will gained through shaping the meaning of the values and engaging in co-creation further deepens that solidarity.

Clearly, a number of concerns could be raised, such as *This was done by academics and graduate students—but what is the potential for regular people to do it? When will people have time?* and *What about potential abuses?* However, the best approach to these questions may be to simply be aware of them, while at the same time recognizing that there are no universal or final answers. There is no basis of certainty that can be established prior to moving forward. In the postmodern world, we must move forward without the confidence of absolutes. Our challenge in encouraging human betterment is thus a question of pragmatic, not theoretic, discourse. Certainly, we need to be aware of what knowledge is available. However, rather than becoming moribund in studies that indicate what has historically been true, we must seek to find ideals that suggest what could be, or ought to be, the human condition. By seeking ideals, building systems on a set of shared values, and continuously reflecting on what we have accomplished, we can widen our horizons to discover better solutions and maintain adaptation to the changing conditions of our environment. Thus, it is not the content of an ultimate solution that matters, but the system dynamics that can enable a self-organizing public dialogue toward a better society.

NOTES

[1] Early Greek democracy is roundly criticized for relying on slave labor and considering only free Greek men to be citizens. Reference to this early democracy is strictly metaphorical and intended to illustrate the concept of citizen participation in public dialogue, not the implementation of ancient Greek governmental and social concepts.

REFERENCES

Asilomar Conference Grounds. URL: www.asilomarcenter.com .

Banathy, B.H., 1996. *Designing Social Systems in a Changing World.* Plenum, New York.

Banathy, B.H., 2000a. *Guided Evolution of Society: A Systems View.* Kluwer/Plenum, New York.

Banathy, B.H., 2000b. The systemic structure of agoras. Personal correspondence to Team E members.

Banathy, B.H., 2000c. Research Team E: The agora project – the new agoras of the 21st century. *ISI conversations*, **March**: 4.

Banathy, B.H., and McCormick, S., January 2000. A guide to the year 'round process'. *ISI Conversations,* **V1**(1): 10 – 11.

Brown, J., Smith, B., and Isaacs, D., 1994. Operating principles for building community. In P. Senge, A. Kleiner, C. Roberts, R. B. Ross, and B. J. Smith (Eds.), *The Fifth Discipline Fieldbook: Strategies and Tools for Building a Learning Organization.* Doubleday, New York.

Chaisson, E., 1987. *The Life Era.* The Atlantic Monthly Press, New York.

Csikszentmihalyi, M., 1993. *The Evolving Self.* Harper Perennial, New York.

Diehl, M., and Stroebe, W., 1987. Productivity Loss in Brainstorming Groups: Towards the Solution of a Riddle, *Journal of Personality and Social Psychology*, **53**: 497-509.

Eberly, D.E., 2000. The meaning, origins, and applications of civil society. In D. E. Eberly (Ed.), *The Essential Civil Society Reader: Classic Essays in the American Civil Society Debate.* Rowman & Littlefield Publishers, Lanham, Maryland, pp. 3-32.

Elgin, D., 1993. *Awakening Earth: Exploring the Evolution of Human Culture and Consciousness.* William Morrow & Co. New York.

Greider, W., 1997. *One world, ready or not: The manic logic of global capitalism.* New York: Simon & Schuster.

Habermas, J., 1990. *Moral consciousness and communicative action.* (C. Lenhardt and S. W. Nicholsen, Trans.). MIT Press, Cambridge, MA. (Original work published in 1983.)

Held, D., 1987. *Models of democracy.* Stanford University Press, Stanford, CA.

Hubbard, B.M., 1998. *Conscious evolution.* Novato, CA: New World Library.

Ray, P., 1996. *The rise of integral culture.* Noetic Science Review, Sausalito, CA.

Roberts, J.M., 1997. *The penguin history of the world (3rd Ed.).* Penguin Group, London.

Salk, J., 1983. *Anatomy of Reality: Merging Intuition and Reason.* Columbia University Press, New York.

Seligman, A.B., 1992. *The idea of civil society.* Princeton University Press, Princeton, NJ.

Ulrich, W., 1983. *Critical heuristics for social planning.* West John Wiley & Sons Ltd., Sussex, England.

Weeks, E.C., 2000. The practice of deliberative democracy: Results from four large-scale trials. *Public Administration Review, Jul/Aug 2000*, **60**(4): 360-372.

Chapter 18

A POST-FORMAL INTERVENTION STRATEGY

The Case of the Texas Educational Model

RAYMOND A. HORN, JR.
Assistant Professor of Education, Penn State at Harrisburg

1. INTRODUCTION

The implication of the phrase "from cradle to grave" is an apt metaphor for the pervasive and intense control of the state of Texas over the education of all children into their college years. For those individuals who become educators, the state control extends to the end of their careers. This control is paradoxical in relation to the state of Texas' often-stated commitment to local control of the educational process. Unfortunately for Texas educators, local control only involves the freedom to comply with the rigidly detailed state mandated educational standards and accountability system.

This compelling and highly politicized educational situation in Texas provides an excellent opportunity to examine a common but extremely important aspect of educational change. That aspect is the enervating pessimism experienced by educational practitioners who wish to make educational changes in the face of seemingly insurmountable state regulation. When the potential of best practice demands implementation in an educational system, the change agents who wish to implement the practice often quickly terminate their change effort when faced with extensive and pervasive internal or external resistance to their initiative. The Texas system, which is so rigidly defined and supported with extensive and effective accountability procedures, allows little room for curriculum, instruction, and assessment that differs from the state plan.

This situation in Texas is a good example of a situation faced by many educators who want to enact change but are constrained by intrusive and controlling external mandates. Many educators in Texas want to promote change in their schools but fail to implement the new theory and practice because of the oppressive externally imposed mandates. How then can a

Texas educator promote change within this rigid and controlling system? How can Texas educators resist the deleterious aspects of this system?

The Texas educational model provides an opportunity to apply post-formal theory as an interventionist strategy that can be utilized by concerned stakeholders to promote best practice, equity, and caring in an educational system that is uncompromisingly controlled by the state. The purpose of this chapter is to present a post-formal strategy that can facilitate effective change initiatives in our schools and that also can be used to ameliorate the deleterious effects of oppressive external influences on social systems. First, a foundation will be constructed by outlining the origins, context and patterns of educational change in Texas. This foundation will provide the context for the presentation of post-formal conversation as an intervention strategy.

2. EDUCATIONAL CHANGE IN TEXAS: ORIGINS, CONTEXT, AND PATTERNS

Educational change in Texas is neither the result of a series of piecemeal change initiatives nor a recent event. A post-formal inquiry traces the origins of this systematic design of the Texas system to the late 1960s (Horn & Kincheloe, 2001). As post-formal processes uncover the origins and context, patterns emerge. These patterns portray emergent properties and phenomenon whose value is contested by both those who favor and oppose the Texas model. This post-formal inquiry will first include a brief review of the origins and structure of the Texas education standardization and accountability model, and then include a presentation of the explicate and implicate patterns inherent in this model. Post-formalism will then be framed as an intervention strategy for those who wish to oppose the state's model or to ameliorate the deleterious effects of the model on their own school building or school district.

2.1 Origins and Context

On the surface, and as presented in the context of the presidential campaign of 2000, the Texas model appears to be a simple system of educational standards that are assessed by a standardized test. A simplistic summary of this system is, if students do not pass the test, than they do not graduate. A related inference, that is usually suggested but not accurate, is that this system of standards and accountability can be implemented fairly quickly, and represents, in educational terms, a relatively quick fix to the perceived problems of an incredibly complex social system. This simplistic statement becomes significantly complex and problematic when post-formally scrutinized.

2.1.1 Modernizing the Educational System

The Texas model is not a recent educational phenomenon. In 1968, during the term of Texas governor John Connelly, a committee was commissioned to research and recommend goals for public education in Texas. This act began a process to modernize public education in Texas. At this time the traditional Texas economy consisted of farming, ranching, and oil and gas production. Modernization of the educational system was necessary if Texas was going to expand its economic base to include the promise of industry based on new technologies. Simply, students had to learn more, stay in school longer, and be prepared for higher education. Proponents of this view argued that this would ensure a labor force consistent with the needs of the emerging technology.

From 1975 to 1979, standards were developed for math, reading, and writing, which became the basis for a criterion-referenced exit-test developed by the Educational Testing Service of Princeton, New Jersey. In the 1979-1980 school year, the first standardized tests were given to 433,520 students in grades five and nine (Texas Education Agency, 1980). The Texas Assessment of Basic Skills (TABS) was designed to be used as a diagnostic tool in meeting the needs of students. However, in the early 1980s, Governor Mark White appointed Ross Perot to chair a task force called the Select Committee on Public Education (SCOPE). "The committee's recommendations were included in House Bill 72, passed by the legislature in 1984, which resulted in increased graduation requirements, a no-pass, no-play measure for student athletes, the requirement for each student to pass an exit-level competency test in order to receive a school diploma, and a maximum five day absence rule per semester. In addition in 1984, House Bill 246 resulted in establishment of a statewide curriculum framework of essential elements of instruction established for each grade level and subject area" (Alford, 2001, p. 110).

2.1.2 TABS, TEAMS, and TAAS

With strong guidance from the business community, the Texas legislature continued the change initiative. In the 1984-85 school year, TABS was replaced by another assessment of minimum skills and a more difficult standardized test, the Texas Educational Assessment of Minimum Skills (TEAMS). TEAMS was administered to students in grades 1, 3, 5, 7, and 9. In 1990, the Texas Assessment of Academic Skills (TAAS) replaced the TEAMS, and the state senate made schools more accountable for student test scores by initiating a school and school campus rating system. Each school and school campus continues to be rated on student test performance disaggregated into categories for African Americans, Hispanics, Whites, and the Economically Disadvantaged. A system of rewards and sanctions accompanied the rating system. The TAAS test continues to be administered to students in grades 3 and 8 with an exit-level test in grade 10. Students who fail the 10th-grade TAAS test have until their senior year to pass the test; otherwise, they cannot receive a diploma. With the advent of TAAS, the test scores of the schools and campuses were published. In addition to the TAAS, end-of-course tests were now required for every course.

With TAAS, the emphasis on minimum skills changed to an idealistic emphasis on critical thinking and higher-order reasoning skills. The test questions were changed to include longer reading passages and mathematical word problems. The content of the tests (TABS, TEAMS, and TAAS) was based on the original standards called the Essential Elements (EEs). In 1997 the State Board of Education changed the standards to the Texas Essential Knowledge and Skills (TEKS), which were supposed to reflect the change in emphasis from regurgitation of facts to higher order reasoning. The TEKS standards are currently being phased in the TAAS test with 100 percent alignment of the standards and the assessment by 2001-2002.

The state legislature in 1999 mandated a new 11th-grade exit test along with the inclusion of algebra 1 and geometry in the math section. Also containing expanded content are the social studies and science tests. Social studies will include U. S. History, and biology, chemistry and physics will be added to the science test. In addition, the legislature required the construction of new 9th-grade and 10th-grade tests.

To monitor student achievement, in 1989 the state established the Academic Excellence Indicator System (AEIS). "This indicator system has been used to rate districts and campuses, and these ratings are the basis for rewards, sanctions, and public information. Districts are rated as Exemplary, Recognized, Acceptable, or Low-Performing on student performance by all sub-groups of Hispanics, Whites, African Americans, and Economically Disadvantged" (Alford, 2001, p. 119). Monetary rewards are provided for schools in the acceptable category if they meet all the sub-group performance requirements. The assessment of a school or campus is essentially focused on the base indicators of TAAS testing results, dropout rate, and attendance rate of the sub-groups. However, the system is more complicated in that there are other indicators such as student performance on SAT or ACT tests, the percentage of students taking end-of-course tests, students' completion of advanced courses, students' completion of the recommended examinations, results of Advanced Placement (AP) testing, results of International Baccalaureate examinations, and the equivalency between performance on exit-level TAAS and the Texas Academic Skills Program (TASP) test (Alford, 2001).

2.1.3 Sanctions for Poor Performance

Districts and campuses can be sanctioned for poor performance. As Alford (2001) notes,

> With the emergence of the AEIS system in 1990, districts and campuses are provided a school report card that is to be mailed to all parents. In addition, all schools must hold a school board hearing to publicize the AEIS report each year. Schools that are designated as low-performing must also hold a public hearing at the school to explain to parents and other interested individuals both the school's rating and the campus plan to improve the scores. (p. 120)

The first year a school is rated as low-performing a monitoring team from the state visits the school to review the campus plan. The second year an intervention team visits the school. The third year a monitor is assigned to the campus. In the fourth year a plan for closing or reconstituting the campus or school is developed (Alford, 2001). Accreditation reviews by the Texas Education Agency (TEA) are part of the District Effectiveness and Compliance (DEC) visits. "The DEC process is used to assess special program implementation, including special education, accelerated programs such as Title I and Migrant, Career and Technology, and Bilingual/English as a Second Language" (Lowery & Buck, 2001, p. 276). The DEC visitation is a source of anxiety afor school districts because of the potentially extensive ramifications for a school's curriculum, instruction, and organization.

2.1.4 The Inclusion of All Stakeholders

At this point it should be apparent that all public school students, teachers, and administrators are directly and significantly affected by this change initiative. This includes special education and Limited English Proficient (LEP) students. Initially, special education students were exempt from the TAAS test. However, beginning in 1996-97, special education students had to be administered an alternative assessment. In 1999, special education data was included in the rating of schools and campuses (Alford, 2001). LEP students can take a Spanish language version of the test, and its purpose is to assess the progress of the students in acquiring minimum skills while learning English (Alford, 2001). However, schools are discouraged from using the test with LEP students from 3rd-grade on (Alford, 2001). Also, LEP students must pass the English version of TAAS to receive a diploma.

Besides low-performing students, special education, and LEP students, economically advantaged and high-performing students are also affected. The inordinate pressure to "teach to the TAAS test" created a situation where repetition in basic content and test-taking skills has lessened the amount of class time for higher order curriculum and instruction. In most schools it has become the responsibility of all teachers in all disciplines, whether or not they are involved in the state testing, to allocate time from their classes to reinforce TAAS related instruction. For students who are well beyond the basic content and skills required by the TAAS, less is not more in their intellectual growth. Some parents of more skilled children are now starting to publicly condemn this perceived "dumbing down" of their children's curriculum and instruction. An additional dimension of the dumbing down process is the control over the cultural capital that can be attained by a disadvantaged child. When their school experience is dominated by the constant repetitious instruction of decontextualized facts, disadvantaged children cannot gain the knowledge and skills necessary to move above their social class. Therefore, all children are served poorly by, what in reality, becomes a transmissional curriculum and instruction situation.

Teachers are obviously affected by the TAAS test because of the need to restructure curriculum, instruction, assessment, and the allocation of school time to meet student performance standards. However, they are directly affected due to the Professional Development and Appraisal System (PDAS). Developed in 1995, the PDAS directly links teacher evaluation to student performance. This linkage is so strong that poor student performance on the TAAS, regardless of whether the students are in the teacher's class or are just in the school, results in a poor evaluation for that teacher. Administrator's, through a computerized profiler system, can access every student's TAAS scores on individual questions and compute student performance percentages for every teacher. Poor student performance on the TAAS can be traced to specific teachers through this system. Teacher effectiveness can be judged within this narrow context of student performance on a standardized test. Other positive teacher activities become subsumed by this highly restrictive focus on test scores.

Administrators are ultimately responsible for the TAAS scores in their building. If scores are low, administrators lose their jobs. Administrators also have the responsibility to accurately record all the TAAS and accountability data with the state. The Public Education Information Management System (PEIMS) is a statewide data management system for public education information.

> School districts must submit organizational, budget, financial and staff data to the TEA through standardized computer files. Other required data includes student demographic and program participation records, student attendance, dropouts, retentions and graduation information. This data is compiled and put together into the AEIS. (Lowery & Buck, 2001, p. 272)

It is also their responsibility to keep abreast of yearly changes in the standards, assessment instrument, and accountability procedures.

In 1990, non-educators were drawn into the model through the establishment of site-based management. As part of the move to create local control over education, councils containing educators, parents and community members were established to overseer the implementation of this model. This responsibility eventually moved to the campus level with campus level teams being charged with the responsibilities for improving curriculum and instruction within the guidelines of TEKS, TAAS, and the other accountability structures (Lowery & Buck, 2001). In most cases, the focus of the site-based management council is primarily on the TAAS scores.

In addition, in 1995 the state approved three types of charter schools: home-rule school district charters, campus charters, and open-enrollment charters. This was an attempt to give parents more options regarding the education of their children (Lowery & Buck, 2001). Also under the requirements of the state mandates, these schools can develop exclusionary tendencies in order to meet the mandates.

Even though they are elected officials, school board members are not exempt from accountability for student performance. The State Board of Education (SBOE) adopted "a framework for governance leadership to be used in continuing education for board members. A structured system of board member training, that includes team building, governance issues, roles and responsibilities of board members and administrators, and other topics based on needs identified by the board members themselves is currently in place" (Lowery & Buck, 2001, p. 275). School board member compliance is monitored by public disclosure of board members' activity in this framework.

2.1.5 Higher Education

Statewide control of education through testing and accountability procedures extends into higher education. The TASP is used to screen student entry into college undergraduate programs, as well as into teaching education programs. Once in college in order to be certified, prospective teachers must pass two state exist tests. One is the professional development version of the Examination for the Certification for Educators in Texas (ExCET) in either elementary or secondary education, and the other is an ExCET in their content area (such as math, science, special education). Regardless of their grade point average or other assessments, they must pass the ExCET to teach.

Principal and superintendency certification programs also terminate with the requirement of passing an ExCET. In other words, no one can be permanently certified to teach or adminstrate in Texas unless they pass the ExCET. This requirement is monitored by the State Board for Educator Certification (SBEC). Established in 1995, SBEC is a 15-member appointed board that oversees all aspects of public school educator certification, continuing education, and the standards of conduct. SBEC ensures accountability by monitoring the Accountability System for Educator Preparation (ASEP). Included in ASEP, besides the ExCET, are the Texas Oral Proficiency Test (TOPT),

Texas Assessment of Sign Communication (TASC), and TASC-American Sign Language (ASL).

In 1998 the educator preparation accountability system was implemented by SBEC. Based on their students' ExCET scores, educator preparation programs were rated as "Accredited," "Accredited-Under Review" or "Not Accredited." If a preparation program is Accredited-Under Review for 3 years, SBEC may send a visiting team to the university and terminate their ability to certify teachers and administrators. ExCET scores are disaggregated into categories of: all students, female, male, African American, Hispanic, and White. To avoid Accredited-Under Review status a certification program must meet the SBEC targets in each demographic group. If one group is low, the preparation program is under review with the potential to lose it's ability to certify teachers and administrators. This creates the potential for exclusionary entrance requirements as well as remediation for African-Americans and Hispanics who consistently score low as a subgroup.

2.1.6 Related Issues

There are other issues that are interrelated with the previously described standardization and accountability structures. The relationship of these issues with the standardization and accountability system magnifies the deleterious effects of the Texas model. For instance, while Texas is raising teacher certification requirements, the state is experiencing a major teacher shortage with predictions of a shortage in excess of 40,000 teachers. In addition, in proportion with the growing minority student population, Texas has a significant problem in providing an equal proportion of African American and Hispanic teachers and administrators. To offset the teacher shortage problem, the state has adopted an alternative certification program that temporarily allows anyone with a bachelor's degree to teach any subject in Texas schools for three years. By the end of the third year, permanent certification must be achieved or the individual can no longer teach. These individuals are paid a substitute teacher salary, which is lower than that of a permanently certified teacher. Besides increasing the labor pool, the cost of teacher reimbursement is lower for school districts.

While more accountability is required of the Texas teachers, Texas continues to be one of the lowest paying states for teacher service. Health care benefits for teachers lag below those provided for other professionals, and in the Houston Independent School District a recent three percent pay increase was offset by a comparable rise in health care costs. As the demand for more teacher professional development increases, the TEA strictly regulates and restricts supplemental pay to teachers for summer work. Another significant issue is that as more pressure is placed on university preparation programs, the state is funding non-university entities, such as educational service centers, to become certifiers of educators; thus, shifting needed funding away from the universities. Finally, despite legal action requiring more equitable funding of education in Texas, funding to schools and universities to facilitate this change remains low. These related issues even extend into the larger community with the indirect effect of the Texas standards on private schools (Harris, 2001), and the school TAAS ratings being included in the real estate industry's market strategies for home buyers.

Another pending support and accountability structure will have significant effect on the teacher preparation programs in Texas. "Texas has initiated a new support system for teachers - the Texas Beginning Educator Support System (TxBESS). One component of TxBESS is the TxBESS Activity Profile (TAP), initially called the Beginning Teacher Activity Profile in Texas (BTAPT). In its current form, the Activity Profile provides formative information to first-year teachers and also results in a rating for teacher preparation programs. This rating may be used in the future to determine whether teacher preparation entities are accredited and whether these institutions can continue to prepare teachers for certification in Texas" (Jasper, 2001, p. 1075). An important implication of this program is that after students graduate and take teaching positions, during their first semester as a teacher they will be evaluated. The evaluative information is used to guide the teachers' professional development, and to be used to judge the effectiveness of their teacher preparation program. Therefore, the preparation programs are responsible for how the teachers teach. If the teacher's administrator requires pedagogical techniques that are contrary to the TEA's expectations, the preparation program not the administrator, will suffer.

Now that an exploration of the origins and context of the Texas model have set the stage, further analysis allows patterns to emerge.

2.2 Patterns

Surface patterns indicate that Texas is finally doing something about its less than adequate educational system. This superficial interpretation based on rising test scores continues with the idea that making teachers, administrators, and students accountable for student achievement is making progress towards a quality educational system. The perception is that this is being accomplished by stopping social promotion, by making teachers work like professionals, by creating an efficient educational system, by forcing out students who don't care about their education, by promoting local control of education, and by allowing the business community to monitor educational policymaking. Because of the rising TAAS test scores, the assertion is that all subgroups of students are learning more and most important they are being prepared for the workplace of the future. Proponents of this view agree that there are problems, but they argue that these problems are a small price to pay for the eventual gain of an efficient educational system.

The previous paragraph essentially expresses the interpretations that are derived from a superficial inquiry into the Texas model. A post-formal inquiry problematizes this interpretive level. One pattern is the top-down centralized control over the whole system by the state. The often repeated mantra of local control is at best interpreted as letting the school districts and university officials have the freedom to meet the state mandated policies as best they can. However, this interpretation is contradicted by the state's immediate reaction to local procedures that are not acceptable to the state. For instance, to avoid losing their ability to accredit students, universities have adopted stricter entrance requirements for their programs. One especially effective requirement is to have the prospective student take a qualifying examination similar to the ExCET. If a target score is not reached, than the student is denied the opportunity to take the ExCET test. Recently the state has made it known that these tests are not desirable, and at best may be called diagnostic tests instead of qualifying or entrance examinations. Even though, state control attempts to subsume local control, the universities mask the qualifying exams by making them exit requirements in a capstone course.

Another pattern that quickly becomes evident is the lack of organized and vocal opposition to the Texas model or any of its components. Only since 1999 has opposition emerged. In federal district court a legal challenge (GI Forum 1999; GI Forum 2000) failed to secure an injunction in the state's use of the TAAS due to its alleged discriminatory effect on minority students. In addition, the Intercultural Development Research Association has consistently criticized the state over issues of discrimination, state funding, and student attrition (Cardenas, 1998; Cortez, 2000; Johnson, 1999). A few scholarly books have recently critiqued the Texas model. Linda M. McNeil (2000) documented the contradictions of Texas school reform, Angela Valenzuela (1999) described the deleterious effects on Mexican-American children, and Raymond A. Horn, Jr. and Joe L. Kincheloe (2001) provided a broad and deep perspective concerning the effects of the model on education from elementary through the educator certification programs. Besides the beginning of occasional parental displeasure as reported in Texas newspapers, there has been an almost lack of criticism or debate. Since Texas is a non-union state for public employees, teachers do not have the collectively negotiated support that teachers have earned in other states. Also, this pattern of silence is indicative of the pervasive power of the state in educational matters, and also over the individuals within the educational community.

Despite the silence, there is a pattern of resistance, but resistance in many ways contextualized by self-interest. The disastrous effects that poor student performance can have on one's professional career have prompted a diversity of creative responses by many educators. Quite simply, cheating has been the recourse of some (Low & Horn2001), while others have sought loopholes in the bureaucratic structure. One initial and effective loophole was to pre-test students and to label as special education those who had no chance of passing. Another was to create an absentee situation for those who would certainly fail the test by suspending them from school for the test day, or suggesting that they not show up that day. An additional strategy deals with "bubble students" who are the students who are closest to passing the test. These students are targeted for intensive instruction on test taking procedures and drilled repetitively to get enough of the questions correct to pass the test. A harsher strategy is to retain the worst students in the ninth grade. In this way the students have another year to prepare for the test, and consequently, they don't hurt the school's current rating. Unfortunately for the students, dropout rates are highest for students who are retained.

The more altruistic educator might try positive reinforcement in the form of pep rallies, prizes, trips, and motivational celebrity speakers (Bertrand, 2001). Standard fare is to pre-test the students to determine those in dire need of remediation and restructure their school day so that they get significant TAAS remediation. Most schools allocate time at the end of the day for TAAS remediation and require all subject teachers to devote the last 15 minutes of every class to TAAS preparation. Chants and slogans as mnemonic devices pervade the schools. Of course, there are many other techniques and strategies devoted to getting that good rating for the school. How is this pattern of teaching to the test and the use of motivational strategies an act of resistance? Whether to save one's job, or to help the children, educator's who utilize these strategies are resisting the state's paradoxical desire to develop critical and higher order skills in the children through the use of an assessment tool that requires the opposite of what these skills demand. The state assumed that after dictating a curriculum, an assessment tool, and accountability procedures to enforce them, instructional procedures would become aligned with the idealistic standards. This not only hasn't occurred, but the situation is exacerbated by the fact that the TAAS test is not aligned with the curricular standards (TEKS). Teachers who care deeply about their students are concerned about the lost instructional time to teach to the test, and are sensitive to the other deleterious effects of the situation, resist as best they can.

The strategies described in the previous pattern contradict what the state wishes to achieve. The state wishes to prepare children for the world of work, as it will be in the 21st-century. As we know, this requires facilitating the development of critical thinkers who can creatively work in collaborative and autonomous situations. However, the actual schooling that occurs better prepares students for the factories of the modernistic industrial age. This rank and sort, drill and fill, or banking model of education is disconnected from what the standards promote and from the real requirements of the technological workplace of the 21st-century—higher order and critical thinking skills.

This paradoxical disparity between the goals of the state and the outcomes promoted by the state's implementation strategy is also evident in the context of higher education. The standards that are tested on the ExCET tests represent research-based best practice. Seemingly, to pass the test students must show evidence of a knowledge of constructivist pedagogy and leadership, of empowering collegial collaboration, of child development, and of current best practice strategies in classroom and community management. In actuality, the state is reproducing industrial age management and instructional strategies by modeling a rigid hierarchical and hegemonic bureaucracy, and by facilitating the use of memorization and repetition in the instructional preparation for a high stakes exit-level test. In addition, by requiring an internship the state ironically negates the best practice that constitutes the test content by immersing the student into educational environments that don't represent best practice mainly because the administrators and teachers must teach to the test to survive.

Some professor's refer to this condition as "ExCET Land."

> In ExCET Land ideal teachers base instruction on their students' prior knowledge and facilitate learning through exploration and interaction with the environment; knowledge is constructed socially in a classroom that is reflective of a cooperative, democratic society. Technology is readily available and is integrated into the curriculum. Teachers foster a positive classroom climate that provides equal access for all learners and they take into account the varied needs and characteristics of students. Instead of producing "cookie cutter," look-alike students, individuality is addressed through use of a variety of instructional strategies and classroom activities that address learning styles and cultural preferences. Teachers build on the learners' knowledge by including the students' experiences and language and by incorporating relevant, real world contexts. Learners develop conceptual understanding as they explore, investigate, and discuss concepts. Furthermore, in ExCET Land teachers do <u>not</u> teach to the TAAS test, the public school exam that their students will be taking, because their students do well without a TAAS-driven approach, and because they believe learning should go beyond the facts-based drill and practice methods commonly adopted by those whose focus is on improving TAAS scores. (Abel, C. D., Abel, C. F., Alexander, V. C., McCune, S. L., & Nason, P. G., 2001, p. 198)

Of course, ExCET Land is an unrealized ideal because of the education system that is fostered by the state's insistence on a high stakes exit-level test as the sole assessment of the standards.

In other words, the state's rigid inability to allow or solicit professional feedback, their reliance on an exit-level test as the sole assessment tool, their modeling of a tightly controlled hierarchical system, their instrumental use of rewards and punishments, and their tight focus on statistical analysis of the outcomes results in a misalignment between what they want and what they are producing.

A final and very significant pattern deals with student attrition or the dropout rate. Prior to the initiation of the Texas model, the state of Texas had a tremendous dropout problem and is attempting to solve this problem. However, despite the state's assertion that rates have declined steadily for a decade, their assertion is invalidated by the inaccuracy of prior dropout data due to improper reporting by school districts and a change in the coding of dropouts (Horn, 2001a). In fact some reports (Johnson, 1999) indicate that the dropout rate has worsened since 1986 when the overall rate was 33 percent as compared to 42 percent in 1999. In Texas between 1995-96 and 1998-99, 53 percent of Hispanic students and 48 percent of Black students were lost compared to 31 percent of White students (Johnson, 1999). "Two of every five students from the freshman class of 1995-96 left school prior to their 1998-99 graduation from Texas public schools" (Johnson, 1999, p. 1). In the legal challenge to the TAAS, both sides testified that only 52 percent of the African Americans who entered high school in the ninth grade in 1995 graduated in 1998 (GI Forum, 1999). In his decision the judge reported that the plaintiffs indicated that "TEA's data shows that there is a very high correlation between the students' scores on the eighth grade TAAS test and their scores on the tenth grade TAAS Exit Test. Districts have an incentive to retain students in the ninth grade who are likely not to pass the TAAS Exit Test in the tenth grade in order to improve their tenth grade exit test scores" (*GI Forum v. TEA*, 1999, p. 18). As mentioned earlier, ninth-grade retention increases the dropout rate. To frame this problem in a larger perspective, from 1985-86 to 1998-99, more than 1.3 million students have been lost from Texas schools due to attrition (Johnson, 1999, p. 2). In Travis county, which is home to the University of Texas one of the premiere flagship universities, in 1998-99 student attrition rates were 60 percent for Blacks, 34 percent for Whites, and 69 percent for Hispanics (Johnson, 1999, p. 8).

The student attrition pattern is the most damaging aspect of the Texas educational model because of the direct deleterious influence that it has on such a large number of individuals over their whole lives. The cultural capital that is denied these individuals has a harmful effect that transcends generations. This pattern of high TAAS failure rates and dropout rates for African Americans, Hispanics, and economically disadvantaged Whites, in relation to the fact that the TEA has held its course and these groups of people have gone unattended over the last decade, indicates a disregard by the TEA and the state legislature for these population groups. This pattern raises the question as to how could human cost of this magnitude become an acceptable part of a change initiative?

An analysis of the origins and context of the Texas model discerned patterns that indeed indicate that the phrase "from cradle to grave" is an apt metaphor for the pervasive and intense control of the state of Texas over the education of all children, not only into their college years, but for many minorities and disadvantaged well into their adult lives. The purpose of this discussion is to explore post-formalism as an interventionist strategy, but intervention implies that there are those who are dissatisfied and are motivated to take action. In Texas, there are teachers, administrators, parents, university educators, students, and many others who are motivated to resist or evolve the Texas educational model. What intervention possibilities does post-formalism offer to them?

3. A POST-FORMAL INTERVENTION STRATEGY: DESIGNING CRITICAL CONVERSATION COMMUNITIES

What is lacking in Texas is critical conversation about their educational system. No where in the web of Texas communities is this kind of conversation taking place. Therefore, the system goes unchallenged, and consequently, only evolves systematically in accordance with the core values of those in control. How then can the development of critical conversational communities that can challenge this control be facilitated by post-formalism?

3.1 Setting the Stage: The Necessity of an Ecological View

An ecological view is inherent in the post-formal process and therefore guides a post-formal view of community intervention. An ecological view sees community intervention as systemic, in that all parts of the community are interrelated and interconnected. This view sees a network of interrelated subsystems nested within the larger suprasystem. An ecological view also recognizes the iterative nature of core values and behavioral patterns throughout the systemic network. As proposed by chaos theory, at every level in a system there are repetitive versions of the same patterns. As in the self-similarity of a fern or cauliflower, so is the repetitive patterning discernable in human systems. For instance, in the Texas model, the characteristic patterns within the controlling bureaucracy are repetitively reproduced in similar versions in schools and universities across Texas. For instance in Texas, the state's rigid focus on student performance on standardized testing becomes the focus of most, if not all, Texas school districts. What is also replicated is the devaluing of those practices deemed secondary or peripheral to this focus. Practice resulting in critical thinking, higher order thinking, and empowerment of others is relegated to the periphery. Schools that persist in maintaining a primary focus on these more complex and emancipatory practices soon realize that they risk being penalized for this persistence because of the potential for low test scores. If more complex pedagogical goals are being realized but student performance, as defined by the Texas standards and accountability test, lags, then the other success cannot offset the penalties imposed on the educators by the state and perhaps the community.

Another aspect of an ecological view is Goerner's (1999) idea of intricacy. Goerner proposes that since all organizations are flow structures, disruptions result in self-organization into more intricate patterns. Recognizing the intricacies of self-organization is significant in that when individuals realize that intervention can force reorganization, than conscious evolutionary change in human systems becomes possible.

Hope is created in those who desire change when they realize that systems are self-organizing, and therefore have the potential to change. This view of the self-organizing potential of human systems also facilitates an understanding of the organizational mind and the orders of consciousness represented within the system. Getting into the "mind" of the system and exploring the different orders of consciousness or the different ways that those in the system view the world is inherently empowering.

A critical aspect of an ecological view is the inclusion of social justice and caring as the centering component in the systemic assessment of the power arrangements pervasive through the relational web of human systems. Banathy (1996) sees our current problems in the context of an evolutionary crisis, a crisis of consciousness. Post-formal inquirers further explicate his concept as also a crisis of conscience (Horn, 2001b).

Another dimension of the ecological view deals with the skills and knowledge needed to see ecologically and benefit from the ecological view. To see broadly and deeply requires a post-formal excursion into the dimensions of human activity systems that ubiquitously pervade the system. Conversation is one of these dimensions. Conversation is ubiquitous in human systems. Knowledge about the types of conversation and the ability to engage in the different types is essential in understanding and intervening in human activity. Obviously the lack of criticism of the Texas model indicates a predominance of certain types of conversation, but also provides an entry point for criticism. Another dimension that pervades human systems is the potential of and for critical reflection. When knowledge and skill of critical reflection are imparted to individuals, the foundation is laid for a continuing skepticism. For instance, as parents of advantaged students protest that the inordinate emphasis on TAAS takes time away from more complex curriculum and essentially "dumbs down" their children, facilitating the parents' ability to critically reflect will allow them to consider the negative effects of TAAS on other children of different socioeconomic, racial, and ethnic groups.

An ecological view requires the inclusion of community. Community building is a dimension of human activity systems that is always present in some form. Knowledge of and skill in building egalitarian communities opens the door for critical concerns to become part of the activity system. To their great credit, many Texas schools have built school communities to answer the TAAS challenge. Through the introduction of knowledge and skill about conversation and critical reflection, these existing communities can be critically enhanced, and therefore, become sites of resistance to the oppressive elements of the Texas model.

Additionally, individuals within a human activity system need to know about human systems. They need to understand that systems can be ideally designed; that a vision of the ideal can be achieved. They need to have the skills associated with systems thinking to achieve their ideal. Also, knowledge and skills about leadership are essential in weaving these dimensions into a coherent whole. Leadership must be developed that is egalitarian and caring, pragmatic, ethical and moral, empowering of stakeholders, and instructionally effective. Ideally, in order to discern the deep and broad patterns within the human activity system, other post-formal processes, as detailed in the earlier chapter, would be introduced and used.

The questions are raised as to who needs to know all of this, and how much knowledge is needed to post-formally intervene? In an ideal world, the initial designers of the intervention would be knowledgeable and skilled in conversational theory, systems theory, leadership theory, postpositivistic strategies, semiotics, affect control theory, and of course have a critical center mediating all of these areas. As the design team becomes larger and more stakeholders become part of the initiative, different levels of knowledge acquisition and skill development in each of the areas would be evident. However, the key is continuous growth in the stakeholders' knowledge base and skill levels. In a less than ideal world, the initial group who is designing the intervention will not have a substantial knowledge of all these areas. However, once again the key is a commitment to growth in knowledge and skill. With this commitment to growth, how then do these individuals design a post-formal intervention?

3.2 Designing Educational Change

How is a post-formal design different from other systems design models? What does a post-formal design have to offer that is different from the other models? This section will answer these questions by critiquing three systems design models and then presenting the post-formal change model.

3.2.1 Idealized Systems Design

Bela H. Banathy (1991, 1992, 1996) has promoted an idealized systems design (ISD) model that attempts to create a consciousness that focuses on

"What Should Be" (1991, p. 27). Basically, this model entails the development of a vision, or "the grand ideal, the underlying philosophy, the inspiration that guide imaging" (1991, p. 27). After this first step an image is constructed, which embodies "a system of core values and core ideas that elaborate the vision and guide the system design" (1991, p. 27). This image is essentially a cognitive map of the desired future and leads to the third stage, which is the design or the model of the ideal system. The essence of Banathy's design is modeling. Models are constructed to view the system from different perspectives. This enhances the likelihood of realizing the vision. Critics of this type of systemic change point to the fact that "few, if any, examples of anyone using ISD to design an educational system exist" (Squire, 1999, p. 634). Kurt Squire (1999) reports numerous barriers to the employment of ISD ranging from agreement among stakeholders concerning the need for change to the large amount of time and energy that needs to be expended. Also, in a "quick fix" society, a process that may take years to culminate in a tangible outcome is hard to sell to those who want to see immediate progress. Is there a variation of ISD or can the ISD process be modified to become more accessible to practitioners?

3.2.2 A Guidance System for Designing New K-12 Educational Systems

Patrick M. Jenlink and Charles M. Reigeluth (2000) propose a guidance system for addressing the needs of a systemic change. Their guidance system is an ISD process that incorporates the ideas of a human activity system (Checkland, 1981) in which "structured sets of activities [act] as a notional system that expresses some purposeful human activity focused on creating an ideal system" (Jenlink & Reigeluth, 2000, p. 8). Activity theory (Engeström, Miettinen, & Punamäki, 1999) also informs their guidance system, in that the systemic change process "is mediated by cultural tools (physical, psychological, and process) such as ideal systems design technology, systems language, systems modeling, design conversation, process facilitation, and related types of tools" (Jenlink & Reigeluth, 2000, p. 8). In this way, discrete and continuous human activity can be deconstructed and aligned to facilitate the systemic design process. They propose the following phases of discrete events: Phase I – Assess Readiness and Capacity, Phase II – Prepare the Initial Core Team, Phase III – Prepare the Expanded Team, Phase IV – Design a New System, and Phase V – Implement and Evolve the New System. There are numerous continuous events in their guidance systems that by nature are ongoing in that the participants continuously engage in these activities. A few are: evaluate and improve the change process, build and maintain political support, develop and sustain appropriate leadership, build trust, develop group-process and team-building skills, engage in reflection, and build and evolve community.

This guidance system is a detailed template for those who desire to engage in systemic change. Of course, some of the previously mentioned barriers to ISD oriented systems apply to this guidance system. Since this change system requires forethought, process, and discipline, it is not appropriate for those who need a quick fix. Jenlink and Reigeluth (2000) also identify other people who would not find this system appropriate. They stress that "the guidance system is designed primarily for people who want to facilitate systemic change in education" (p. 3). The significance of their statement is that their change model is *only* for people who: recognize that systemic change requires going beyond the classroom or school building to include district-level changes; recognize that the participation of all groups in the community is required; realize that systemic change is difficult and want some guidance; and, realize that systemic change requires a change in the collective consciousness and worldviews concerning how social systems function (Jenlink & Reigeluth, 2000, p. 3). Obviously, people who do not want to promote wide spread change, do not want to allow participation of all stakeholders, do not want external help in the process, and do not want to change the way they think about the world will not find this model attractive.

As in all ISD models, guiding beliefs or core values are integral to any systemic change effort. Jenlink and Reigeluth identify twenty-nine guiding beliefs that are essential if this system is to be employed successfully. In fact, they succinctly state that "this guidance system is not likely to be appropriate for facilitators and stakeholders who do not hold these beliefs" (Jenlink & Reigeluth, 2000, p. 4). Undoubtedly, various individuals could contest each of these beliefs; however, some are inherently conflictual with the beliefs of certain individuals. For instance, the first belief is expressed as caring for children and their future while two other related beliefs are creating a socially just system, and being socially responsible for all members of the community. In relation to the Texas educational situation, the dropout rates for minority and disadvantaged children that have been tolerated for decades, and the restrictive practices resulting in the decreased involvement of minorities in education are not indicative of a caring attitude about children. Individuals who consciously contribute to these deleterious outcomes for children will resist any change that opposes their own interests.

In addition, the holistic view required by ISD would be deemed inappropriate by administrators and teachers who need to "own" and "control" their "territory." A school district that is organized like a feudal system with the principals functioning as medieval Lords ruling their kingdoms in a state of sublime autonomy, would view any intrusion into their domain as a threat. Another belief that is problematic for certain kinds of educators is inclusivity. Allowing substantial participation of all types of stakeholders challenges many practices and beliefs of those who find security in strictly regulated hierarchical systems. Also, the equitable participation of a school's community in educational decision making is in many cases opposed by the buffering functions that are inherent in schools (Ogawa, 1998). Another downside for some would be the fact that inclusion opens the door for significant stakeholder empowerment, which contradicts the traditional hegemonic and hierarchical organization of schools.

The last reason for resistance to this change model is self-criticality. One characteristic of the Texas change initiative is the lack of criticism and critical conversation about the initiative. This critical silence is a pervasive societal phenomenon in Texas in which centralized decision making is not criticized, whether on a state or local level. The other beliefs of this guidance model that could be equally problematic for some individuals are: ideal systems design, capacity, capability, creativity, co-evolution, facilitator, process orientation, context, time, space, participant commitment, respect, responsibility, readiness, community, vision, wholeness, language, conversation, democracy, and culture. All of these contain elements that would problematize an authoritarian view of education in some way.

It is apparent from this detailed list of beliefs that this guidance system for change is not for everyone. This type of ISD requires the cooperation of a number of like-minded individuals including those who hold power and are willing to share power. This leads to the consideration of what does an individual or a small group of concerned individuals do to initiate change in a system that is resistant to large scale change of the ISD type? One potential answer lies in the Opportunity Initiated Systems Design (OISD) model.

3.2.3 Opportunity Initiated Systems Design

OISD (Squire, 1999), which is posed as a hybrid between systemic design and evolutionary change, allows individuals or small groups to exploit opportunities that can be manipulated to create deep systemic change (p. 634). Specifically, this theory of systemic design "could be used by researchers, teachers, and practitioners who have visions, opportunities, and projects in their system that they want to use as a springboard for system change" (Squires 1999, p. 637). Because "a practitioner should be able to use this theory to guide the development of a project so that it may have deep, far-reaching systemic effects…it needs to be friendly to the practitioner" (Squire, 1999, p. 637).

The Six phases of this model approximate the systemic models of Banathy, Jenlink, and Reigeluth. Initially, self-knowledge of the system needs to be developed. This includes creating a descriptive statement of the system and its environment as well as the identification of leverage points for change, key stakeholders, and environmental forces. The second phase involves the creation of a vision of a preferred system state. The act of visioning an ideal system is grounded in the principles of ISD but differs from ISD in that the OISD vision is not one in which all stakeholders may agree. Instead, it is the vision created by the change agents, who desire to move the system towards this vision. Also, phase two consists of a further analysis of the current system in relation to the ideal vision of the change agents, and the potential opportunities to foster change. The third phase focuses exclusively on initiating change. In this phase, the change agents attempt to capitalize on opportunities that are manifested through external or internal imposed changes or disruptions to the normal functioning of the system, In addition, through a deconstruction of the current system, certain practices or conditions may be identified as opportunities for change. Finally, in phase three, the change agents may try to take action to create opportunities to move the system toward the change agent's ideal vision. Phase four consists of an evaluation of the potential of the opportunity. Here the term potential implies not only the viability of the opportunity, but also whether the opportunity has the "potential to spur systemic design and development" (Squire, 1999, p. 643). Also in this phase, other members of the system can be brought into the OISD process. With the inclusion of other stakeholders, the original change agents will know share ownership in the initiative. In the fifth phase, the original proposal "is revisited, modified, and defined to satisfy the system requirements revealed in Phase 4" (Squire, 1999, p. 644). In this phase, "the opportunity really begins to stop belonging to the core design team and really becomes owned by the larger system" (Squire, 1999, p. 644). The final phase is full implementation of the OISD initiative.

As in the guidance system of Jenlink and Reigeluth, this model has great potential to initiate systemic change, but also is limited in its use and effectiveness. Squire identifies potential weaknesses or risks in the model; however, these conditions are only weaknesses or risks when a systemic perspective is required to validate the action taken by those using the model. For instance, Squire writes that "there certainly is a risk that opportunities may lack sufficient character or size to be systemic in influence," or that "the impact of a change may remain narrow, without rippling through the peer subsystems, without affecting peer systems, and with leaving the suprasystem unchanged" (1999, p. 646). In addition, "OISD risks being too local in addressing the needs of the system, advancing a limited perspective of the system's purposes and needs" (Squire, 1999, p. 646). These potential outcomes of OISD activity are risks and weaknesses only in relation to the goal of promoting systemic change. As an interventionist strategy, the weaknesses and risks of OISD arise out of the need to make the promotion of systemic change the primary concern of the change initiative, rather than the need to disrupt or confront situations that are socially unjust and uncaring. In phase two, Squire cautions that "if at all possible, this phase should include buy-in from top leadership, namely, the building principal and the superintendent" (1999, p. 641). Without a doubt this is a necessary but not a sufficient condition for determining the viability of taking action. In some cases, these individuals may be the primary resistors to change or the actual agents of socially unjust policies. In this case, their compliance or assistance in a change initiative will not come through "buy-in" but through the development and subsequent exercise of power by the change agents.

Phase four is an essential but conditional part of an interventionist strategy. The ideal is certainly the opportunity for systemic design and development; however, any opportunity for action should not be dismissed because of its initial lack of systemic potential. A primary ideal of a post-formal interventionist strategy is that any action is better than no action, and that any action, no matter how limited or constrained, has the potential to enfold as a future site of leverage. Squire (1999) lists eight criteria to be used in the evaluation of the opportunity's potential for systemic change. The identification of conditional criteria severely limits the applicability of OISD as a model that can initiate change in situations contextualized as socially unjust or uncaring.

Considering that there are levels of change and levels of need for change, systemic change is one level and the change models more closely linked to ISD have the potential to achieve significant change on this level. However, there is also the local level characterized as one classroom, or one school building within a school district, or one school district within a statewide system. Initiation of change on a local level that effects other levels of a school system is the ideal, but pragmatically, social injustice and uncaring policies need to be immediately challenged regardless of their ability to impact the ideal. Therefore, the criteria listed by Squire are problematic in relation to a post-formal intervention strategy that values an immediate challenge to injustice. For instance, Squire identifies compatibility of the proposed project with the vision, values, and purposes of the larger system as one way to judge the viability of an opportunity. Obviously, if the locus of the injustice lies within the vision, values, and purpose of the larger system, than compatibility is not possible. Another OISD criterion is whether the opportunity is of a sufficient level of complexity so that it will have an impact on at least two levels of the larger system. If the action of a teacher or group of teachers is limited in its ability to disrupt or confront inequity outside of their classroom, grade level, or school, the action still needs to be taken.

The distinction that I am making between the previous systemic change models and a post-formal model is not a distinction of merit between the two, but a distinction between the larger systemic view and the more narrow local view. Both views are inherently emancipatory, and both views need to be employed when social justice and caring are the ideals driving the change initiative. Squire (1999) comments on the implications of OISD, which are inherently compatible with post-formal intervention, by proposing that OISD has the potential to hone leadership within a system, challenge people's mindsets about education, and leverage the naturally occurring self-organizing processes (p. 647). A post-formal intervention strategy can act as a scaffold to the OISD process and to more ideal guidance systems. However, the primary center of all activity in a post-formal intervention is the promotion of social justice and caring. This is the center and the bias; post-formal intervention does not allow a neutral posture by the change agents.

3.3 Post-Formal Intervention

Unlike the previous change models, the primary purpose of a post-formal intervention (PFI) is to promote social justice and caring through systemic or local change initiatives. The ideal is to effect systemic change. However, immediate and on going intervention is required in the very real and pervasive oppressive policies and practices that result in immediate and ongoing harm to those who are marginalized and result in collateral damage to some members of the dominant culture. The Texas standardization and accountability model not only continues to directly affect those who are marginalized in a deleterious way through dropout rates and through control over their access to cultural capital, but also affects those who are part of the dominant culture by diminishing the time in which students can engage challenging curriculum and instruction. Confining change initiatives to idealized models does nothing to address or ameliorate the immediate human cost associated with current and on going reproduction of oppressive dominant culture. Post-formal intervention is a change strategy that can address the immediate nature of oppression in human systems and also lay the foundation for idealized systemic change.

3.3.1 Phases of Post-formal Intervention

Post-formal intervention recognizes the need for the sequential development of change by organizing change initiatives into interactive phases similar to those of the idealized systemic models. Phase one consists of identifying the core team who will facilitate the change initiative, the beginning of the process of capacity building in post-formal skills and knowledge, and the development of knowledge about the system in which the change will occur. The purpose of phase one is to start the change process. In some idealized models, the first phase evaluates the systems potential for change and a decision is made to continue the change process or to abandon the change because of the system's lack of readiness and capacity. If the chances for systemic change are slim and the system is abandoned, the decision that is made is based on the primary desire to promote systemic change. This amelioration of oppression is seen as a secondary goal. A post-formal intervention views changing oppressive structures as the primary goal, even if the effects will be contained within a subsystem of the local system.

The core team must consist of individuals who are committed to a critical examination and intervention because of their concern for issues of social justice and caring. Before beginning the initiative, the core team needs to establish a foundational understanding of post-formal skills and knowledge with the further understanding that this professional development is an ongoing process. The OISD model identifies phase one as development of system self-knowledge. The development of this knowledge and knowledge of post-formality can be accomplished simultaneously. Recognizing that the three components of phase one are interactive and interrelated is also the realization of the dynamic nature of a post-formal intervention. Once again in a post-formal intervention, when phase one begins, the change process begins.

Inherent in a post-formal intervention is the act of questioning. Asking questions promotes the development of critical reflection, and facilitates a crystallization of personal and collective vision. Questioning is an ongoing and dynamic process; however, certain questions need to be initiated in phase one. A sampling of seminal questions would be what are my concerns? Should I take action? Who shares my concern and might join me in taking action? What do I need to know to better understand my concerns? What skills and methods can I employ in gaining a better understanding and in making a significant intervention? What are the systemic, cultural, economic, and political aspects of my system that sustain the areas of my concern and those aspects that have the potential to facilitate my intervention? These questions not only lay the foundation for the subsequent phases but also continuously reoccur in the later phases.

Phase two can be viewed as a discrete activity in the progress of the core team's activity toward intervention, or it can be viewed as a continuous activity that occurs in all stages. Phase two is characterized as an envisioning process. A vision of how the system *ought to be* in relation to *how the system is* needs to be developed. Also, as in the envisioning phase of the OISD model, an idealized vision of the potential opportunities for change is developed. The process of developing a vision by post-formal change agents is similar to and compatible with the vision building process reported by Squire. Squire suggests that the vision in an OISD initiative needs to be a fuzzy vision of the potentially powerful opportunities for change. The vision needs to be fuzzy because "the goal of this phase is not to generate a vision that everyone can agree on; rather the goal is for a smaller subset of stakeholders that is interested in a particular opportunity to develop their vision for a system and possible opportunities that will drive the system toward the desired end state" (Squire, 1999, p. 641). Thus, "an individual or, possibly, a small group could 'dream up' an indealized system, develop it to the fuzzy image stage, and then begin to approach key stakeholders for buy in" (Squire, 1999, pp. 641-642).

Besides the envisioning process, phase two is also characterized by the expansion of the core team, continuing capacity building in post-formal skills and knowledge, and the continuing development of a knowledge base about the team's system. Primary questions asked by the team would be: What are the values that guide the current system? How are these values sustained?

What is the ideal vision that should guide the system? What are the potential opportunities that may occur or could be fostered that will facilitate the development of the ideal vision?

Phase three in the OISD model is entitled, "Manifestation, Identification, or Creation of an Opportunity and Its Definition" (Squire, 1999, p. 642). In phase three of the post-formal intervention model, each of these origins of opportunity take on the following meaning. Opportunity to promote change may be manifested by an external action. In relation to the Texas educational system, these opportunities could be a critical presentation of the Texas system through diverse sources such as a segment on the television show *60 Minutes*, a scholarly book, parental criticism, a change of administrators, or poor TAAS results. Also, identification of opportunity could come from a metacognitive critique that uncovers schoolwide attitudes, decision making processes, interpretations, and knowledge that can be utilized as an entry point for critical conversation into the system. A metaemotional critique could uncover individual and collective emotions, a pervasive feeling tone, or isolated groups and individuals who have the potential to serve as leverage points for change. The creation of opportunities implies a seeding process that initiates conversation about the concerns and vision of the core team. Passing around books and articles appropriate to the teams' goals, speaking out at meetings, questioning, engaging others in conversation are some of the possibilities. As in all phases, phase three is characterized by the expansion of the core team, continuing capacity building in post-formal skills and knowledge, and the continuing development of a knowledge base about the team's system. By this phase the team should have detailed knowledge about the system's culture, gatekeepers, and power arrangements. Knowledge of the structure and operation of the system is imperative in taking advantage of opportunities as they present themselves. Questioning becomes pragmatic in relation to how events that arise and components of the system can be appropriated for the teams' use.

In phase four a strategic plan is designed that will seek to aggressively create opportunity for change. After a thorough post-formal inquiry into the existing system, objectives and targets need to be developed. Aspects of current programs, people, artifacts, attitudes, knowledge, and socio-cultural rules need to be targeted as potential opportunities for intervention. Raising questions that cause stakeholders to explore their ideal beliefs and assumptions about education in relation to the current state of the system is another powerful intervention strategy. An implementation time frame needs to be developed along with an assessment component. This is followed by the fifth phase in which the plan is implemented. As phase five begins, processes identified in the previous phases continue.

3.3.2 Accommodating Linearity and Nonlinearity in Human Activity Systems

The previous six phases are not to be construed as steps in a process but as markers in an action to disrupt and confront oppression. A post-formal intervention must be understood to be as much an intuitive action as a pre-meditated action. A well-developed plan of action supported by appropriate envisioning and capacity building is the ideal. However, when an opportunity arises, action must be taken. In an ideal situation, intervention by the change models more closely aligned with ISD can linearly impact a system and activate the change potential in the nonlinear aspects of the system. However, when all is less than ideal, a post-formal intervention has the greatest chance to activate change in a nonlinear system, which is the nature of human systems.

A PFI recognizes the linearity and nonlinearity of human activity systems and of the change process. Besides the characteristic of sequential development, linearity in change processes is also characterized by the need for an external force to initiate the change because "change from a linear perspective is not a potential within the system" (Goldstein, 1994, p. 10). Conversely, nonlinearity implies development that arises out of seemingly random or chaotic conditions, and implies that there is an inherent potential for change that can be activated under the right conditions (Goldstein, 1994).

As we have seen in the six phases, PFI recognizes the need for the sequential development of change by organizing change initiatives into phases similar to those of the idealized systemic models. This recognition of linearity implies the potential to act upon the system or externally "push" the system in question to recognize the need for change and to take action. Like the ISD models, this aspect of linearity, results in discrete phases that can be employed. In addition, PFI recognizes the nonlinearity of human activity by trying to create the right conditions that will activate the inherent potential of the system for change. Along with the guidance system model, PFI recognizes that within and through the linear phases characterized by discrete activity, there are dynamic and continuous activities that in actuality represent a seeding and nourishing process.

These continuous activities are the ongoing and recursive development and critique of: post-formal skills as described in the previous chapter; the knowledge base of the change agent, which includes all types of knowledge such as knowledge of the issue, of the place, of the culture, and of one's self; the vision and guiding principles; the implementation plan; and, the questions that are asked and need to be asked. Once again the purpose of this continuous activity is to activate the potential by disrupting the status quo or the equilibrium of the system that sustains the system's oppressive traits. The more individuals question, the more they know, the more skilled they become, and the more willing are they to act from a moral center of justice and caring.

Disruption of a system's equilibrium is essential. Jeffrey Goldstein (1994) reports that "a system that predominantly seeks equilibrium is a system that resists change. Such systems allow neither creative development nor innovative response to their environment" (p. 15). According to Goldstein, the promotion of change requires the identification and exploitation of noncon-sensus and far-from-equilibrium situations. Far-from-equilibrium situations are indicators of potentiality for change, or indicators of leverage points. These situations "are the conditions that facilitate an organization or work group in coming up with creative, new solutions to the challenges it faces" (Goldstein, 1994, p. 15). Post-formal skills and knowledge have the potential to become these conditions and to capitalize on other far-from-equilibrium conditions. Introducing post-formal conversation and other postpositivistic techniques into the culture of a school is akin to seeding the school for far-from-equilibrium opportunities. If a school receives a bad TAAS rating, the equilibrium of the system becomes disturbed. As the school struggles to regain equilibrium, post-formal processes can act as attractors that capitalize on the nonlinearity of the system to promote change that is just and caring.

Goldstein (1994) defines an attractor as "a pattern that defines the behavior of a system when it is in a particular phase of development" (p. 59). Post-formal processes that become part of a system can become patterns of activity within the system. When far-from-equilibrium situations develop that create the potential for change, these post-formal patterns increase the probability that concerns about justice and caring will mediate the behavior of the system as the system strives to regain equilibrium.

Another important aspect of the self-organizing quality of nonlinear systems is the relationship between far-from-equilibrium conditions and the specific kinds of nonlinearity in a system. In other words, the intervention must match the system's nonlinearity. In educational systems, it is common to have teachers not participate in a proposed change even when offered a small monetary stipend. On the other hand, it is also common to have teachers work assiduously to incorporate a change for no discernable material rewards. The distinguishing nonlinear characteristic between the two examples is the teachers' concern for the well being of their students. If the change is viewed as another quick fix that will fade away and not affect the well being of their students, not even a moderate monetary reimbursement will motivate many teachers to give up their personal time. However, if the change is perceived as significant, many will commit their time and energy to the project without financial reimbursement. For post-formal interventions to be effective, they must match the system's nonlinearity, and in the previous case that match was the teachers' concern for their students. Unfortunately, many educational leaders in schools are adept at manipulating nonlinear aspects like this to get teachers to comply with interventions that actually perpetuate the oppressive status quo or equilibrium. Teachers need to develop the post-formal skills and knowledge to critically reflect on interventions in their system and take appropriate action.

3.4 Implementing A Post-Formal Intervention Strategy

Individuals can post-formally intervene on many levels. Intervention can occur in classrooms, schools, school districts, or in the larger state and national educational community. Interventions can utilize the potential of the Internet, involve written interactions in traditional media, or capitalize on the interpersonal dynamics of public meetings. However, the place where the potential and possibility of intervention is greatest is in an educator's school. A hypothetical situation involving the Texas situation will now be provided that explains in more practical terms the possibility of post-formal intervention.

3.4.1 Background

The fictitious Bowie Independent School District (Bowie), located in a rural part of Texas, consists of approximately 1000 students in grades K-12. Bowie is a typical Texas school in that it closely approximates all of the state averages. For instance, 46 percent of the households have an income of less than $25,000, 65 percent of the adults in the community hold a high school diploma, and 35 percent of the students who started with the graduating class of 1999 did not graduate. The school represents the demographic diversity of Texas in that 14 percent are African American, 35 percent are Hispanic, 48 percent are White, and the remaining 3 percent represent Asians and Native Americans.

Bowie is typically poor as represented by the fact that on average each school in the district has about 65 percent of its students on the free and reduced lunch program. As in many other Texas schools, elementary teachers conduct periodic clothing drives and fund raising activities to provide clothing, shoes, and teaching supplies for the more needy students. One manifestation of this poverty is the student attrition rate. Bowie is very comparable to most school districts in Texas in that 69 percent of the Hispanic students, 60 percent of the African Americans, and 34 percent of the White students dropout of school.

As are all public schools in Texas, Bowie is very greatly concerned about the TAAS test scores of it's students. The exact time when Bowie did become concerned was when the TEA gave them one year to raise the scores or Bowie would be closed and the students bussed to a neighboring school district. The loss of their local identity was too much to bear, and TAAS became the number one priority in Bowie. New administrators were hired, teachers were reassigned and fired, and money was allocated to increase achievement on the TAAS test. For our purposes, we will focus on the Travis Elementary School.

3.4.2 Problems

The principal and three teachers realized that the problem was more complex than just increasing the TAAS scores. They feared that teaching to the TAAS would deprive their students of educational activities that have great potential in breaking the cycle of poverty. According to these educators, preparing children for the real world instead of for achievement on a standardized test was a critical priority. Another concern dealt with equity issues. These educators were very much aware that starting in grade eight students who would do poorly on the TAAS would probably drop out. If remediation wouldn't work to get a student's pretest scores into an acceptable range, then the likelihood of the student leaving school was greatly increased. Therefore, they wanted to make sure that every child had the knowledge and test taking skills that could possibly be provided by their school.

They also knew that many of their colleagues did not share their concern for equity. Some of these colleagues felt that since their jobs were on the line it was too risky to play around with "important curriculum and instruction." After all, "they are paying us to get good TAAS scores, and what else we can work in is fine, but must be secondary to the TAAS." The four educators knew that since the TEA's evaluation system of teachers directly tied teacher evaluation to student performance on the TAAS, it would be difficult to argue with their colleagues' insistence on focusing on the TAAS. And obviously, if the scores didn't go up, the principal would be fired.

3.4.3 Significance of the Problems

The larger significance of these complexly related problems is systemic in nature. From a teacher and administrator point of view, their pay raise and possibly their job security is tied to student performance on the TAAS. In addition, teacher morale is affected by the perceived necessity to teach to the TAAS and abandon those activities deemed important by the teacher. Many teachers understand that by teaching to the test they are complicit in the reproduction of existing class structures because the African American, Hispanic, and economically disadvantaged White children do not gain the cultural capital necessary to move to a higher socioeconomic level. Another significant context for educators who wish to make more ideal changes is that if they deviate from the repetition and memorization that characterize teaching to the TAAS, then they will put themselves and their school at risk in relation to student performance on the TAAS. Despite any other measure of success, if the scores fall for any reason, then they will be held accountable. In Texas, the ultimate measure of success is student performance on the TAAS.

Therefore, based on this context how do these educators intervene in a system that is resistant to, or at best doesn't reinforce more ideal forms of curriculum, instruction, and assessment that are socially just and caring?

3.4.4 Intervention Possibilities

In this case, the core team consists of the principal and three other teachers. The principal is a doctoral student in an educational leadership program at a regional university, and through her coursework has developed a knowledge of conversation theory, systems theory, critical theory, and other skills and knowledge associated with other postpositivistic paradigms. Pragmatically, the principal lays the foundation for raising student achievement on the TAAS. However, in these efforts to build a foundation, the principal, recognizing the high degree of motivation in the faculty concerning the TAAS problem, includes conversation about the critical issues that would be raised by a person with a post-formal view. Also, along with professional development aimed specifically at the TAAS problem, the principal pairs activities designed to build post-formal capacity. These initial actions by the principal and supported by the other members of the core team would not constitute an intervention phase but the beginning of continuous capacity building in post-formal knowledge and skills.

At the same time the core team would become knowledgeable about the systems aspects of their school, their school district, and the TEA mandates and accountability system. In addition, they would develop at least a fuzzy vision as to how their school and school district should ideally function. As their insight into their system and their problem grows, any opportunities to increase the core group or to take action would be seized. For better opportunity identification and intervention targeting, the workings of the school and community could be reduced to strategic components such as teacher professional development, parental involvement, funding acquisition through grants and increased local budget appropriations, student involvement, a reorganization of the faculty, and the school's vision and mission.

Teacher professional development would entail devising activities that achieve the purposes of raising TAAS scores, developing criticality in the faculty, and developing research-based best practice techniques in the faculty. Parents could be invited into the school and organized to participate in the evolution and achievement of the vision and operational goals. While in contact with the school, every effort would be made to develop critical awareness and critical skills in the parents. Bringing money into the school adds potential to what can be done in curriculum and instruction, the improvement of the facility, and the improvement of the health and welfare of the students. In addition, this will result in the enhancement of the principal and core team members as effective and caring educational leaders. This enhancement will make others more respectful and susceptible to the vision of the core team.

The strategy of utilizing the students as peer tutors and classroom collaborators helps achieve better test scores and aids in the development of a more critical view of others in the students. However, the most important activities involving the students have to be the promotion of the ideal vision and the development of post-formal skills in the students. The same activities can relate to the faculty. Besides promoting the vision and evolving the vision, the principal and core team need to break up the traditional organizational structures of the faculty. Balkanized and individualized faculty (Hargreaves, 1994) need to be restructured into interdisciplinary teams that transcend grade level and subject departments. Space needs to be appropriated for community building activities between faculty, administration, students and parents. Finally, an idealized community vision needs to evolve and can only evolve if in every activity the appropriate questions are asked by the core team and less than ideal ideas, habits, and attitudes are problematized.

Also, metacognitive and metaemotional patterns will play out as the core team's knowledge and skill grow. Faculty knowledge and emotions concerning the problem and the problem's significance become leverage points in developing a critical perspective about the change initiative. In every activity to increase faculty knowledge is a hidden opportunity to problematize their attitudes, feelings, and perceptions attached to the old values, beliefs, and knowledge. In all the activities that occur, the core team needs to elicit this old knowledge and emotion and provide opportunity for the other faculty, parents, and students to critique their thoughts and feelings in relation to the ideal vision. Pragmatically, this needs to happen repetitively before change occurs in individuals. Also, members of the core team need to be intuitive about when to intervene and how hard to push. Their intuition is based on sensitivity developed through post-formal analysis of their own beliefs and practices, of the system, of the stakeholders, and of the progress the stakeholders are making toward the ideal vision. In a school, a small act of kindness might generate a large return in moving the school toward an ideal vision.

As these examples indicate, disrupting a system and stakeholder's equilibrium is a series of discrete and continuous acts. If these acts are conducted in a just and caring manner, contextualized by a pragmatic sensitivity to the place and the time, they can become far-from-equilibrium points that can be utilized to ideally change the system. Essentially, one view of this post-formal process is that the original core team's primary responsibility is to set things in motion, scaffold when necessary, be inclusive, and above all, raise questions. Questions are important because they generate other questions as well as answers that represent agreement with the change initiative or represent objections that can be used as potential leverage points in moving the initiative to fruition.

Another situation in some Texas schools that promotes resistance to idealized change is that when a school achieves exemplary status the stakeholders no longer are receptive to change. The feeling is that change is unnecessary because the school has met the state requirements to an exemplary degree. This view is reinforced by the subconscious fear that any change in the system could result in lower test scores. If the administrator is new to the school and wants to move to a more idealized vision, how does this administrator intervene in this system?

The same change process would enfold in a similar manner. What is significantly different is that the motivation of job security now works against the change agent. Therefore, the principal and other members of the core team must be more politically aware in their development of a supportive network, in waiting for and recognizing far-from-equilibrium points where the equilibrium is disrupted, and in applying external pressure on the system and individual stakeholders when appropriate. In this situation, it is even more important for the principal to develop a deep understanding of the local system and of the post-formal process. Fortunately for the principal, the TEA continuously changes the standards, the test, and the related accountability structures. This continuous change in the rules of the game guarantees an on going tension and level of anxiety that creates the potential for far-from-equilibrium experiences. For instance, the TEA is now moving to a new test, "son of TAAS," which will incorporate all of the TEKS. Educators predict that when this test is first given the student scores will be low throughout the state because schools will not know how to teach to the new form of the test. Also, another pattern of instability in the Texas educational system is the rapid turnover of faculty and administration. Due to low teacher pay, teachers move from school district to school district more frequently than in many other states. This turnover of faculty and administration disrupts the equilibrium of a system. For an administrator in this situation, the challenge is to develop a change culture that transcends change in individuals and capitalizes on the systemic disruptions

4. CONCLUSION

The bias of this chapter is clearly that critical conversational communities need to be established if social justice and caring are to be promoted in our educational systems. The educational system in Texas was used as the context for the promotion of a post-formal intervention strategy. As presented, there are other strategies that have the potential to achieve the same goal. The uniqueness of post-formal intervention is that a few individuals can utilize it on a small scale, and yet have the potential to evolve it into a larger systemic change. Additionally, post-formal intervention strategies allow change agents to take immediate action in promoting socially just and caring change.

The essence and potential of post-formal intervention is succinctly expressed in this quotation by Robert F. Kennedy:

> It is from numberless diverse acts of courage and belief that human history is shaped. Each time a person stands up for an ideal, or acts to improve the lot of others, or strikes out against injustice, he or she sends a

tiny ripple of hope, and crossing each other from a million different centers of energy and daring these ripples will build a current which can sweep down the mightiest walls of oppression and resistance.

REFERENCES

Abel. C., Abel., R., Alexander, V., McCune, S., & Nason, P. (2001). Anatomy of an ExCET. In Horn, R. A. & Kincheloe, J. L. (Eds.), *American standards: Quality education in a complex world—The Texas case.* New York: Peter Lang Publishing, pp. 191-206.

Alford, B. (2001). The Texas accountability system past, present, and future through one educator's lens: A continuing journey toward system improvement. In Horn, R. A. & Kincheloe, J. L. (Eds.), *American standards: Quality education in a complex world—The Texas case*. New York: Peter Lang Publishing, pp. 107-130.

Banathy, B.H. (1991). *Systems design of education: A journey to create the future*. Englewood Cliffs, NJ: Educational Technology Publications.

Banathy, B. H. (1992). *A systems view of education: Concepts and principles for effective practice*. Englewood Cliffs, NJ: Educational Technology Publications.

Banathy, B. H. (1996). *Designing social systems in a changing world: A journey toward a creating society.* New York: Plenum Press.

Bertrand, L. (2001). Promoting student success through the mastery of the Texas assessment of academic skills. In Horn, R. A. & Kincheloe, J. L. (Eds.), *American standards: Quality education in a complex world—The Texas case.* New York: Peter Lang Publishing, pp. 131-140.

Cárdenas, J. A. (1998). October. School-student performance and accountability. *Intercultural Development Research Association Newsletter*, **25**(9): 1-2, 17-19.

Checkland, P. (1981). *Systems Thinking, Systems Practice*. New York, NY: John Wiley & Sons.

Cortez, A. (2000). Why better isn't enough: A closer look at taas gains, *IDRS Newsletter*, **28**(3): 1-2, 6-9, 12).

Engeström, Y., Miettinen, R., & Punamäki, R. (1999). Perspectives on activity theory. New York: Cambridge University Press. ö ä

GI Forum, et al. V. Texas Education Agency et al., (1999). No. SA 97 CA 1278EP (W.D. Tex., November 8, 1999, Plaintiffs' Post Trial Brief).

GI Forum, et al. V. Texas Education Agency et al., (2000). No. SA 97 CA 1278EP (W.D. Tex., January 6, 2000, Memorandum Opinion).

Goerner, S. J. (1999). *After the Clockwork Universe: The Emerging Science and Culture of Integral Society*. Edinburgh, UK: Floris Books.

Goldstein, J. (1994). *The Unshackled Organization: Facing the Challenge of Unpredictability Through Spontaneous Reorganization*. Portland, OR: Productivity Press.

Hargreaves, A. (1994). *Changing Teachers, Changing Times*. New York: Teachers College Press.

Harris, S. (2001). One private school's response to educational standards. In Horn, R. A. & Kincheloe, J. L. (Eds.), *American standards: Quality Education in a Complex World—The Texas Case*. New York: Peter Lang Publishing, pp. 319-328.

Horn, R. A. (2001a). Is Texas failing to equitably educate minorities? In Horn, R. A. & Kincheloe, J. L. (Eds.), *American standards: Quality Education in a Complex World—The Texas Case*. New York: Peter Lang Publishing, pp. 159-174.

Horn, R. A. (2001b). The moral and ethical implications of educational standards: Commentary and questions. In Horn, R. A. & Kincheloe, J. L. (Eds.), *American standards: Quality Education in a Complex World—The Texas Case*. New York: Peter Lang Publishing, pp. 369-380.

Horn, R. A., & Kincheloe, J. L. (Eds.). (2001) *American Standards: Quality Education in a Complex World—The Texas Case*. New York: Peter Lang Publishing.

Jasper, W. (2001). Assessment of teacher preparation programs in Texas: The TxBESS activity profile. In Kincheloe, J. L. & Weil, D. (Eds.), *Standards and Schooling in the United States*. Santa Barbara, CA: ABC-CLO Publishers, pp. 1075-1082.

Jenlink, P. M., & Reigeluth, C. (2000, July). *A Guidance System for Designing New K-12 Educational Systems*. Paper presented at the meeting of the International Society of Systems Science, Toronto, Canada.

Johnson, R. (1999, October). Attrition rates in Texas public high schools still high. *Intercultural Development Research Association Newsletter*, **26**(9): 1-2, 8-15.

Low, R., & Horn, R. A. (2001). The administrator's caper. In Horn, R. A. & Kincheloe, J. L. (Eds.), *American standards: Quality Education in a Complex World—The Texas Case*. New York: Peter Lang Publishing, pp. 141-148.

Lowery, S., & Buck, J. (2001). Three decades of educational reform in Texas: Putting the pieces together. In Horn, R. A. & Kincheloe, J. L. (Eds.), *American standards: Quality Education in a Complex World—The Texas Case*. New York: Peter Lang Publishing, pp. 267-280.

McNeil, L. M. (2000). *Contradictions of School Reform: Educational Costs of Standardized Testing*. New York: Routledge.

Ogawa, R. T. (1998). Organizing parent-teacher relations around the work of teaching. *Peabody Journal of Education*, **73**(1): 6-14.

Squire, K. D. (1999). Opportunity initiated systems design. *Systemic Practice and Action Research*, **12**(6): 633-648.

Texas Education Agency. (1980, November). *Texas Assessment of Basic Skills: Statewide and Regional Results as Reported by the Commissioner of Education*. Austin, TX: Texas Education Agency.

Valenzuela, A. (1999). *Subtractive Schooling: U. S.-Mexican Youth and the Politics of Caring*. Albany: State University of New York Press.

Chapter 19

THE EXPERIENCES OF LONG-TERM PRACTITIONERS OF BOHM'S DIALOGUE

MARIO CAYER
Department of Management, Faculty of Administrative Sciences, Laval University

1. INTRODUCTION

This chapter introduces Bohm's dialogue through data gathered from a research study conducted on the experience of long-term practitioners of Bohm's dialogue. The research was not intended to develop a theory or model, may it be explanatory, predictive or prescriptive. The purpose was rather a qualitative description of a diversity of experiences, understandings, and meanings of Bohm's dialogue. In other words, to provide access to different realities about Bohm's dialogue through individuals who are among those with the most experience in this practice.

The chapter will address four important topics: the effects of practicing Bohm's dialogue, the side effects of the practice of dialogue, the difficulties encountered, and the facilitator's role. To illustrate each topic, participant voices are used, because, who better than they can describe their own experience.

2. EFFECTS OF THE PRACTICE OF DIALOGUE

In many of his texts, Bohm mentions that the effects of the practice of dialogue appear after a few years of sustained practice. In this respect, the participants in the research, except for one, were long time practitioners of Bohm's dialogue. Still, it is important to underline that the nature of this work does not allow for generalization of the results.

Some of the questions used to gather data on the effects of dialogue were: "What does dialogue do for you? What changes have you witnessed in your life through participation in dialogue? Did you learn something about yourself through your experience of dialogue?" Sometimes, other formulations were used (mostly with the first participants in the study): "Can you identify any values, beliefs, assumptions, attitudes, behaviors that have been challenged by the practice of dialogue? How has your experience or understanding of society changed through dialogue?" Did dialogue trigger any questioning about yourself?

From the answers to these questions, four categories (with sub-categories) were created: improved capacity to communicate with and relate to people, experiencing the transpersonal, development of a more subtle awareness, and personal growth.

3. IMPROVED CAPACITY TO COMMUNICATE WITH AND RELATE TO PEOPLE

Many participants state that dialogue has positively influenced their relationships and their way of communicating. Two themes were included in the same category since, usually, good communication leads to good relationships and, good relationships are usually based on good communication. This category was divided in four sub-categories: improved capacity to deal with conflicts, arguments, etc., improved listening skills, developing respect and consideration for people, and improved capacity to empathize.

3.1 Improved Capacity to Deal with Conflicts, Arguments, etc.

Some participants mentioned that the practice of dialogue had an influence on their way of dealing with conflicts, arguments, etc. The following comments from a participant show how, through the practice of dialogue, the anger he experiences during deep disagreements now seems to dissipate more quickly because of his desire to better understand the nature of the conflict:

> One other thing too, I suppose, is that my upset and anger at times of deep disagreement seem to dissipate more quickly. I begin to move into a mode of wanting to understand the nature of the conflicts whereas before I'd just sit with it and stew with it for long periods of time. I still do that a little bit, but not nearly so long as I used to.

In the same line of thought, another man mentions his greater ability "to see through the conflict" more quickly[1]:

> We used to sometimes have conflict for a couple of days. My sense now is that it is hard for us not to see through that a lot more quickly. Perhaps

> some conflict will come up over an issue for 15 minutes or half an hour, but it doesn't go on much longer than that.

For another participant, dialogue seems to have helped him develop the capacity to withstand conflict in relationships:

> The thing that amazes me is that, on one level, nothing much seems to have changed. And, on another level, even though in many of my relationships — they're fraught with difficulty: interpersonal difficulties, arguments, that sort of thing — the same relationships have carried on, being able to withstand some pretty extreme blows through arguments, battles, disagreements. We've been able to ride through those, to survive them without everything breaking up or falling apart, or so it seems to me. I don't know about the other people. For me it's sort of, even though I've come to points of deep frustration when a part of me wants to say: "To hell with all of this. I'm going elsewhere, and I'm just going to leave all this", I realize that that isn't necessary. There's still something more I can do, further that I can go, that sort of thing. The fragmentation that would have probably taken place quite readily when deep disagreements were reached hasn't occurred since I've been doing dialogue.

3.2 Improved Listening Skills

Many participants mentioned that the practice of dialogue improved their listening skills. As one woman summarizes it:

> . . . I think in all my relationships, I've been able to learn to listen more deeply, articulate myself more clearly and reach a shared understanding. So dialogue has had a great effect on my life . . .

While some participant mentioned having improved their listening skills through dialogue, others indicated how this came about:

> But basically what real dialogue in a group setting has showed me was how my judgments interfere with my perception, and how, through dialogue somehow, I've been able to see beyond a lot of my judgments, and listen and talk to freely and feel a great affection for people I ordinarily would not be as interested in conversing with.

Here is a good description of the shift dialogue allowed this man to make by teaching him to take a closer look at what people are saying rather than passing judgment too quickly:

> Since I've been in the dialogue groups, I've learned to be more patient and more tolerant with views that are quite different from my own, and taken a willingness to take a closer look at what people are saying, rather than being quickly judgmental.

For others, dialogue taught them to be less impulsive, to suspend their feelings in order to be better listeners:

> In my relationships I would say that I have become better at listening, and also I have at least tried to suspend my feelings in situations, where they earlier probably would be the dominating way of relating to a specific situation.

In more complex terms, another man also makes the relationship between his listening skills and a greater capacity to suspend his reactions, his judgments:

> Well, I think chiefly it's allowed me to have a better or quicker insight into the differences between reactivity and response and that . . . and also there's a challenge to my state of capacitance, my state of ability to absorb without reacting. But at the same time not to suppress, not to ignore, but to allow a feeling state to grow into awareness . . . So, in a sense, it's a challenge to reactivity and impulsiveness and listening, being able to listen to the other and not care, really, whether they're an asshole or not. So, it's a very good way of seeing that judgment about another is a projection of one's own state of fear about oneself, unexamined functions within oneself. So I think it's a very useful thing to do.

3.3 Developing Respect and Consideration Towards People

The comments below emphasize the development of attitudes, which are, in my view, the very foundation of respect for others. One participant directly addresses the topic of respect:

> For me, one of the important things with dialogue is that it is where respect can develop among the participants and, generally speaking, in our contacts with one another not in dialogue (and sometimes in dialogue too) but, our so-called normal contacts with one another, respect is not present . . . Whereas in dialogue, I think that one of the most important things that can arise is this feeling of respect. It can happen. I think it does happen.

One participant states that the practice of dialogue allowed him to develop an attitude that makes him acknowledge the interests and points of view of other people, which he now takes into account when he has to make decisions:

> Well, in terms of relating to people I've found a change. I find my natural inclination and my strength of personality is to be single-minded and singular in direction and, as a result, I've been quite successful in management and quite effective in getting clear about what needs to be done and having it done. The dialogue I found to be particularly helpful because it, by nature, includes all the participants. The method I had before didn't take the people into account too much whereas . . . it's so obvious from a dialogue that all people have different interests and different viewpoints and different understandings. I was always concerned with the experience of other people, but in many ways in a more manipulative way in order to get them to do what I wanted them to do . . . Whereas now I find my interest is because their interest is part of the whole equation not, not to persuade them, to make it mine but actually to take theirs into account. So it's affected my relationship with people quite a lot.

Through the practice of dialogue, another participant has become more respectful and accepting of others:

> And maybe also I think it's made me somewhat less demanding on people. I realize that people have got their way of doing things and I have my way of doing things and, you know, see if you can do it together without trying to change people. I think that's a big thing in everybody's life, trying to change people and to realize the futility of that is a worthwhile thing.

This last comment shows that one basic condition for being able to respect others resides in the capacity to let go of the need to do things your own way or be right all the time or having others agree with you.

3.4 Improved Capacity to Empathize

Although respect and empathy are, in a way, intertwined, the following comments illustrate how, through a change in the sense of the self (strengthening or expansion of the self), participants, through the practice of dialogue, become more empathetic and are able to bring out what is interesting and worthwhile in people:

> I've so often felt that everybody has something interesting and worthwhile in them to bring out, to show, to share. But so often, because of the structures that we're in, there's only certain things that are allowed

> to come out, that are supposed to come out, that are admissible for any given individual. And in dialogue that starts to loosen up and all kinds of things can come out. And the process has confirmed so far for me my suspicion that people in general are remarkably interesting and alive at some level, however dead they may be in the process of becoming, because of the context that they're in. There's a pulsing, living, quality that wants to get out, and dialogue really can show that very powerfully. So that's more of a confirmation than a change. That's something that you see in people when you are in dialogue with them. But seeing that, the more I've seen it, it has deepened whatever that means to me. I mean, to see that that quality is there in people means something. I can't say exactly what, but it has a very significant meaning for me. And so in that way there has been a change because it's deepened that meaning . . . So seeing, confirming that's there . . . that has deepened that meaning, which I think in some way makes me more relaxed around people in general, because I'm less inclined to give all my attention to the surface presentation and to be really sort of confident, instinctively confident that whatever is going on top, that there's a living, interesting human being in there somewhere.

Self-empowerment, strengthening of the self led to more empathy for this woman participant:

> . . . first the self-knowledge is tremendous, and is different and also it's very empowering, self-empowering because, before I was doing dialogue, when I had problems, I was never able to solve them by myself. I always had to go to some kind of professional who claims that they can solve problems or they can help you to help yourself. . . . And because of dialogue I found something very immediate to deal with my psycho-emotional state. Through that I understood other people go through the same, so I also, and I know they do, so it's enlarged my knowledge about that all human beings are more or less the same. It is the same suffering going on.

For others, their experience of dialogue led to a very different sense of self than the feeling of strengthening mentioned above. In fact, some participants stressed feeling some kind of expansion of the self, which also makes them more empathetic:

> . . . I feel the dualistic distinction embedded in our culture between me, on the one hand, and the rest of the world, on the other, has been diminished. I am much more able to see myself in others than I had before, also more able to own up to what it is in me that engenders undesirable or paradoxical results, including good results, obviously . . .

> So I am talking about an attitude of inclusiveness, saying non-separation might be going too far. But the chasm has been reduced.

4. EXPERIENCING THE TRANSPERSONAL

The participants' answers to the question: "What are the most significant or profound experiences you have had with dialogue?", illustrate that the practice of dialogue triggered profound and significant experiences for some. It was also obvious that these experiences were not a common occurrence but could happen occasionally during dialogues.

Words to describe such experiences do not come easily to the participants. I encountered the same difficulty when it came to labeling this category. Many words could have been used to qualify these experiences: mystical, spiritual, transcendental or sacred. I chose the term transpersonal because it seems to have fewer connotations than the others and, yet, expresses what people have described: an experience beyond the personal.

To describe his experience, one man talks about a feeling of profound connection, of connectedness, of communion:

> We had talked a great deal over the weekend but the pace of the conversation had slowed down so much that you felt a profound connection to each person who was speaking. They were speaking very slowly and very deeply from the depths of their being, we felt that . . . each one . . . and it was some kind of thinking together but it was more profoundly a feeling together of communion. A feeling of compassion for each person there, a sense, a very high sense of energy. It was, I do not know how else to describe it in terms of words, it stayed with me until this day, I remember it as the experience that stands out the most of all my group experiences in dialogue.

This participant conveys a similar experience when she says that the group behaved as one organism rather than separated organisms; she adds that a feeling of both peace and aliveness accompanied this experience:

> But there came a point where it felt like I tapped into, or we all tapped into this collective humanness on a very deep level where all distinctions disappear and we were like one organism rather than separate organisms and that we were moving as one, like all the cells that make up a hand are working together to be able to do something. All of a sudden they were working together and it was a great feeling of peace and tranquility yet stimulation, really a . . . I don't want to use the word excitement because

> that's too frivolous, but a profound sense of aliveness and stimulation, yet a real peacefulness.

Finally, another participant talks about those special moments in dialogue like, for example, the awareness "of the miracle of being together, living the beauty of life":

> . . . the dialogue has been rather intense. It seems that we will reach a certain point and it just seems that all of the opinions and arguments and ideas are really not that important and that suddenly there will be a kind of silence, a kind of quiet which is not a silence that we have planned. But, suddenly there seems to be a silence where we will realize that . . . well, it's even hard to put it in words. As I try to put it in words it seems kind of trite but I guess the closest I can put it is that we will suddenly be aware of the miracle of being together, living the beauty of life.

5. DEVELOPMENT OF A MORE SUBTLE AWARENESS

For many participants their awareness has become increasingly subtle following the practice of dialogue. This seems to manifest in at least two ways: increased clarity and understanding, and awareness of one's assumptions. In this last category, special attention will be paid to assumptions related to oneself, meaning self-knowledge.

I should have introduced the word *insight* in the title of this category. The fact of seeing a situation with a clearer view, of suddenly being aware of an assumption can, in my view, be considered as insights. Moreover, some participants mentioned that their most profound experience in dialogue had been an insight. This effect of dialogue has probably not been emphasized enough.

5.1 Increased Clarity and Understanding

Many participants mentioned that the practice of dialogue increased clarity and understanding. Some of them seem to better understand what is going on, experience less confusion and answers are clearer. It is with a touch of humor that a participant shares that his participation in dialogue makes the "mental communication channel" cleaner and that creates space for things to happen:

> R: What does dialogue do for you?
> P: Nothing.
> R: Nothing?

> P: No, it doesn't do anything for me. I mean it's just, it seems it doesn't, I mean, it makes (pause) I mean I think it's almost like a dharma, or a duty or something, it's like asking what does cleaning my house do for me? Well it makes me a bit tired, but also I have the satisfaction of seeing everything clean. So it creates a space of clarity and in that space things can happen. So, to me dialogue is cleaning the mental communication channels. I think that's a very worthy task.

Other participants have difficulty describing the phenomenon of clarity resulting from the practice of dialogue. They will simply state: "It's increased my awareness or opened up levels of understanding for me that I didn't have before." or " It's just that what is happening is clearer — maybe not always at the moment but shortly thereafter."

On the other hand, one woman gave a good description of the enhanced clarity and understanding:

> . . . I can give you examples of what happens to me at dialogues. And that happens more and more. That my psycho-emotional energy level goes up and there's more clarity in my understanding of what life is about . . . what human beings are about. I find some answers for questions I always ask myself, which I couldn't before answer. And now, somehow, the answers are coming to me without much thinking about it, like they're obvious answers.

5.2 Awareness of One's Assumptions

Another way the development of a subtler awareness manifests is through an awareness of one's assumptions. Some participants mentioned that the practice of dialogue allows assumptions to surface and allows them to acknowledge some of them. This is an important point: we know that assumptions play a major role in our perception of things, our behaviors and our emotions. One man describes how dialogue confronts us with our assumptions:

> . . . I find that dialogue will unfailingly throw your assumptions right up in your face for you to see, if you're willing to truly engage in dialogue. It will show you the limit of your flexibility, and the limit of your mental agility, the limit of . . . the nature of your personality. And it will put in front of you the opportunity to see that. I wouldn't say change it, but at least to see it, bluntly, uncompromisingly. Not because somebody's sitting there stuffing it down your throat but because the circumstances

> themselves have conspired to put it, just like your hand is in front of your face, it's clear as a bell. This is who I am. And that's not always pleasant.

Another participant talks in a different way about the influence of dialogue in making her aware of her assumptions. First, she states that dialogue made her aware that her assumptions are often shared by many of the others and, according to her, this shows the collective nature of many of our assumptions. Secondly, this quotation also reveals that for this participant the process of challenging one's assumptions, process inherent to dialogue, does not take place solely during meetings of dialogue groups but also spreads to daily life. Finally, she gives a good description of how the practice of dialogue also allows her to be aware of how her communication was determined by a lot of reflexes put on her by her family, the media, and society:

> P: What dialogue showed me is that our society really has very fundamental thought patterns that are similar, and that we can really tap into this collective movement, not the creative movement, but what I call more an unconscious conditioned movement, each culture has their own shared meaning, both destructively and creatively. And so I felt like dialogue really helped educate me about how societies act and how their behaviors are just an expression of some of these fundamental assumptions in the culture.
>
> R: Do you have an example of that?
>
> P: Well, America. I mean, we have this fundamental assumption that we are very powerful and the greatest country and somehow we have the right to the greatest standard of living. That thought drives us to work crazy hours and justify really destructive habits, drinking and smoking and driving cars all over and getting stressed out. But this idea that we're Americans and we have to live and be successful and have money, that really drives a certain incoherent behavior that many people don't question because that's we think the way it is.
>
> R: That's the way we have to be . . .
>
> P: That's the way we have to be, never realizing that . . . it's just ideas that, progress and capitalism and wealth and power are so important.
>
> R: You could have learned that in a course or at school. How did dialogue make you realize these things? With people in a dialogue group you question these assumptions or try to become aware of these things?

> P: This brings up another distinction. The dialogue process, I think, is for me something that's happening all the time, whether I'm sitting with a group of people or not. What I mean by that is just attentiveness to my assumptions, and my reflexes to want to say something or do something and just slowing that whole thing down. And the dialogue process, when I'm sitting in a group together, really helps me see that my communication has been determined by a lot of reflexes. So, having an understanding of the collective nature of these conditionings because, I see someone who looks very different from me, who dresses, looks, acts, talks very differently from me in ordinary life, when we sit in a dialogue group and really start digging into what's meaningful and what our problems are, I see that we share a lot of these, such as, "I work too hard", "I don't have enough time for myself, my family", and so on. So I see we share this assumption that has been put on us by the media and our families, television, the whole past of our society.

Among our assumptions, those relating to our selves are very important. It seems that, for many participants, the practice of dialogue allows them to be aware of certain assumptions relating to themselves. In other words, many participants think that dialogue allows them to better know themselves:

> As a therapist I have spent an awful lot of time and money exploring myself in the spirit of "physician, explore thyself." No question, I have learned a lot about myself and about human beings in general. But dialogue added a dimension that seemed to augment and mirror what I had known before or begin to know it in another way . . . It is likely that dialogue has refined some insights that had their origin before and provided a vehicle in which to test them.

One woman mentions that, for her, the practice of dialogue accelerates the process of self-knowledge:

> I think dialogue groups speed up the whole process for me. So I would just miss the quickening of self-discovery, or self-knowledge. Because, I do it anyway, with what I can. It doesn't have to be a dialogue group. I catch myself dialoguing with anybody, anywhere, rather than having a polite conversation. So that will go on anyway. But in dialogue group, because there are more people than one or two or five there are usually more than 20 people. The whole process just speeds up.

6. PERSONAL GROWTH

Other effects of dialogue can be included in the category "personal growth". These effects are: being more creative, more open, and less fearful.

6.1 More Creative

Some participants mentioned that the practice of dialogue makes them more creative, while others rather talked about the many insights they had through the practice of dialogue.

One man believes that the practice enhanced his capacity to think creatively:

> These days it seems to make me feel smarter. It's strange. I tend to think better and more creatively. I'm not sure whether this is entirely true, but if I sort of look back at the process of my thinking and my ability to think new thoughts — new to me — I seem to be better at it now than I used to be. I don't know whether that would be objectively true.

Another participant provides a good example of how, concretely, creativity resulting from the practice of dialogue helps him face situations that previously made him feel insecure:

> I would say that I feel more comfortable in speaking, not knowing exactly what it is I want to say and being more spontaneous and creative in certain situations where before I would feel fairly insecure and not knowing what I was going to say next, or not being able to solve a particular problem.

6.2 More Open

Some participants will use expressions like "more broad-minded", "more liberated as a person" while another talks about dialogue "as a process of opening".

One person says that despite the fact that it is sometimes difficult to cope with being challenged, the end result is nonetheless refreshing because she feels an opening; she is not limited anymore by the search for answers and the security she gained from them:

> But some of the ideas I had have been really challenged and that's sometimes difficult to swallow. But it's so refreshing to come out on the other end of that because there's a real openness. After the second David

> Bohm seminar I attended, I remember feeling so open and like I didn't want any answers because that was ending something rather than beginning something. And it was really a powerful moment because so much of my life I've been looking for answers, and so dialogue makes it difficult because these answers I felt security with no longer are relevant. But to realize that, there's a great feeling of openness and simplicity in a way.

The process of opening can also manifest through an opening to change. Dialogue helped this participant realize that living implies being open to change. In this respect, through the practice of dialogue, a person wouldn't limit herself to only one way of doing things, namely her way:

> It makes you realize that to live is to always be open to change. Living and learning are coexisting. Of course the older one gets the easier it is to get into a rut and you know, having your own way of doing things and I think dialogue counteracts that to a degree. It's very good in that respect.

6.3 Less Fearful

Other participants mentioned that their practice of dialogue had rendered them less fearful. For example, one woman states that dialogue made her a lot more outspoken. Another one says it allowed her to be less fearful in front of a group; and another one states that dialogue has changed her; she used to be timid and frightened and is now confident and alive:

> P: They [colleagues and friends] say I've changed. People keep telling me I've changed so much over the years that I used to be this very timid, insecure, frightened person and now I'm confident and alive. So they see some changes.
>
> R: And for you it's related to dialogue?
>
> P: Definitely. I wasn't, I'm not doing anything else.

I would like to end with a comment from a participant illustrating how the practice of dialogue affected his life. This comment is special not only because it did not really fit into the previous categories but mostly because the impact on this participant's daily life is very concrete and profound:

> Well . . . the changes for me have been quite profound. When I first engaged in dialogue, I had a career situation in the community that I helped establish. I had an ongoing situation that would provide for me, and my family's general wellbeing. But taking the nature of dialogue back into that community, I started to discover the limitation of what was happening there. What I saw to be inquiry, other people found to be quite dangerous and more like betrayal. My exploration of dialogue led to me withdrawing from that. The way which I earned my income changed and my general activities changed quite substantially. It had at the same time an impact of a similar nature on others that I knew as well. In that way it was an agent for quite substantial change. And I have to say that it still is undermining in some ways because I find it more difficult to attribute value to some of the activities than previously.

Table 1 summarizes the different effects of the practice of dialogue.

Table 1. Effects of the Practiced of Dialogue

Improved capacity to communicate with and relate to people
Improved capacity to deal with conflicts, arguments, etc.
Improved listening skills
Developing respect and consideration towards people
Improved capacity to empathize
Experiencing the transpersonal
Development of a more subtle awareness
Increased clarity and understanding
Awareness of one's assumptions
Personal growth
More creative
More open
Less fearful

7. SIDE EFFECTS RESULTING FROM THE PRACTICE OF DIALOGUE

The issue of side effects was not addressed during the first interviews. Actually, for some participants, this question was asked at the second stage of the research when they had to answer supplementary questions and send their answers back. In this case, four participants did not answer the question.

The question was formulated slightly differently as the interviews unfolded: "What are the drawbacks in practicing dialogue? Are there any counter-productive effects in practicing dialogue?", "Do you think there are any drawbacks in practicing dialogue?", "Has the practice of dialogue held any unexpected consequences for you?, I would like to know if you have experienced any "side effects" from the practice of dialogue. Could you describe these "side effects"?

The notion of side effects, by definition, implies an unpleasant or unwanted effect. However, this does not mean that the effect, regardless of its nature, does not have its importance or its reason to be in the process of dialogue. It often seems that the transformation process, is accompanied by unpleasant effects:

These kinds of things very definitely occur. If they did not occur to someone who was involved in dialogue in any sustained way, I would seriously question the nature of the group that they were involved with or the nature of their participation and so on.

In fact, for some participants, the side effects they identified do not have to be eliminated since they are inseparable from the transformation process generated by the practice of dialogue (for example, the suffering accompanying the enhanced awareness of violence and suffering in the world). On the other hand, this is not true for all side effects. For example, even if some participants mentioned, as a side effect, the tendency to neglect practical things, this would not be presented as a tendency inseparable from the transformation process resulting from the practice of dialogue!

The side effects will be presented under four themes: confusion and identity crisis, addiction and turning dialogue into a religion, neglect of practical things, and pain coming from awareness.

8. CONFUSION AND IDENTITY CRISIS

In the participants' comments about side effects, confusion and identity crisis seem to be related. They are sometimes presented separately and sometimes associated, in the sense that the identity crisis is seen as a more pronounced form of confusion.

One man underlines the risk of confusion associated with the search for meaning in dialogue. He even stresses that the risk can have either beneficial or awkward consequences depending on the circumstances:

> I suppose one of the side effects of the endeavor to make sense of everything that's happening for everyone can lead in turn to a sense of some confusion in one's personal life, which may be very fruitful or it may be an awkward side effect.

While some participants talked about their personal experience, others described what they observed in other participants, namely a shattering of the sense of reality and a questioning of one's identity:

> Personally, I have not experienced it, but I have seen it in several people in the sense that their reality picture has been turned upside down, or their identity has been questioned, or their language has been cut to pieces and put together again, so that the language doesn't mean the same thing as it used to. And this could be very profound experiences, like going through psychotherapy, and if you don't have anybody to talk to about it, you have two options, either you suppress it or you become very depressed, or even a third one, you become very restless. In all three reactions, there's an element of suppression.

The next quotation brings to light the relationship between confusion and identity crisis; the latter being considered as the outcome of too much disorientation, or disappointment with oneself, or agitation. Here are examples of how this confusion manifests itself and this, based on the personal experience of the participant (and confirmed, according to him, by the experience of other participants). We can see how the confusion, resulting from a "substantial dialogue encounter", spreads to very concrete actions and this, to a point where usual habits such as entering one's home or driving down familiar streets can seem strange and unfamiliar:

> With regard to the third question about secondary, usually unpleasant or unwanted side effects, most definitely, I would say after any substantial dialogue encounter, I feel raw at the edges, at least raw at the edges, uncertain about where I stand with all kinds of things. . . . this thing of disorientation, and sometimes it is literally disorienting, sometimes deeply so, for me. . . . I know many people share this experience. Walking back into your own home, even if there's no one there can be a strange experience, driving down familiar streets can seem unfamiliar. . . . this feeling of disorientation is in fact sometimes carried over into that context [workaday world], but I would not say that it's furthered by that context or aggravated by that context. You know, some other components of this are just agitation or disappointment with yourself or disorientation to the point of identity crisis.

The next comments give a good illustration of the feeling of confusion and how a deep questioning of behaviors and values can lead to challenging one's own identity:

> But I think I prefer the awareness from what's really going on than feeling superficially happy. I stopped enjoying going shopping and buying clothes or trying to look better. I do less and less of this, even though in the past, it used to be very important to look the best, as a woman. And I really enjoyed buying clothes. Now I can't find so much motivation. I don't have a very good reason to do that because I see all

> the action behind it. All the conditioning and it stops me. Also I stopped having desires. Sometimes there is nothing I want. (Laughter) It really worries me, like there is no desire. I stopped having desire to have a relationship or have sex or go out and buy myself something beautiful. I don't have the desire to make lots of money. I'm a very strange member of society. I don't actually enjoy living in a big town with all the shops. Sometimes I desire being in empty space in the countryside and going for a walk in the forest. That's not very strong desire; it would be like sensual pleasure for the body. But I enjoy less and less all the Western society and Western culture. I don't desire to go and watch the next great movie because it's just as the one before. And I don't enjoy . . . Sometimes it worries me. So I feel sad. I don't have desire because there is no energy, which drives me. I used to enjoy so many things. And now those things, which are available, I don't want them. There is something I want but I don't know what it is. I think my secret dreams and desires, and I sort of investigated (?) a little bit, maybe the only desire I have is to . . . for the human race to become sane. And that brings me excitement and energy . . .

9. ADDICTION AND TURNING DIALOGUE INTO A RELIGION

During the interview, some of the participants talked about the risk of transforming dialogue into a cult. They also mentioned the danger of becoming addicted to dialogue. One man's comments stressed the fact that some participants are in such a need of what dialogue can offer, that, when they are in contact with dialogue, they make a religion out of it:

> . . . the world is a bit of a desert at this point and that when you begin to bring this [dialogue] into being, people get so hungry for it they think it's the complete and only answer. So they pour themselves into it as if it were the only thing, and they make a kind of little religion out of it, or whatever, when it isn't. That's not what it is. That's not what it's for. It's in practice, what I think, to be the antidote to some of that. But, because there's so little space in our world for this kind of thing, people will rush to it.

The same participant adds that others are so desperately seeking "the kind of cosmic orgasm, the ultimate in sense of harmony" that they can become addicted to dialogue. (He is quick to add that dialogue is not about "cosmic orgasm".) Finally, he underlines the tendency to "create in-groups and out-groups" which leads to exclusion or inclusion, to comparison between groups to know which one is more advanced, etc., and this results in isolation between dialogue groups.

10. NEGLECT OF PRACTICAL THINGS

Some participants mentioned that one of the side effects from the practice of dialogue is neglecting practical things. There seems to be two causes to this. One man who says that sometimes certain persons only pay attention to the conversation and this to the point that they forget other very practical dimensions describes the first one. In that case it seems that the neglect happens when dialogue is unfolding and this, to the detriment of more practical activities:

> P: I do notice sometimes people becoming quite intellectual and insensitive to practical things in their own environment. So, people are so busy talking that they don't notice that they've missed all sorts of things immediately around them. And I suppose, at times, I become susceptible to that. Conversations going on late into the night really to the detriment of all sorts of others things that has to happen, for example, at a practical level. I've become aware, I've become sufficiently aware of that, I think, to be able to deal with it, but certainly at times it's been a problem. People talking so much they can't wash the dishes . . . whatever.
>
> R: So, you're saying it's the . . .
>
> P: It's counterproductive . . .
>
> R: Yes!
>
> P: . . . since the intention of the whole dialogue is that people . . .
>
> R: . . . being attentive.
>
> P: . . . should be aware, be attentive to everything, but in fact become aware only of an exchange of ideas and concepts rather than a general awareness. And I think that's quite a challenge for the dialogue to not be purely intellectual, conceptual . . . or to be purely emotional and experiential, but to be involved with the general sense of both thought and feeling and the immediate circumstances. So that, I think that's a fairly easily perceptible unfortunate side effect.

According to another man, the neglect of practical things occurs when the questioning or challenging, inherent to dialogue, is extended to situations requiring decision-making or action. In that case, there seems to be a risk of getting stuck in the questioning (ponder, ruminate) and avoid decision-making or action:

> I suppose the only one I can think of is the joke about how many dialoguers does it take to change a light bulb (laughter). I think actually sometimes that does tend to happen to me where I'm asked a question that demands at a practical level an answer and to be dealt with. . . . I often find that I will consider them from sort of a dialogian point of view or something and never actually get around to dealing with what is required. And that is a problem, I think. It is quite a specific activity for a specific purpose and probably shouldn't be generalized too far before it becomes incoherent.

In other words, some people become over-involved in the process and this can affect daily activities and even decision-making.

11. PAIN COMING FROM AWARENESS

Inquiry leads to awareness and the awareness of one's own values and conditioning also carries some pain or sadness. While describing why dialogue can bring out pain, one woman does not see this as a drawback:

> I think it can be quite painful. I mean, you know, it raises many questions in one's self. I've had to look at aspects of myself perhaps which I might have preferred not to. My motivations and so on. Why I do things, why I say things in the way that I do. I don't think that's a drawback, quite the opposite.

Another participant illustrates, in a very concrete way, how the practice of dialogue manifests in her daily life and how it is a source of sadness. It seems that being conscious and refusing the social conditioning conveyed by the media, such as movies and television, keep her from feeling jolly and happy all the time:

> The drawbacks for me, like in every-day life, I find when I watch TV, and I watch it with proprioception so I don't only watch the play or the movie or the news, I don't just hear the news, but I hear what is behind it. Sometimes it's so horrible that it can send me to a tremendous sadness, or shock. Sometimes I can't watch TV if I don't feel strong enough, I don't want to watch TV. Because if I see so many movies and all about the

same subjects, adultery or killings . . . There's so much violence and also in newspapers. Like the whole world is excited about violence. It is sort of terrifying. So being too aware is difficult to feel jolly and happy all the time. But I don't mind not feeling happy, so that's O.K. But if I was really after happiness and feeling jolly and jumping up and down and being full of enthusiasm, I would be a very unhappy person with all this awareness.

It is interesting to note how this participant reframes the issue of happiness by mentioning that it is OK for her not to feel happy. Table 2 illustrates the side effects from the practice of dialogue.

Table 2. Side Effects of Dialogue Practice

Side effects resulting from the practice of dialogue
– Confusion and identity crisis
– Addiction and turning dialogue into a religion
– Neglect of practical things
– Pain coming from awareness

12. DIFFICULTIES ENCOUNTERED IN THE PRACTICE OF DIALOGUE

One important dimension of the participants' experience relates to the difficulties and challenges encountered in their practice of dialogue. This dimension was explored by asking the following question: "What difficulties do you encounter in the practice of dialogue?" Answers to other questions also shed light on this issue. Some participants answered the question in a very personal way while others answered in a more general way, more impersonally.

The great diversity of difficulties encountered by the participants made it hard to group them into categories. Sometimes, it seemed that each difficulty should have its own category, which obviously would have resulted in an overabundance of categories. For this reason, the difficulties were grouped into four categories: 1) coping with inconsistent intentions, 2) coping with paradoxes, 3) logistic and physical issues, and 4) hindrances to dialogue. Some of these categories are complex. For example, the category ''coping with paradoxes" includes seven different paradoxes and often, many difficulties are involved in the unfolding of the same paradox. To better grasp the nature of the difficulties encountered, the reader will need to pay attention not only to the larger categories of difficulties but, more importantly, to their constituent parts.

Before describing the difficulties, it is important to underline the space given to difficulties in dialogue. Some participants mention that difficulties are an integral part of the practice of dialogue and should not be avoided nor considered as something to get rid of, but rather they should be seen as material for observation:

> And I don't make that list [of the difficulties that I encounter] with a sense of "get rid of that", but it's a factual thing, it's there to be observed and looked at.

Also, by acknowledging these difficulties and facing them, the individuals and the group can reach a greater sense of coherence:

> . . . we were talking about the difficulties, right? So, the reason that a certain level of difficulty tends to be . . . you could say even "welcome" in a dialogue or "fruitful," I mean in the sense that it's precisely in coming up against something difficult and then seeing what's happening. That is, in a sense, when you feel that you're touching something, something deep and something that binds you to people and that makes you, gives you a sense of coherence precisely in seeing the incoherence. In other words, the clear perception of the incoherence of consciousness is a tremendous energizing perception.

The participant goes on to state that certain difficulties would not be "fruitful" and, if not successfully handled, they could eventually result in the breaking up of the group, which is something that actually happened to his group.

It does seem that certain difficulties would be an integral part of dialogue and that it is not possible, nor desirable, to avoid them without losing the essence of dialogue. On the other hand, it seems that other difficulties are not an integral part of dialogue, in the sense that they could represent a threat to the integrity and continuance of the group.

13. COPING WITH INCONSISTENT INTENTIONS

Despite the fact that the answer to "Why do they practice dialogue? shows a great variety of motivations among the interviewees, some of the intentions remain, nonetheless, a source of difficulties for many, because these intentions are not consistent, or are not coherent with the spirit of dialogue. In a very general way, it would be possible to characterize these difficulties by saying that they originate in the lack of intention, on the part of some participants, to engage in dialogue:

> I think there are many situations where people have no intention to engage in dialogue. And, it may be very difficult to propose dialogue to someone who really has defenses or has very strong opinions . . . Normally in those situations where I participated, dialogue works well when people come there for that reason. . . . they're interested in it.

It may seem strange that some people would come to a dialogue group without the intention of engaging in dialogue. A more detailed analysis of the interviews sheds light on this apparent contradiction. I grouped into two major themes the difficulties related to intentions: the consumption attitude and propagandizing. A third theme, lack of background, is indirectly related to intentions.

13.1 The Consumption Attitude

Some people participate in dialogue with a consumer's attitude, which is characterized by wanting to get something out of dialogue in exchange for something else. Dialogue is then considered as some consumer goods. One of the participants described this difficulty by qualifying it as an acquisitional tendency, meaning the tendency of certain people to participate in dialogue with the intention of acquiring something:

> So that's another difficulty in dialogue . . . the acquisitional tendency, I want to get something out of this, I want to go somewhere with this. Or maybe I will become better, or smarter or all of this. It's all related in the same pattern of . . . acquisition and self-enhancement. And again I'm not judging that movement but when the question is what are the difficulties, those are the things, those are the difficulties.

Another participant makes reference to this type of difficulty when he mentions that some people see dialogue as a commodity to buy or not to buy, depending on what they will find:

> One of the major difficulties that I encounter in the dialogue practice is that many people approach the dialogue group with the attitude as if the group is already formed as a commodity. That is, they tend to see it as something that they're going to buy on the basis of whether they like what they see or they're going to reject because they don't like what they see.

Other difficulties relate to the intentions of people coming to dialogue looking for individual therapy, entertaining, distraction, and searching for a mate.

13.2 Propagandizing

Some of the participants have difficulty with persons who use dialogue groups for propaganda, for "selling" an ideology or a belief system:

> Well, there are people who come into it to sort of propagandize a particular point of view. And as gently as possible one tries to let them know that this is not the nature of dialogue. It's one of the main things.

The object of the propaganda can also be a person. When this happens, people find themselves involved with individuals who participate in dialogue in order to be recognized as "teachers" or in order to "sell" their gurus:

> That is the difficulty of when I think that people are trying to take on the role of teacher, guru really . . . Or even keep quoting their own gurus and bringing them in all the time, it can be very tedious . . .

Although these intentions are judged as divergent, the persons whose behaviors are influenced by these intentions do not always recognize them. One participant states that it is a priority that these motives be brought to light as soon as possible:

> That's the problem with it, it can be a terrible waste of time unless these hidden motives [looking for sexual contact, having conversation in a nice place with some bright people . . .] are brought out and seen for being totally inadequate reasons for being there.

I also included in this category the difficulties related to a lack of preparation because the way a person prepares before joining a dialogue group could be, in most cases, related to his or her intentions.

13.3 Lack of Background

The fact that some people join a dialogue group without knowing what dialogue is, represents a source of difficulties for some participants:

> . . . sometimes people were brought that weren't informed and it went terrible. They came and they thought this was a debating club, they were very frustrated, they often left angry or were angry while they were there. And then, sometimes after these same people would be given information

> about what we were doing and they said, "Oh! This is what you were doing." Oh! They had no idea and they would come next time and they would be perfectly fine and enjoy it.

One woman expresses what she feels personally and how she behaves in such a situation:

> In the group dialogue, when there are new people coming, sometimes I experience lots of frustration because some people don't want, or they choose not to get familiar with the material available like some tapes of the previous dialogues or books. They just read *Dialogue: A Proposal* and then they want to start a dialogue but they're missing out on the ideas, like the felts and thoughts and memory and so on and the connection. So sometimes I wish everybody did their homework but they don't. Sometimes I find that difficult and frustrating. That's why I tend to be quite sharp with them . . .

14. COPING WITH PARADOXES

A careful analysis of the interviews revealed a pattern that seemed to be repeating itself: there was a tension, a movement back and forth between two situations apparently in opposition. In fact, many participants mentioned the difficulty in choosing between two behaviors or attitudes, which apparently cannot manifest at the same time. It seems, in certain cases, that neither behaviors are satisfactory despite the fact that they represent apparently the only two possible choices. In other cases, the two behaviors should take place together! I included the diverse manifestations of this pattern into one category. These manifestations take many forms: holding back vs. indulging, the individual vs. the collective, the personal vs. the impersonal, inclusion vs. exclusion, listening vs. speaking, content vs. process, thought vs. feeling.

14.1 Holding Back vs. Indulging

Some participants underlined the difficulty of finding a balance between repressing their emotions and expressing them. One woman expresses this by saying that people are either too polite or too psychotherapeutic:

> Sometimes feeling like it's a thin balance between repression and indulgence. And sometimes I feel like we can indulge in our emotions and it kind of gets more psychotherapeutic. And other times I feel like we repress things and it gets more cerebral and intellectual. So the balance, I

> think, is in between and that's when we flow into that place together where we're really moving cautiously, patiently and things are being revealed as we go. But to get there we do a lot of bouncing back and forth between a kind of emotions on ice, repressing and just trying to talk about things and not really being totally honest, being too polite . . .

The next quotation is a good example of the difficulty of expressing anger or frustration in a way that is not antagonistic:

> And another thing is the question of how polite you have to be . . . sometimes it may be appropriate if you are really put off or annoyed or angered by another person to express it, but not in a way that would make another person want to leave the group, that kind of thing. Not in a way that would be antagonistic.

It seems, however, that there is a tendency, in dialogue groups, to repress emotions, to hold back: "I think the tendency is to be more repressive and too polite with dialogue." The participants mention this kind of repression in a variety of ways. One will mention the conversations " . . . where there's obvious "withholds" in the group, and they're unconsciously held or even consciously held . . . " or will mention the situations where people do not take the risk of saying they don't like what is going on: "So they're there, but they don't really want to be there, or they're there and they don't understand what's happening and they don't like it but they don't say it."

Very few examples were given where expressing emotions had been a source of difficulty. On the other hand, some participants clearly indicated their difficulty (or that of the group) in dealing with emotionally charged situations.

14.2 The Individual vs. the Collective

In their interviews, some participants explicitly underlined this tension between the individual and the collective, the difficulty to reconcile the individual and the collective. It seems that the participants are attracted to one of the poles, namely the focus on the individual:

> I have noticed in my participation that, overall, the power of coming back to the personal, individual experience seems to overpower the focus on the art of thinking together and talking together and communicating together.

14.3 The Personal vs. the Impersonal

Although this paradox seems similar to the previous one it is, in fact, different. Here, the difficulty resides in keeping a balance between two opposing ways of dealing with the world, one being the personal and the other, the impersonal. One man describes this difficulty in the following way:

> Either one is deeply personal and sobbing about experiences, in trying to get the group to sympathize, or one is so abstract and lost in an impersonal universe, that you're not taking other people into account. I think that's the most difficult thing, somehow to remain in touch with your abstract flight and at the same time be completely cognizant of the personal difficulties and idiosyncrasies that we all have.

Another man uses the expression "personal and philosophical" to describe the same difficulty. The expression "concrete and abstract" could also have been used. As you will see below, many participants said that dialogue was often very abstract, very intellectual. On the other hand, others mentioned that certain persons who dominate the group often do it coming from their personal history.

14.4 Inclusion vs. Exclusion

One of the difficulties often mentioned by the interviewees and also mentioned during the dialogues in which I participated, relates to the tension between the tendency to want to include as many people as possible in the dialogue and the tendency to exclude those who hinder the flow of the dialogue. One man summarizes this situation:

> It seems contrary to the spirit of the activity to omit anybody, to ask anyone not to participate, to keep out anybody, but, on the other hand, there are difficulties with the inclusion of everyone who's interested to participate. The question is how to resolve that sort of situation.

The issue of inclusion and exclusion can take many forms. For example, certain persons find it difficult to integrate newcomers in the group and prefer that their group be closed to new people, at least for a determined period of time. In fact, one of the groups I participated in, as an experiment, did not accept new participants for a period of one year.

14.5 Listening vs. Speaking

This relates to the tension between being taken over by one's emotions and repressing them. This difficulty is not characterized by its emotional content but rather relates to the tension between doing (talking) and letting things happen (listening and not expressing). Here is an example:

> . . . I find a challenge just in, in a very simple way, in terms of how much to share insights and experiences and knowledge that I have and how much not to. To what extent should I bring out a situation and to what extent should it be left to be brought out itself?

This difficulty also relates to the balance between "me" and the others in the sense of taking one's place by talking or leaving space for the others by listening:

> Well, I used to feel very shy and find it difficult to talk and now I find it much too easy so I find it quite difficult not to get so involved, in a sense. But I'm talking all the time, and I don't want to talk all the time. I want to give space to other people. And I find it difficult to notice when I'm getting very verbal.

The last quotation shows how this tension can manifest in the same person. But this tension also manifests in the whole group between those who talk a lot and those who do not talk. Some participants have mentioned their difficulty in dealing with persons monopolizing time and space with their unceasing verbiage while others mentioned their difficulty in taking their place in the group and expressing themselves.

14.6 Content vs. Process

The participants did not mention this in a specific way, but it was at the heart of a whole dialogue in which I participated, and it also had an influence on the unfolding of another. Content vs. process manifests the following way: one or some participants want to talk about a topic, a content, while others want to talk about what is going on right now, the process. One interviewee relates to that difficulty when he mentions:

> Then the conversation goes off in some very irrelevant tangent of the weather or of Peruvian testosterone levels, or something. And what's

> going on in the room is like a big elephant in the middle and no one's talking about it.

In fact, in one of the groups in which I participated, the tension between content and process led to a person leaving the group. That person wanted to talk about a specific content rather than deal with what was happening.

14.7 Thought vs. Feeling

This dichotomy, this polarization between thought and feeling happened in almost all the groups in which I participated. An interviewee who participated in numerous dialogues in many countries corroborates this observation:

> What very often happens in the dialogues that I've been involved in all over the world is . . . this question about thought and feeling, this almost inevitably comes out. And comes up in such a meaningless way because people haven't really considered it very carefully and there's a lot of anger around it.

This difficulty carries a "charge" when it manifests. In a classic way, some people feel that participants in a dialogue should express themselves from their gut-feelings because they are the real thing, they can be felt contrary to thought which is pure abstraction. On the other hand, other participants feel that intellect or rational thought should guide our interventions because they would not be biased or warped by our emotions. One man mentioned that, in his group, the feeling advocates were called "the love group" while the intellectuals were called "the hate group"!

15. LOGISTIC AND PHYSICAL ISSUES

Many of the difficulties mentioned by the participants relate to psychological dimensions. However, one category of difficulties groups more concrete aspects together. It deals with the physical and logistic aspects of dialogue.

A first aspect relates to the difficulty of gathering enough people. One must remember that, in accordance with Bohm's proposition, a dialogue group includes 20 to 40 persons. The nature of dialogue would make the gathering of such a great number of people difficult:

> In very practical terms, gathering enough people to make it possible. To gather a large group of people is not the easiest thing in the world when one hasn't got a particular philosophy or goal or . . . for people to give up

> their time to travel some distance and be regular and committed if it's to be an ongoing group.

But the difficulties do not stop with the gathering of people: these people must also sustain the practice over time. That aspect seems to represent a real challenge for many of the groups:

> The most difficult thing has been how hard it has been to sustain a dialogue. When we started that dialogue in . . . in 1986 we had 18 people but at the end of the year we had only 5. We tried to start a couple of groups even later, and it was always that we would get a few people and we were not able to sustain it over time. I have been part of many groups that have started and stopped.

It is interesting to see that among the five ongoing groups in which I participated for this research, only one had more than fifteen participants. The other groups were made of a rather stable core group, which other persons joined for periods varying in time.

Another difficulty resides in finding a suitable meeting place that can accommodate a group of 20 to 25 persons and finding a suitable frequency. The more participants in the group, the more difficult to draw up a schedule that suits everyone. Finally, another difficulty brought up by a woman relates to the weekend format, where it can be exhausting to be inactive and sit two days in a row, especially if you are an active person: " . . . five to six hours just sitting. I need to get up, breathe and take a walk."

16. HINDRANCES

Most of the difficulties, obviously, do not help the unfolding of dialogue and rather hinder the flow of dialogue. It also seems that some represent escape strategies (probably not deliberate) adopted by one or many persons or by the whole group when facing situations where people do not know how to react (for example, dealing with violence, grief, frustration, etc.). In the present category we will find: flight into abstraction, domination, lack of questioning of assumptions, lapses in attention.

16.1 Flight into Abstraction

Flight into abstraction was often mentioned in the interviews and also in the groups. This happens when individuals move away from what is happening in the here and now. The focus is on the cerebral dimension of dialogue:

> But I found myself very often quite uninterested in what was going on. The quality of inquiry was often quite cerebral and abstract . . . here is a real danger for dialogue to be conducted primarily in the head and a lot of its transformative potential would get lost in the process of exploring large existential questions but failing to pay attention to the stuff that is actually happening within the group.

One man even mentions that intellectually articulate people often dominate the group and, as a result, participants who do not function with this type of intelligence are hindered from contributing to the group:

> . . . that is the tendency of intellectually articulate egoists to take over. Bohm points out an inborn structure in dialogue that helps to undo this, but in practice I feel the problem is seldom sufficiently addressed. The result is a put-down and discouragement for those who bring other equally important sub-cultural dimensions to the dialogue process.

The last quotation also relates to the fact that some participants can dominate the group. This was also brought up by other participants.

16.2 Domination

Domination of the group by one or many persons usually takes the form of monopolization of time of speech. The underlying reasons can vary. The situations mentioned by the persons interviewed include: "psychotic" individuals who monopolize the group in order to talk about their problems, individuals who are not clear on what exactly dialogue is and take the floor to talk about their interests, individuals who take advantage of the space offered by dialogue to "sell" their belief systems. And finally, there are people who simply dominate the group because they want to talk about their own thing without taking into account the other participants:

> . . . a couple of people simply monopolized and told stories that meant a lot personally to them but were wholly unresponsive to what the group was up to. In other words, it's the perfect situation where someone can just dominate.

The domination of the group by one or many persons put some of the participants in contact with the difficulty of *holding back vs. indulging,* in the sense that they do not know if they should openly express their frustration and discontent or if they should repress their reactions.

16.3 Lack of Questioning of Assumptions

Another difficulty arises when the group does not go beyond the superficial level, in the sense that the participants do not question the underlying assumptions. One man, states that this difficulty could have its source in the fact that people are so focused on understanding the message that they forget to question the underlying assumptions:

> I think you can listen to somebody for a long time and just be absorbed in taking in their meanings and lose sight of the importance of questioning the assumptions behind what you hear and also lose sight of questioning the assumptions within you that give rise to various feelings.

A woman mentions that the lack of questioning could come from the fact that the group is moving too fast or is not taking the necessary time to explore the beliefs and feelings behind what is said or not said. The hurried pace of exchanges becomes a difficulty for another participant:

> I did not mention then that I think one of the things that's been frustrating for me is the speed very often of the dialogue groups. So very often I find that dialogue moves into a discussion with people putting forth points of view so quickly that we lose the spirit of really inquiring and slowing down the thought process.

In my own observation of dialogue groups, the speed in which dialogue moves was often brought up. One way, often suggested, to deal with this consists in demanding silence after each intervention. Despite the fact that silence was often proposed, none of the dialogue groups in which I participated accepted to experience it; there were always people opposed to it. This observation is corroborated by the statement of a participant who expresses her difficulty with such a demand and sees it as an escape from a tensed situation:

> Another thing that constantly comes out . . . that I find difficult, that is when people demand silence. "Let's have silence." Well, you know, silence is a very important part of the whole thing but you can't demand it. It may come or it may not come. And usually, when people say let's have a period of silence, what's going on in the brain? And this I find, almost impossible. I really feel so angry when this happens I feel like just saying, "What on earth are you talking about?" Or they'll say, "Well we've come to a difficult part." Where really, there's a lot of tension which one has to be able to stay with which isn't easy . . .

16.4 Lapses in Attention

Constantly paying attention to the present moment represents another difficulty. This could be considered as a meta-difficulty since it manifests when people are not attentive to the fact that they are using abstraction as an escape or they are forgetting to explore the underlying assumptions, etc. One man mentions this in terms of "lapses of attention":

> . . . the frequent lapses of attention, the slipping off the cutting edge of the present moment into issues of the past or future, then and there, getting oneself seduced or lulled back into thought. Greater tiredness seems to make these lapses more frequent.

Another man clearly describes how this "lapse of attention" triggered some sort of automatism, of "program" which, at a certain point during the dialogue, took control:

> . . . somehow I did not participate in a way that I wanted to. That somehow I had fallen into this role of being too submissive in the dialogue and not saying enough about what's on my mind. So, that way, it is such a challenging environment that anything can kind of elicit programs that exist in you. To be alert, to be awake, to be learning and so on. It's quite challenging. I am really listening well but it slips over into being quiet out of some kind of a program.

Table 3. Difficulties Encountered in the Practice of Dialogue

Difficulties Encountered in the Practice of Dialogue
Coping with inconsistent intentions
Consumption attitude
Propagandizing
Lack of background
Coping with paradoxes
Holding back vs. indulging
The individual vs. the collective
The personal vs. the impersonal
Listening vs. speaking
Content vs. process
Thought vs. feeling
Logistic and physical issues
Difficulty to Difficulty to gather enough people
Difficulty to sustain a group over time
Difficulty to find a suitable place
Difficulty of being seated and inactive for many hours
Hindrances
Flight into abstraction
Domination
Lack of questioning of assumptions
Lapses in attention

In fact, this difficulty is at the heart of dialogue because, according to Bohm, dialogue is a space to develop greater attention. Table 3 summarizes the difficulties encountered in the practice of dialogue.

17. THE FACILITATOR'S ROLE

Bohm, very briefly addressed the topic of the facilitator role, in his works on dialogue. He proposed very general guidelines but no details on how to carry out this responsibility. Considering that many of the participants had started groups and acted as facilitators themselves, they were, thus, very concretely confronted with the difficulty of defining and assuming the role of facilitator. For many participants, the situation was rather different when David Bohm was alive since he usually started the groups and also participated in them. Actually, this topic was included in this study following a request from a participant and the issue of facilitation was the object of many comments. It is in such a case that the learning aspect of this study takes all its meaning.

One must also understand that the topic of facilitation is very controversial and very political within the dialogue community. All participants do not share the same point of view on the role of the facilitator. But, contrary to other topics addressed in the scope of this study where many points of view prevailed, the role of facilitator is a topic that can lead dialogue in a certain direction. According to the participants, the vision conveyed by the facilitator in a dialogue, the nature of his or her interventions and behavior, can induce a type of dialogue that respects or strays away from Bohm's vision of dialogue. Therefore, the topic of the facilitator's role is a very delicate and, as it is often the case with controversial topics, it seems that people use the same words to describe different realities. For example, almost all the participants mentioned that the facilitator should make his or her role obsolete as soon as possible. Participants' responses show that independently of the vision of the facilitator's role, the meaning of "as soon as possible" can vary greatly.

In this section, quotes from the participants will prevail even more since their comments are sometimes difficult to interpret. For many of them, their reflection is still in progress and the political nature of the topic did not encourage very clear comments. The numerous quotes will enable the reader to judge whether or not my interpretation makes sense and will also help the reader make his or her own interpretation.

The question asked in order to explore this topic was: "In your view, what should be the facilitator's role: a) when a new group starts? b) when participants have some experience with the process?" I also asked occasionally about the qualities required of a facilitator.

18. THE ROLE

Let us start with the elements that bring together the vast majority of participants regarding the facilitator's role. First, almost everyone agrees that the facilitator is essential at the start of a group. On the other hand, there is a great variety of views on the nature of that role. Secondly, almost all the participants agree that the facilitator should make his or her role obsolete as soon as possible, and this by seeing that the role is taken on by the group members rather than being simply abandoned. Yet, the appropriate moment for doing so does not reach consensus.

It seemed that the many opinions about the facilitator's role followed a continuous progression, from not being given much importance to being given a very important place. Therefore, I divided this progression scale into five positions: the procedure-oriented facilitator, the teacher, the bearer of the proposal, the modeler and the interventionist.

18.1 The procedure-oriented Facilitator

For some the main role of the facilitator is a role in logistics, in following procedures: finding people interested in dialogue, finding a room, drawing up the schedule, etc. I am not saying that for these participants the facilitator's role is limited to that, but, for them, that seems to be the most important aspect. One woman says:

> Of course, when you come into a group, and say it's a fairly new group where people don't know one another, somebody has to take the responsibility of communicating certain procedural things such as how to sit in a group, in a circle, to, as far as possible, allow one another space to talk. What else is there to say? I don't really know that there's a lot more to say beyond that. I mean there are certain other things. I mean you could talk about suspension, you can talk about proprioception but, to me, those are all things that should come after the dialogue. They should come out of the dialogue and then reflect back on what is actually happening. Whereas if the facilitator imparts that knowledge before hand, I think it might make people a little uncomfortable.

The comments from this person illustrate very well the difficulty in interpreting. The participant goes on by giving the following comments that hide a more important facilitator's role:

> . . . If a situation develops which can be rather ugly, then somehow it's good for the facilitator to be able to diffuse it in some way. But, at the same time, it's those ugly situations, which are showing something. So to

> me the purpose of dialogue is not to make us all feel good; but, rather to reveal what is actually happening in each one of us and in the group in general. So the facilitator, I think, has to be quite sensitive to know how much space to give a situation and when it's becoming locked into itself and not moving into the group where people are. So, in that sense, a facilitator can help perhaps. And perhaps also maybe winding up, you know. I mean, you know, ending the sessions, so to say.

Finally, she resumes the idea of procedure: "More really, I look upon it as a procedural thing, but not necessarily interfering too much or trying to tell the others what to do." Another participant had this to say:

> Place, time and length of meeting is still about the most difficult things to organize and once the group has settled on a regular time and place a lot of the facilitator's job is complete. The rest is done when the group has some understanding of the intentions and significance of dialogue as an activity.

Here again, the position taken up by this participant is hard to grasp. Just before, in answer to the question on what it takes to facilitate a dialogue, she said: ". . . being able to ask questions gently, not being defensive and not arousing other people's defensiveness if possible and being prepared both to take risks and support and encourage others to take risks." This could lead to believe that the facilitator plays a specific role. However, this interpretation could be wrong if we consider that this comment could apply to all the participants in a dialogue group.

18.2 The Teacher

Other participants see the facilitator as the person responsible for introducing new participants to David Bohm's ideas about dialogue and the functioning of thought as well as the person responsible for the procedural tasks:

> I think the facilitator should be familiar with the ideas of dialogue the way David Bohm conceived it. I think that's the major role. On the other hand, I have experienced dialogue without a facilitator and it was as good. So I'm not really sure if dialogue needs a facilitator if I really think about it. I prefer that it doesn't, some people prefer it does.

When asked to specify what the role of the facilitator should be at the start of a group, the participant replied:

> In the beginning, I suppose, the facilitator is necessary to start, to bring the basic idea of what it is about, how it is different from discussion and so on. At the beginning it is necessary. Later on, as the dialogue unfolds, I think the facilitator is not necessary, in the same sort of role. We need more an organizer, who organizes what to do, where it takes place. But yes, at the beginning there should be somebody, it doesn't have to be one person, it could be a few people, who are experienced with the dialogue for some years and they want to introduce it to other people because they feel it's important for them as well as for the other people who don't know it yet.

Another participant is more specific about the role of the facilitator as a teacher. According to him, if the role of teacher is well carried out, the role of facilitator will not be necessary:

> And I also differentiate between the facilitator's role and a teacher's role which is necessary to introduce some of the ideas around this activity to people who are new to it. . . . My ideal of the facilitation role is that the facilitator, as quickly as is humanly possible, inspires the other participants to share the task of facilitation so that eventually the task of facilitation merges, it becomes part of everyone's participation. I tend now to think that if we can begin with some good teaching, that is a form of teaching that clearly presents the intentions of dialogue and then allows some general discussion of those intentions along with some reading so that everyone is able to set out on the same path, then no facilitation is needed.

18.3 The Bearer of the Proposal

One participant sees beyond the role of teacher in the facilitator's role without, however, asking the facilitator to model the behaviors, as it is the case in the next category. (In fact, one interpretation of some of the comments from this participant could be that the role of the facilitator is to model behaviors, but he also proposes another role.) For him, the facilitator not only presents the dialogue proposal but also is able to demonstrate why this proposal is important and deserves attention. This demonstration is made by using the group situation rather than being theoretical or conceptual. According to him, the participative nature of dialogue does not require that the facilitator master suspension and proprioception skills to demonstrate how this works but rather that he or she be ready to get involved in the activities in the company of others. Still, according to him, it is in such an involvement that the learning nature of dialogue reveals itself, and not in a context where the facilitator already masters these abilities and teaches them to the others.

His comment being very long, I took the liberty to underline key elements. As I mentioned at the beginning of the section, it is important that the reader be able to make up his or her own mind:

> I think, that the facilitator's job, at first, is to explain something of the situation, that there is a difficulty in communication and that it may relate to the nature of thought and the movement of consciousness, and that we are coming together to see how we can talk and to explore whether this is in fact, the case. So, I think the facilitator makes a proposal of a particular kind and many other people could make other proposals, so, part of the job of a facilitator is to make clear why he is making the proposal, so that other people will take it seriously. He makes a serious proposal and engages with others as to why, why this is a serious proposal. Hum . . .
>
> R: I would like to explore this a little bit. In your view, do people have to accept this proposal? Is it a kind of requirement . . .
>
> P: I think that if people do accept the proposal, then we can go further into it. If people don't accept the proposal then there are two possibilities: either people put the proposal into abeyance and they say: "Well, maybe he's right, maybe he's wrong, he seems like he's quite an intelligent sort of person, so I'll stay around a while and see whether this has value or not". And I think it's fine that certain people may not accept the proposal fully but they're happy to put it in abeyance to see what happens. Then I think we can proceed. If the person rejects the proposal, then I think there's some work for the facilitator to do in the immediate experience of the conversation to demonstrate why the proposal is a serious one to be pursued. I don't think that can simply be brought about in conceptual and theoretical terms. I think the facilitator is then required to demonstrate something of it within the group, with the assistance of everyone else present. So, that in a way takes us further because I think then, part of the role of the facilitator is that he has to accept that the dialogue itself is part of his facilitation role. He can't simply come and do it. He can come and make a serious proposal but eventually it has to take form within the dialogue itself for it to be a clear facilitation role. So the facilitator is dependent upon the dialogue for his ability to facilitate, and that very dynamics, I think, is quite helpful in bringing out what's happening. But, I think it's also the clue, as one goes on, to see that the facilitator is not separate from the dialogue, but is a participant in the dialogue who happens to play an unusual role at the start of it. So, he has to make it possible for others to do the same thing that he has been doing, because, in the end, that is inherent in the nature of the dialogue that we're all

dependent on the dialogue for it to go further, but we all have to take some personal responsibility for it. So I think this is unusual because I think traditionally the facilitator is separate from the thing he's facilitating. A teacher is separate from, the people he's teaching for example.

R: So are you saying that it's necessary for the facilitator to master suspension, proprioception of thoughts? Is it necessary for a facilitator to have the skills and also not only to have these skills but also to use them when it's appropriate or . . . all the time? Maybe all the time is too strong but . . .

P: Again, I think that depends a little on everybody who's present, as to what is possible. But I think what the facilitator has to do is to make the serious proposal and enable it to stick. So, I may be a facilitator who hasn't mastered the skills of suspension, but I may still be able to make the proposal of suspension serious enough that I engage with it further and others do further. So I might not be the master in suspension. But I can take that activity further in myself and draw others in the end to demonstrate features of suspension that I don't know. So, in that sense, it is, it is a really quite a participatory activity. If I've just mastered the whole thing and am getting other people to understand it, then we aren't in a learning situation of the type that is present in dialogue. So, in a way, the chairman doesn't have to be able to do everything. But I do have to be able to take the proposal that I'm making seriously myself. I think that, that is more important than the actual developed skills. In that sense, I think someone who has not developed great ideas about dialogue, but has caught the spirit of it, and the seriousness of the proposal, could set up a dialogue group and it could move ahead very effectively. And I think another person with greater skills but wasn't prepared to engage in the immediate situation would be a less effective facilitator, interestingly. So, the participatory nature of this is quite fundamental. (Silence) . . . But inevitably, if we proceed over a period of time, as a facilitator, I'm going to become more and more engaged into the elements and I'm going to become clearer about the nature of suspension or proprioception, of the necessity of various features which it's become apparent are important. But still, I would say, I have to find that in the group I'm in. It's not enough for me to say: "Well, of course when we were in a dialogue in Sweden, this happened." That's very interesting but what's happening here is the significant thing about this dialogue, and how to go further in this one is a challenge. So, that's why, in some of the earlier questions I've talked about something like courage, because, fairly quickly, one's past experience is less important than one's courage in the immediate situation to go further.

18.4 The Modeler

Some participants see the facilitator playing a role of modeling the behaviors and required skills for dialogue. In this respect, he or she must have already mastered the art of asking questions, of suspension, and of proprioception of thought. Obviously, it implies that he or she also have a deep understanding of Bohm's work on thought and dialogue. One participant mentions:

> In my view the most important aspect of facilitation is the introduction into the dialogue of perceptions in the present of how our individual and collective thought is lacking proprioception.
>
> A woman adds:
>
> I think a degree of modeling may well be necessary, at least in the beginning of the life of a dialogue group, to illustrate the power of a well-timed and worded intervention to make manifest what had been latent.

Another participant states that, according to what type of group we are dealing with, the role of modeler can be important at the beginning:

> In some ways the dialogue facilitator at first, the outset, models the space of dialogue, not so much by . . . sort of it leads by example as opposed to by direct overt direction.

18.5 The Interventionist

At the other end of the spectrum we find the interventionist facilitation. This happens when the role of the facilitator goes beyond modeling behaviors. In such a case, the facilitator can intervene directly to comment on the behavior of some of the participants, make observations, have participants show respect for the spirit of dialogue, etc. I am well aware that among those who favor a more interventionist approach, others could qualify some interventions suggested from participants as incompatible with the spirit of dialogue.

One participant mentions that, from the perspective where there is a facilitator, he or she has the task of informing participants if they are not behaving accordingly to the spirit of dialogue. According to him, the facilitator assumes a "negative function":

> Well, if there is a facilitator, he or she should make . . . should aim at making themselves unnecessary in a short while. And I can only see it at

> most as a negative role to point out that this particular approach is not a dialogical approach. Unless you have a particular subject matter, which is one way of doing dialogue, the facilitator can point out the fact that you're really off the track . . . But all the facilitator can do basically is the negative function, which eventually becomes unnecessary.
>
> R: What do you mean by negative function?
>
> P: Well to point out that you're not following the . . . that this manner of relating is not consistent with dialogue or that if you have a specific subject what you are saying is not really to the point. I can't think of any positive roles.

A participant gives specific examples of the type of interventions that would be appropriate in a context where there are many conflicts:

> Although in some setting with high conflict, it's often useful to have some interventions that engage the participant, so that the facilitator might learn to say things like, "I noticed for three times, you know, Joe and Fred have talked and gotten into a conflict around what they said. Can we understand what it is that allows this to keep happening this way and can we explore this a little more? Can we understand, Joe, what goes on in you? And Fred, what goes on in you? And in others, what goes on in your bodies and in your consciousness and minds and hearts, as you watch this?" Not so much with an intention and tone to fix, but to discover and explore and inquire. Deepen the inquiry, so they can learn to make moves that will deepen, by engaging patterns, they can learn to make moves that will broaden, by naming patterns. They can bypass difficulty that'll allow the flow to keep going. Like they can notice that three or four people have tried to speak and have not been able to, and they can finally say something like, "Jane, you get a chance to finish," or, "Joe, you get to speak." A deeper intervention might be, "I noticed that Jane's been trying to speak all day and I've noticed no one noticing," or, "I mentioned before and we agreed and then nothing happened." You know, what is it about how we're operating that leads us to systematically exclude pain? Or to systematically not notice that we're excluding Jane, while doing it?

Finally, at the other extreme, we find the following comment where the facilitator must not only monitor the tone and the energy but also use his or her authority to have the spirit of dialogue respected:

> I think the facilitator should be able to monitor the tone, should be able to see where the energy is, where it's stuck or see what kind of level of

> energy is and . . . be able to do something about it. So if the energy is incredibly stuck, physically stuck, (I mean people have gotten rigid) perhaps the facilitator could get people running around the room a couple of times. Or, if the whole thing becomes so personal that it's just become one person dominating the group, the facilitator could perhaps step in and change the direction. If the person continues to be obstructive and starts using the group, it should be pointed out, in no uncertain terms, that this is isn't free therapy and if the person wants to continue doing so, then they have to contribute $100 to each member of the group.

I will end with one last comment. A participant states that the type of facilitation must be contextualized. In other words, he stresses the fact that the type of facilitation should vary according to the characteristics of the participants:

> . . . the majority of the groups that David's [Bohm] worked with over the years, which is groups that pay some nominal or sometimes large but mostly nominal amounts of money to come, because they know about him and they know about dialogue and they're interested in new consciousness and they're from a pretty specific band of humanity. You know, white, typically white-middle-class-spiritually-oriented people. Sometimes intelligentsia. But, that's the band, you know, to be specific. In that group, facilitators are somewhat less important . . . I think that this need for facilitation will vary depending on the degree of self-selection, the degree of prior familiarity, the degree of, you know, if people are sort of on a spiritual search, then recognize that this could be a root for it, they'll leap to it and they won't fight it very much.

Figure 1 illustrates the five ways of conceiving the role of the facilitator and the relative importance given to this role. In addition, according to my interpretation, the greater the importance, the greater the time before the facilitator should become obsolete.

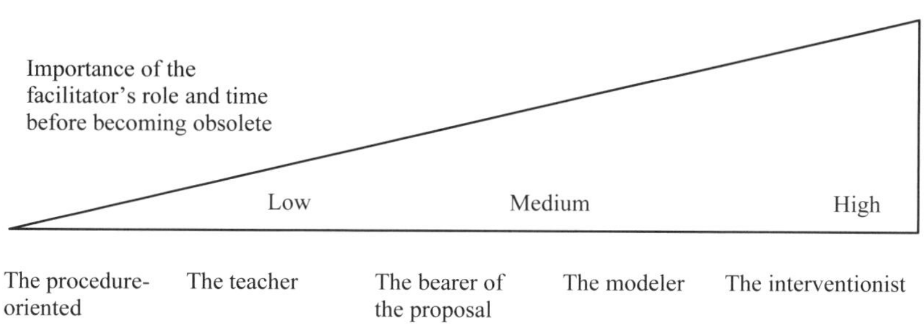

Figure 1. The continuous progression of the importance of the facilitator's role

19. WHAT IS REQUIRED TO BE A FACILITATOR

It seemed important to present comments from the participants regarding what is required to be a facilitator, namely prerequisites, skills, attitudes and qualities. This should help to better understand the role of the facilitator or, at least, provide avenues to explore. However, few participants formally answered this question[2]. This information is nonetheless very precious. It was not possible to include it in a specific framework, as was the case for the role of the facilitator. Thus, I will present it as an enumeration by simply quoting from the participants.

What brings together most participants is the prerequisite concerning the necessity of understanding Bohm's theory on dialogue and the nature of thought. Another prerequisite that was mentioned is having a substantial experience in dialogue. One participant makes these points very clear:

> . . . what I would say is the first prerequisite of being a facilitator of dialogue, Bohm-style, and that is, you must do the dialogue, right? I find it impossible to believe that an individual could just start facilitating without having done a substantial amount of actual dialogue as a participant, and without quite a clear understanding of the theory.

Some participants spontaneously talked about the pertinence of having some experience as a facilitator in other types of groups, namely group dynamics. Is this kind of experience useful in order to act as a facilitator in a dialogue group? There seems to be ambivalence on this issue.

One woman, judges both useful and useless her experience in different sorts of groups, while another talks about the potential risks of manipulation on the part of an experienced facilitator "versed in the play of group dynamics":

> A gifted facilitator, versed in the play of group dynamics, could convert the role to enhance his own standing and authority at the expense of that of the group. It could make the group more passive and inhibit learning opportunities. The role of the facilitator in dialogue is a very complex issue and we need to be sensitive both to its enabling and its inhibiting propensities.

Another participant elaborates on the irrelevance and sometimes on the relevance of certain types of interventions generally made in other contexts. I did not leave anything out from the quotation (even if it a repetition of some previous comments) because it emphasizes the vision of this participant on the role of the facilitator and in relation to the characteristics of dialogue:

> Because there are many people who are excellent facilitators in different arenas, that bring their skill to bear in Bohm-style dialogue and it doesn't fit very well, it's a bit mechanical, it's a bit formulaic, sometimes very formulaic, and so on. So, at the same time, I would have to say that sometimes people who have that skill in general, of being a facilitator in other contexts, can use it extremely well in Bohm-style dialogue, but that loops us back to what I would say is the first prerequisite of being a facilitator of dialogue, Bohm-style, and that is, you must *do* the dialogue, right? I find it impossible to believe that an individual could just start facilitating without having done a substantial amount of actual dialogue as a participant, and without quite a clear understanding of the theory. Because, as I mentioned a moment ago, that often the way that people will engage and intervene as a facilitator in other contexts is very inappropriate, which is sometimes just to clarify the meaning of a person's point, not that that's inappropriate in Bohm-style dialogue, but sometimes to allow the feeling to get conveyed, to allow the misinformation to sustain itself, is often the point of dialogue, and not to patch up, not to clarify, which often means to reformulate in ways the first individual may not have meant. That kind of facilitation is certainly not useful in dialogue. To facilitate so that this business. . . . often one of the key points of a facilitator is to make sure that each person understands the other person, that everybody understands one another. Of course, in a way that's desirable in dialogue, but understand *what* about the other person, right, about the other people, about the group as a whole? What is it we're looking to understand, you see? Is it the personal experience of another individual? Well, perhaps, but perhaps not. That could change from moment to moment. Are we looking to understand how to achieve our objective, which is often a crucial part of typical facilitation? Well, if so, that's pretty rare in dialogue, because more or less we have no objective, except to try to understand more deeply what the thought process is. And if a facilitator works toward that end, so be it. Although I've certainly seen that become mechanical too, that Bohm's theory of thinking and so on becomes the reference point. And there may be a place for that, done skilfully, but we must also beware of facilitators who have become reflexive about that. So that's another point. That could occur even within the "Bohm world" of dialogue. That kind of mechanical activity could recur there.

Concerning prerequisites and skills, one man mentions that not only must the facilitator have a grasp of Bohm's work on thought as a system but: " . . . the required background is giving value to paying attention to the nature of this thought process." Moreover, he states: ". . . the most important aspect of facilitation is the introduction into the dialogue of perceptions in the present of how our individual and collective thought is lacking proprioception." And he adds: " . . . the facilitator must be considering these issues in his or her own daily life."

One woman also talks about the fact that the facilitator is required to be attentive to what is going on and he or she must manage to share this awareness:

> To me facilitation requires an ability to be attentive and sensitive to what is going on, moment to moment, as the dialogue unfolds and people become engaged (or fail to engage) with each other. And through questions and comments, have that become a shared awareness.

According to another participant, one of the most important skills of a facilitator is being able to help the group understand that paying attention to meaning is meaningful, and, from that, to "help the group understand the dynamics of the flow of meaning". In order to do this, the facilitator should be able to listen while having in mind that he or she must "shepherd" the group towards certain levels of meaning:

> I would say the same thing as to participate, listening. But as a facilitator, there is some listening in a slightly different way, which is, in a certain sense, you are like a bird perched to comment in a way that in some way tries to make the thing, what would we say, your comments based on your listening would be oriented toward shepherding, for lack of a better term, the drift of the group towards certain levels of meaning, towards certain levels of understanding, towards certainly revelatory things, revelatory meaning only revealing, not necessarily something profound, but revealing something not seen. And, I guess, one of the deeper skills, I think, would be, in succeeding at those things, in helping the group to understand that paying attention to things that it might normally not pay attention to, is useful, is worthwhile, is meaningful. That attending to meaning is meaningful. That would be an essential skill of a facilitator, although they may never use those terms, and so on and so forth. And then to be able to, in some way, help the group understand the dynamics of the flow of meaning. This obviously is crucial and central to dialogue. But I don't know that there's a formula for that, I don't know that I could

> even describe the qualities about that. Again, this is terrain that I ponder a lot, but I don't have many answers for, in terms of responding to your question.

Another skill mentioned by a participant is that the facilitator should be able to challenge someone with respect and without attacking him or her:

> He or she shouldn't challenge . . . there is a way of challenging a person but without attacking them and . . . to show that you may disagree with the point of view but you respect the person and you are asking him or her to examine it from a different angle, so the person does not feel demolished if they change their mind. To give the person a feeling that indeed a change of mind will be a way of growth.

The ability to head the dialogue in the direction of a common inquiry has also been mentioned by a participant:

> I suppose another one is the abilities of the facilitators to either head the dialogue in the direction of a common inquiry or not. If there isn't the ability, then the dialogue can seem to go on forever without it coming to a deeper point.

One woman makes up a list of the qualities a facilitator should have:

> The qualities for a facilitator: patience, good humor, openness and I think humility and compassion (which I am still working on) a willingness to learn, not being hung-up on "what I know", being able to ask questions gently, not being defensive and not arousing other people's defensiveness if possible and being prepared both to take risks and support and encourage others to take risks.

Finally, a participant mentions that an important skill for a facilitator is the capacity "to bring a group of people together and help them grant one another the permission to build a safely dangerous setting."

Table 4 lists the prerequisites, skills, qualities, and attitudes required to be a facilitator as mentioned by the participants. (I tried to stay as close as possible to the comments of the participants or to quote them without overloading the text.)

Table 4. Prerequisites, Skills, Qualities, and Attitudes Required to be a Facilitator

Requirements
• a clear understanding of Bohm's theory on the nature of thought and dialogue.
• a substantial experience in dialogue as a participant.
• the capacity to give value to paying attention to the nature of the thought process.
• the capacity to introduce into the dialogue perceptions in the present of how our individual and collective thought is lacking proprioception.
• the ability to consider in his or her own daily life the issue of individual and collective thought lacking proprioception.
• the ability to be attentive and sensitive to what is going on, moment to moment, as the dialogue unfolds and, through questions and comments, have that become a shared awareness.
• the capacity to help the group understand that paying attention to meaning is meaningful.
• the capacity to help the group understand the dynamics of the flow of meaning.
• the capacity to listen while keeping in mind that he or she must "shepherd" the group towards certain levels of meaning.
• the capacity to challenge someone with respect and without attacking.
• the ability to head the dialogue in the direction of a common inquiry.
• the following qualities: patience, good humor, openness, humility, compassion, a willingness to learn, not being hung-up on "what I know", being able to ask questions gently, not being defensive and not arousing other people's defensiveness if possible and being prepared both to take risks and support and encourage others to take risks.
• the capacity to bring a group of people together and help them grant one another the permission to build a safely dangerous setting.

NOTES

[1] We must underline that this participant recognizes the fact that the development of this ability is probably not uniquely a consequence of practicing dialogue. He states: "So it is hard to really separate the individual learning that has been going on anyway, even prior to dialogue. My sense is that they each feed on each other."

[2] There are two reasons for this. The first one is that the question on requirements was not asked to all the participants. The second reason is that some participants did not send back their answers to the supplementary questions.

SECTION V

REFLECTIONS ON CREATING THE FUTURE

Chapter 20

CREATING OUR SHARED FUTURE

Reflections on Conscious Evolution of Society

PATRICK M. JENLINK
Stephen F. Austin State University and International Systems Institute

1. INDRODUCTION

In this closing chapter, first, I briefly review the purpose of this work and highlight the learning experience offered in the chapters of this compendium. Then, I explore evolutionary systems as future creating. I examine evolutionary design as a new intellectual technology and the need for a new order of *creative surge* in creating the future. Next I examine the evolutionary journey of our species. I then present role of design conversation in the conscious evolution of our species. Finally, I examine conscious purposeful evolution Today, we face a critical evolutionary predicament, as well as a crucial opportunity in the life journey of *Homo Sapiens Sapiens*, to enter into the evolutionary design space, create evolutionary design communities, and use the power of design conversation in creating our shared future.

2. A REVIEW OF THE LEARNING JORNEY

Developing this Compendium, the editors had two *purposes*. The first purpose was to introduce the learner to design conversation as the means of collective communication for which we must be both students and consumers as members of a changing, global society. The second purpose of the Compendium is to demonstrate–and develop an appreciation for the empowering and liberating quality of conversation as a medium and means of communication for collective creativity and societal change.

The path of the *learning journey* began by exploring design conversation. This exploration set in place a historical perspective of design conversation, providing an examination of the emergent and developing characteristics of disciplined conversation. We next explored the different perspectives of design conversation that are sources of methodological ideas and practices of dialogue. We examined different modalities of design conversation in a variety of settings, and provide examples of conversation events. We also explored practical applications of design conversation that illuminated different approaches to systems design that enable individuals, groups, and communities to initiate, engage in, and guide the disciplined inquiry of design.

3. EVOLUTIONARY SYSTEMS DESIGN

Evolutionary systems design in the context of human activity systems is a future creating, disciplined inquiry. People engage in design in order to devise and implement a new system, based on their vision of what that system should be.

There is a growing awareness that most of our systems are out of sync with the new realities, particularly since we crossed the threshold into a new millennium. Increasingly, the realization of postmodernity challenges past views and assumptions grounded in modernist and outdated modes of thinking. Those who understand this and are willing to face these changing realities call for the rethinking and redesign of our systems. Once we understand the significance of these new realities and their implications for us individually and collectively, we will reaffirm that systems design is the only viable approach to working with and creating and recreating our systems in a changing world of new realities. These new realties and the societal and organizational characteristics of the new millennium call for the development of new consciousness, new thinking, new perspectives, new insight, and–based on these–the design of social systems that will be in sync with those realities and emergent characteristics.

In times of accelerating and dynamic changes, when a new stage is unfolding in societal evolution, inquiry should not focus on the improvement of our existing systems. Such a focus limits perception to adjusting or modifying the old design in which our systems are still rooted. A design rooted in an outdated image is useless. We must transcend old ways of thinking and engage in new ways of thinking, at higher levels of sophistication. To paraphrase Albert Einstein, we can no longer solve the problems of society by engaging in the same level of thinking that created them; rather we must equip ourselves to think beyond the constraints of science, we must use our creative imagination. We should transcend the boundaries of our existing system, explore change and renewal from the larger vistas of our transforming society, envision a new image of our systems, create a new design based on the image, and transform our systems by implementing the new design.

3.1 A New Intellectual Technology

Evolutionary systems design in the context of sociocultural systems is "coming into its own as a serious intellectual technology in service of human intention" (Nelson, 1993, p. 145). It emerged only recently as a manifestation of open-systems thinking and corresponding soft-systems approaches. The epistemological and ontological importance of systems design is recognized when situated within the complex nature of social problems in society and in relation to the teleological issues of human purpose (Nelson, 1993). As an intellectual technology, systems design enables us to align our societal systems, with the "new realities" of the postmodern information/knowledge age. Individuals who see a need to transcend existing systems, in our case educational systems, and design new systems that enable the realization of a vision of the future society use systems design. This vision of the future society is situated within the societal and environmental context in which these individuals live and from which they envision new systems decidedly different from systems currently in existence. Evolutionary systems design brings to the foreground a requirement of cognizance in systems philosophy, theory, and methodology. As an intellectual technology and mode of inquiry, evolutionary systems design

> seeks to understand a problem situation as a system of interconnected, interdependent, and interacting issues and to create a design as a system of interconnected, interdependent, interacting, and internally consistent solution ideas. (Banathy, 1996, p. 46)

The need for systems knowledge and competencies in relation to accepting intellectual responsibility for designing the inquiry system as well as applying the inquiry system to resolve complex social problems, sets systems design apart from traditional social planning approaches. From a systems perspective, the individuals who comprise the sociocultural system are the primary beneficiary or users of the system. Therefore, these same individuals are socially charged with the responsibility for constantly determining the "goodness of fit" of existing systems in the larger context of society and our environment, and engaging in designing new systems that meet the emerging needs of humanity.

3.2 The Challenge of a Creative Surge

The challenge before us is developing a new order of *creative surge*, a new and consciously renewed generative order that can guide the conscious evolution of society and or our species. The challenge to evolutionary systems design is to develop evolutionary epistemological understanding and pedagogical strategies that construct the knowledge base necessary to foster such a "creative surge," necessary to holistically understanding the complexity of globalization and evolving new social and cultural realities. Fostering a new order of "creative surge" is not concerned with the outward side of development and evolution as a sequence of succession, but "with the deeper and more inward order, out of which the manifest order of things can emerge creatively" (Bohm & Peat, 1987, p. 151).

What is really needed to create a genuinely new order of *creative surge* is the state of mind that is continually and unceasingly observant of the fact of the actual order of the social and cultural medium in which one is working. Otherwise, one's efforts are foredoomed to failure, because the order of what is done will not correspond to the actual nature of things. And this will make conflict of *some* kind inevitable. Indeed, no really creative, evolutionary transformation can possible be effected by human beings, either in nature or in society, unless they are in the creative state of mind that is generally sensitive to the differences that always exist between the observed fact and *any* preconceived ideas, however noble, beautiful, and magnificent they may seem to be (Bohm, 1998, p. 19).

The enfolded order in our species, the diversity of experiences and beliefs and perspectives and all that makes us decidedly human and that defines us, offers a vast range of potentiality, which can be unfolded. The way it is unfolded depends on many factors. Our consciousness, the way we think, and so on are among those factors. The implicate order that distinguishes one culture from another, one group of *Homo Sapiens Sapiens* from another, implies mutual participation of everything with everything. No thing is complete in itself, and its full being is realized only in that of participation. The implicate order provides an image of how this participation might take place in our conscious evolution various ways.

Consciousness, as Bohm (1998) explains, "therefore, is really our most immediate experience of this implicate order" (p. 106). A conscious awareness of the implicate order "would help us to see . . . that everything enfolds everything. To see that everybody not merely *depends* on everybody, but actually everybody *is* everybody in a deeper sense" (p. 110). It is through a new order of creative surge, in a creative act of perception, one first becomes aware of a new set of relevant differences, and one begins to feel out or otherwise to note a new set of similarities, which do not come *merely* from past knowledge. This new order of creative surge "extends into science, culture, social organization, and consciousness itself" (Bohm & Peat, 1987, p. 210). In scope of evolutionary systems design, this new order of creative surge gives rise to a system of new orders that constitutes a system of new kinds of structure within and across society, a new order of consciousness by which we evolve our species.

The caution before us, at this moment in our evolutionary history as a species, is to recognize that a society, as Bohn and Peat (1987) explain, "that has gone beyond its first creative surge hangs on to the habitual orders that are contained in its customs, taboos, laws, and rules and that are held in its unconscious infrastructure" (p. 206). It is because these orders are "fixed and limited, they will be bound eventually to become inadequate, in the face of every-changing reality" (p. 206). As *Homo Sapiens Sapiens* civilizations grow more complex in structure, what is most significant is the activities that define society itself, within and across nation-states and globally. These activities may "lead eventually to decay, more or less independently of the institutions, the will, and desire of the people who make up the society" (Bohm & Peat, 1987, p. 206).

The need for a new order of *creative surge* becomes more apparent as we move into the new millennium. We must recognize our potentials as a species and we must recognize our responsibility to consciously engage in the evolutionary history of humankind.

The new order of *creative surge*, of evolutionary creativity, operates in the evolutionary design space, enabling participants to engage in "conscious, self-guided evolution" (p. 318). Whereas collective consciousness is the force that triggers the emergence of an evolutionary generation" (p. 317), wholeness is the core purpose of emergence and creativity is the core evolutionary force that yields novelty.

4. AN EVOLUTIONARY VIEW OF HUMANKIND

According to Banathy (2000) we, as a human species, are entering the fourth seminal event of our socio-cultural evolution: the Fourth Generation of *Homo Sapiens Sapines* (HSS). Our species has reached an evolutionary crossroads. Three milestones mark the evolutionary journey of our species into higher levels of complexity, each leading us to the threshold of the fourth milestone where we now stand poised. The first, lasting some six million, is marked by our humanoid ancestors, the First Generation, entering on the evolutionary scene, as the people of the *Human Revolution*—the Cro-magnons. Following this milestone, our species, some 10,000 years ago, entered the Second Generation of *Homo Sapiens Sapiens* (HSS), the Farmers, marked by the *Agricultural Revolution*. Their journey toward the second crucial event marks an evolutionary transformation, which, as Banathy posits, was the greatest event of our evolutionary history–*Homo Sapiens Sapiens* started the revolutionary process of cultural evolution. At the third milestone of our evolutionary journey, some five hundred years ago, our species entered into the Scientific-Industrial Revolution of the Third Generation of HSS. Today, humankind has arrived at the threshold of the fourth major event, "the revolution of conscious evolution," when it becomes our responsibility to enter into the evolutionary design space and guide the evolutionary journey of our species. As Salk (1983) explains, a conscious awareness of human evolution "places upon human beings a responsibility for their participation in and contribution to the process of evolution" (p. 112).

Today, we have arrived at the threshold of the second revolution: the "revolution of conscious evolution," when it becomes our responsibility to enter into the evolutionary design space and guide the evolutionary journey of our species (Banathy, 2000). Escalating societal complexity, global terrorism and oppression, cultural and economic globalization, technological advancements, and the destruction of environment mark this threshold of our second revolution.

With the emergence of evolutionary science in the middle of the last century, we became increasingly conscious of evolution. The science of evolution became the container of much of what the science of life is about.

The explosive knowledge base of evolutionary science is manifested in a host of disciplines and fields of study. We now know how evolution has worked in the evolutionary design space, what have been its operating principles, and how the various life forms of our species have been tested in that space. In one phrase, we have attained evolutionary consciousness (Banathy, 2000).

Some forty thousand years ago, the human revolution brought forth a state of self-reflective consciousness. The emerging *Homo Sapiens Sapiens* became aware of their existence: they knew that they know. Today the revolution of evolutionary consciousness has brought forth another kind of reflective consciousness: we are now aware of our evolutionary history: we now know how we have become what we are and that the burden of this knowledge means that we must now take responsibility for our own continued evolution (Banathy, 2000).

Previous generations of our species were shaped by evolutionary forces across time, and the evolutionary design space we lost "the innocence of ignorance," the innocence of not being responsible for designing our future. We now have the responsibility for consciously guiding our conscious evolution by entering into a new evolutionary design space. Our knowing of how we have become, our evolutionary consciousness, provides humanity a springboard of conscious purposeful evolution, marking the third crucial event of the evolution of our species: the revolution of conscious evolution: the emergence of the Fourth Generation of *Homo Sapiens Sapiens* (Banathy, 2000). We now stand at the threshold of a second human revolution: the revolution of conscious self-guided evolution.

4.1 Evolutionary Epistemology

The conscious evolution of our species, a responsibility we are now confronted with as we stand poised on the threshold of our Fourth Generation, as Banathy argues, is possible to the

> extent we acquire the power of evolutionary competence. Evolutionary competence has several interacting components, including evolutionary consciousness, the knowledge and understandings of the '*what*' of evolution, and the knowledge and understanding of '*how*' evolution works. Such knowledge becomes the basis for developing evolutionary epistemology, which we can apply as we engage in conscious evolution. (2000, p. 244)

The evolutionary epistemology, proposed by Banathy (2000), is viewed as only one possible approach that might lead to many different futures for society. The epistemology is premised on three essential strategies: "*transcending* the existing evolutionary state, *envisioning an image* of a desired future state, and *transforming our systems by bringing the image to life by design*" (p. 293, italics in original). This evolutionary epistemology takes direction from his in-depth examination of our evolutionary journey, and is informed by his compendious and well-articulated presentation on conscious evolution. Hubbard's (1998) work with conscious evolution, representative of Banathy's comprehensive examination, is instructive as a definitional referent for conscious evolution,

> Conscious evolution is a meta discipline; the purpose of this meta discipline is to learn how to be responsible for the ethical guidance of our evolution. It is a quest to understand the processes of developmental change, to identify inherent values for the purpose of learning how to cooperate with these processes toward chosen and positive futures, both near term and long range. (pp. 57-58)

Banathy explains conscious, self-guided evolution as process of giving direction to the evolution of human systems and developing in these systems the human capability and organizational capacity to:

- nurture the physical, mental, emotional, and spiritual development and the self-realization ethics of individuals and their systems;
- extend the boundaries of social and economic justice and genuine civic participation;
- increase cooperation and integration among society systems;
- honor and integrate diversity by creating healthy and authentic communities; and
- engage in the design of social and societal systems that serve the common good. (Banathy, 2000, p. 235-236).

Our conscious evolution, guided by systems design, requires a cognizance of the ever increasing complexity in *Homo Sapiens Sapiens* evolution as a species, as well as an understanding of the need for a revolution in conscious evolution—the need to accept responsibility for our own cultural evolution. As Banathy (2000) argues, what stands before our species is the need now to evolve an evolutionary epistemology that will guide the creating process of the Fourth Generation of HSS. Jantsch (1981) is supportive here, in his position that as a species "We are evolution and we are, to the extent of our power, responsible for it" (p. 4). The call for conscious evolution, as the taking of responsibility for our self-guided evolution as a species, will require us to enter the evolutionary design space and engage our creative potential—to engage in evolutionary systems design. This marks the transition from evolutionary consciousness to conscious evolution.

4.2 Evolutionary Perspectives

Evolutionary systems design, as Banathy explains, calls for evolutionary perspectives and organizing perspectives. Evolutionary perspectives include collective consciousness, wholeness, and creativity. Evolutionary perspectives are generic to all evolutionary guidance systems (EGS). Herein, collective consciousness creates the shared view as "well as generates and maintains internal consistency" (Banathy, 2000, p. 317). Wholeness is a "meta-quality of an evolutionary system" (p. 317), organizing the evolutionary markers into an evolutionary system. Creativity operates in the evolutionary design space, enabling participants to engage in "conscious, self-guided evolution" (p. 318). Whereas collective consciousness is the force that triggers the emergence of an evolutionary generation" (p. 317), wholeness is the core purpose of emergence and creativity is the core evolutionary force that yields novelty.

Organizing perspectives include purpose, focus, underlying values, and underlying assumptions, which "are *specific* to any human activity system that enters the path of conscious evolution" (Banathy, 2000, p. 318). The purposes of an EGS are unique properties of the evolutionary designer community that designs the EGS—its users and benefactors. These purposes, when shared by the design community, generate the individual and collective commitment needed to engage in evolutionary inquiry and self-guided conscious evolution

5. THE FUTURE ROLE OF DESIGN CONVERSATION IN CONSCIOUS EVOLUTION

Evolutionary design is a creating activity, which brings forth a potential-driven, intended novel socio-cultural system in the *evolutionary design space*. This design space is created through design conversation that is dialogic in and which embodies a participatory, inclusive consciousness. Space in this sense is understood as public, a space that ensures individual as well as the collective voice an opportunity to be heard. It is this space that gives way to the *creative surge* of a new order of consciousness and creativity.

In this space, alternative design ideas are proposed and examined for their viability and for their "*goodness of fit*" with their social and cultural environment, which becomes their life-space. Evolutionary design is also a process that carries a stream of shared meaning by a free flow of conversation – the consciousness evolving dialogic flow of ideas and that animates design conversation – among members of an evolutionary designing community, who seek to create their own Evolutionary Guidance System (EGS) (Banathy, 2000). An EGS guides an evolutionary design community toward its envisioned ideal future system. A combination of generative dialogue and strategic dialogue comprises a comprehensive method of intentional social communication in an evolutionary designing community. (The root meaning of conversation is "to turn to one another.") Members of an evolutionary designing community turn to one another without reserve and in truth and openness, accepting and honoring each other. Before the design community engages in the substantive task of evolutionary design, it involves itself in generative dialogue. This involvement will lead to the creation of a "common ground" as the community focuses on the harmonizing of ideas, values and world views of its members and creates a flow of shared meaning, shared perceptions, a shared world view, in a social milieu of friendship and fellowship. At this point the community is prepared to move on and engage in the strategic dialogue of evolutionary design inquiry.

Recently, the idea of Conscious Evolution has gained wide-ranging acceptance through the works of Banathy (2000), Chaisson (1987), Csanyi (1989), Csikszentmihalyi (1993), (Elgin (1993), Hubbard (1998), Laszlo (1987), ad Salk (1983). We can attain Evolutionary Consciousness as we learn to understand: how evolution itself has evolved, how evolution has worked in the evolutionary design space, what are the necessary conditions of the emergence of a new evolutionary system, what are the rules of evolution, and what systemic organizational arrangements can generate the emergence of a new evolutionary system. The attainment of evolutionary consciousness is one of the prerequisites of engaging in Conscious Evolution. Others include: (1) the acquisition of evolutionary *competence* by *evolutionary learning* (2) the attainment of competence in social *systems design.* (3) the *willingness to engage* in *Conscious Evolution*, and (5) most significantly, the *mastery of consensus-building dialogue.*

Banathy (2002) poses a critical question: How does Evolutionary Design Work*?* To answer this question, let us assume that we are working in an evolutionary designing community; be it our family, our neighborhood, a system that we live in and work as a community. We also assume that we have developed evolutionary competence. The first question we ask, as an evolutionary designing community, is: What kind of society do we wish to have? Now we envision that society by creating an ideal image of it. That ideal image becomes a magnet that pulls us and guides us toward our evolutionary future. The second question we ask is: given, that ideal societal image, what system can we design as our evolutionary system that can make a contribution toward creating the envisioned society? We now create an image of our intended evolutionary guidance system.

As Banathy (2000) explains, the image we create should define all the key dimensions of the human experience; including the social-action and social-justice dimensions; the moral/ethical dimension; the wellness dimension: including the spiritual, the emotional, the cognitive, and physical/health; the economic; the aesthetic; the learning and human development dimension, the scientific and technological dimensions, the communication dimension, the societal/polity dimension, and the relationship with nature. And there is the integrative dimension, which integrates all the above dimensions and, thus, creates the wholeness of the human experience. These dimensions are built on explicitly stated and collectively defined values and beliefs shared by an evolutionary design community. The dimensions defined become the key markers around which our evolutionary guidance system is created. The mode of communication in developing all the above dimensions is design conversation, an integration of generative and strategic dialogue.

6. CONSCIOUS PURPOSEFUL EVOLUTION

A conscious awareness of human evolution "places upon human beings a responsibility for their participation in and contribution to the process of evolution" (Salk, 1983, p. 112). Embracing the teleological imperative of consciously evolving our species, the conscious purposeful evolution of Fourth Generation *Homo Sapiens Sapiens* must be understood in relation to the need to direct our evolution toward a greater complexity that marks the future of our species. Our ability to evolve toward greater complexity will, in large part, determine the nature of our humanity, and its potential for sustaining our species' evolution on a global society level. The sustainability of purposeful evolution as a species will require, as Elgin (1993) states, that we find

> a common ground of human experience and a shared vision of evolutionary potential that transcends the differences that now divide humanity. If we cannot find a universal and familiar ground of experience, we will not be able to develop a vision of healthy social evolution that draws out our energy and enthusiasm. (p. 272)

The common ground of human experience must be conceived of on a global society level and in relation a common experience of conscious self-guided evolutionary design. In this sense, conscious evolution is "the potential of giving direction to our own evolution and the evolution of the systems we inhabit, the evolution of our communities and our society by purposeful and deliberate design" (Banathy, 2000a, p. 317). As such, conscious evolution is a complex purposeful activity, which "provides a sense of direction of cultural and societal processes by illuminating those processes with guiding images" (Banathy, 1996, p. 322).

The challenges before humankind are the outcomes of a level of conscious that we achieved as we have evolved as a species, the consequences of which have positioned us at the third event of our evolution: conscious evolution of the Fourth Generation of our species. The question looming before us is whether we have the capacity to determine the evolutionary path that will be our next evolutionary step in consciousness. Salk (1983) believes we have the capacity

> By imagining ourselves inside the process of evolution and by imagining the process of evolution working inside our minds, we may discover how to deal with the problems as well as the opportunities arising from the uniqueness of the natural world, and we may learn how to empathize in a way that might influence the direction of evolutionary choices. (p. 14)

Conscious purposeful evolution of our species is, as Laszlo and Laszlo (1997) state, "an ongoing requisite since the ability of human societal systems to evolve, and even to survive, depends in a great measure on their ability to adapt to changing realities" (p. 16).

Evolutionary systems design is concerned with conscious evolution of our species on a global societal level, and embodies as a design referent the evolution of democratic civil society as an appropriate moral code to guide our choices. This code embodies the traditions of democracy including equality of voice and inclusive participation. The code also embodies the beliefs of civil society as public spheres that mediate between the city state of government and the individual citizen, thus providing a space and place in which citizens may make public the social issues and problems of the day, and engage in social action to address these same issues and problems. Equally important, this code is guided by conscious purposeful evolution. This code guides the evolutionary design process, and therein the participants in design conversation to take "into account the wisdom of tradition, yet [be] inspired by the future rather than the past; [the code] should specify right as being the unfolding of maximum individual potential joined with the achievement of the greatest social and environmental harmony" (Csikszentimihalyi, 1993, p. 162).

6.1 Stewards of Conscious Evolution

The responsibility of *stewards* is to create knowledge bases for evolutionary inquiry, develop resources for evolutionary learning and explore suitable approaches, methods, and technologies. Individuals committed to the creation and evolution of evolutionary design communities, on a local, state, national, and global level are the primary social actors in the conscious evolution of our species.

Members of the *stewardship communities* are scholarly practitioners who embody both the epistemological understandings and ethical imperatives of evolutionary systems design. Additionally, stewards are the social agents of democratic civil society. As such, stewards embrace the importance of civility. Civility, the mutual recognition of each individual's innate human dignity and membership in society is, as Edward Shils (1991) has argued, at the heart of civil society and, in his words, "at bottom the collective consciousness of civil society" (p. 35).

Stewardship, as explained by Banathy (2000a),

> means assuming shared responsibility for the advancement of conscious evolution . . . the kind of stewardship development described here could become a critical component of the evolutionary inquiry movement. Beyond the development of resources for evolutionary learning, through geographically distributed groups of evolutionary stewards, they could enhance the building of a network of evolutionary designing communities . . . (p. 362)

Such stewardship would necessarily begin with identifying individuals who are willing to accept the responsibility for creating communities of conscious evolutionary design, fostered by a shared belief in the potential of humankind to evolve beyond current limitations and constraints of societal systems.

6.2 *New Agoras* Of The 21st Century

The Agoras of the City States of the Classical Greeks were public spheres where democracy was lived by citizens who made collective decisions about issues affecting their daily lives. *New Agoras* is a metaphor for social action contexts in which people can make collective decisions about their future. People in the settings of their families, neighborhoods, community groups, organizations, and institutions have the potential to organize themselves as evolutionary design communities. Participants in the *New Agoras*, through evolutionary systems design and pubic discourse, collectively enjoin to establish a new public sphere that can sustain a meaningful actionable design dialogue among individuals within and across *New Agoras*. These *New Agoras*, communicatively linked, would serve as the infrastructure for democratic civil society and a system of public spheres animated by evolutionary conversation and guided by evolutionary design with purpose of self-guided evolution of the society—cultural evolution of our species, *Homo Sapiens Sapiens*.

The re-enchantment of participative democracy and a socially active citizenry, through the creation of *New Agoras*, portends a great potential for accepting the responsibility of the third event in our species evolution: the revolution of humankind's conscious evolution and the emergence of the Fourth Generation of *Homo Sapiens Sapiens*.

Banathy (2000), in his work with conscious, self-guided evolution, has advanced the idea of *New Agoras* as public spheres for intentional public discourse, a democratic dialogue that would catalyze social actions focused on guided evolution of society, and our species—*Homo Sapiens Sapiens*. Having arrived in our evolution as a species at a level of conscious awareness, and now being responsible for guiding our own evolution, Banathy offers as an ideal for public spheres *New Agoras*,

> where we can not only (re)establish true democracy, but also bring it alive as a shared culture (a democratic culture) and (re)constitute a method and procedure by which our institutions could serve us and establish arrangements by which we can govern ourselves (establish a cultural democracy). Most significantly, we need to hold up democracy as both a guiding idea and a process by which to work in the evolutionary design space. (p. 358)

The citizenry of the twenty-first century will participate in a new participative democracy, creating "a *democratic culture* as their evolutionary culture. People of the *New Agoras* will organize their lives and build institutions that reflect the core ideas of democracy. Toward this ideal state, they will build a *cultural democracy* that will manifest a democratic culture" (Banathy, 2000, pp. 398-399). The *New Agoras* will become new communities of conscious evolution wherein self-guiding activities will be designed and ideals will become realized through social action. Democratic civil society will be fostered through citizens conjoining as evolutionary design communities. The traditions of the Agora of ancient Athene, those of equality of participation in democratic dialogue, will serve as a guiding value and practice in the evolutionary design of new systems. Relatedly, as designing systems, Banathy (2000) characterizes the *New Agoras* "as *ethical systems*. They are *ideal seeking and creative. They are capable of building consensus* by disciplined conversation. They create themselves *as authentic, healthy, and nurturing communities*. In their lives they *manifest democratic culture* and *build cultural democracy*" (p. 317, italics in original).

7. FINAL REFLECTIONS

The challenges we are confronted with, escalating societal complexity, global terrorism and oppression, cultural and economic globalization, technological advancements, and the destruction of our environment are critical markers of our evolutionary path thus far. As a species we are capable of such beautiful dreams and such terrible nightmares. We are now faced with the responsibility for determining the future of humankind. Whether we embrace our dreams or give way to our nightmares, is our choice. Conscious evolution is offered as a modest beginning for creating the future of our species by engaging in evolutionary systems design.

In the life journey of our species there never has been more need for creative invention than it is today, when it becomes our responsibility to enter into the evolutionary design space and create our own future. This necessity has brought forth well-formulated design theories, models, principles, and methods that we can now engage in the service of evolutionary design. This necessity has also brought fourth rich and comprehensive resources produced by dialogue research and scholarship; such as: dialogue approaches, methods and tools; which we can now learn and employ as the means of collective, consensus building evolutionary design. The editors and authors of this Compendium are pleased to offer their contribution to the advancement of the dialogue movement.

REFERENCES

Banathy, B.H., 1991. *Systems Design of Education: A Journey to Create the Future*. Educational Technology Publications, Englewood Cliffs, NJ.

Banathy, B.H., 1996. *Designing Social Systems in a Changing World: A Journey Toward a Creating Society*. New York: Plenum Press.

Banathy, B.H., and Jenlink, P.M., 2003. *Dialogue as a Means of Collective Communication.* Kluwer Academic/Plenum Publishers, New York.

Bohm, D., 1998. On Creativity. (Ed. Lee Nichol) New York: Routledge.

Bohm, D., and Peat, F.D., 1987. *Science, Order, Creativity*. Bantam Books, New York.

Chaisson, E., 1988. *The Life Area.* The Atlantic Monthly Press, New York.

Csanyi, V., 1989. *Evolutionary Systems ad Society*. Duke University Press, Northern Carolina.

Csikszentmihalyi, M., 1993. *The Evolving Self; A Psychology for the Third Millennium.* Harper Perennial, New York.

Elgin, D., 1993. *Awakening Earth*. Crown, New York.

Hubbard, B.M., 1998. *Conscious Evolution: Awakening the Power of Our Social Potential.* New World Library, Novato, CA.

Laszlo, E., 1987. *Evolution: The General Theory*. Hampton Press, Cresskill, NJ.

Laszlo, E., Laszlo, A., 1997. The contributions of systems sciences to the humanities. Systems Research and Behavioral Sciences, **14**(1): 5-19.

Margulis, L., 1999. *Symbiotic Planet; A New Look at Evolution.* Basic Books, New York.

Martin, D., 1999. *The Spirit of Dialogue*. International Communication for the Renewal on Earth, New York.

Nelson, H., 1993. Design inquiry as an Intellectual Technology for the Design of Educational Systems. In C. M. Reigeluth, B. H. Banathy, & J. R. Olson (Eds.), *Comprehensive Systems Design: A New Educational Technology*. Spring-Verlag, Berlin. With NATO Scientific Affairs Division, pp. 145-153.

Salk, J., 1983. *Anatomy of Reality: Merging of Intuition and Reason*. Columbia University Press: New York.

Shils, E., 1991. The virtues of civil society. *Government and Opposition* **26**(2): 3-20.

Toynbee, A., 1947. *A Study of History*. Oxford University Press, New York.

Index